THE PHYSICS OF GRAPHENE

Leading graphene research theorist Mikhail I. Katsnelson systematically presents the basic concepts of graphene physics in this fully revised second edition. The author illustrates and explains basic concepts such as Berry phase, scaling, Zitterbewegung, Kubo, Landauer and Mori formalisms in quantum kinetics, chirality, plasmons, commensurate–incommensurate transitions, and many others. Open issues and unsolved problems introduce the reader to the latest developments in the field. New achievements and topics presented include the basic concepts of van der Waals heterostructures, many-body physics of graphene, electronic optics of Dirac electrons, hydrodynamics of electron liquid, and the mechanical properties of one-atom-thick membranes. Building on an undergraduate-level knowledge of quantum and statistical physics and solid-state theory, this is an important graduate textbook for students in nanoscience, nanotechnology, and condensed matter. For physicists and material scientists working in related areas, this is an excellent introduction to the fast-growing field of graphene science.

MIKHAIL I. KATSNELSON is a professor of theoretical physics at Radboud University. He is a recipient of the Spinoza Prize and the Hamburg Prize for Theoretical Physics. He is an elected member of the Academia Europaea, the Royal Netherlands Academy of Arts and Sciences, and the Royal Society of Sciences at Uppsala, and is a knight of the Order of the Netherlands Lion.

THE PHYSICS OF GRAPHENE

SECOND EDITION

MIKHAIL I. KATSNELSON

Radboud University

CAMBRIDGE
UNIVERSITY PRESS

University Printing House, Cambridge CB2 8BS, United Kingdom

One Liberty Plaza, 20th Floor, New York, NY 10006, USA

477 Williamstown Road, Port Melbourne, VIC 3207, Australia

314–321, 3rd Floor, Plot 3, Splendor Forum, Jasola District Centre, New Delhi – 110025, India

79 Anson Road, #06–04/06, Singapore 079906

Cambridge University Press is part of the University of Cambridge.

It furthers the University's mission by disseminating knowledge in the pursuit of education, learning, and research at the highest international levels of excellence.

www.cambridge.org
Information on this title: www.cambridge.org/9781108471640
DOI: 10.1017/9781108617567

Second Edition, First published 2012

Printed in the United Kingdom by TJ International Ltd, Padstow Cornwall

A catalogue record for this publication is available from the British Library.

ISBN 978-1-108-47164-0 Hardback

In memory of my teacher Serghey Vonsovsky and my friend Sasha Trefilov

Contents

Preface to the second edition

First of all, I still agree with everything that I wrote in the preface to the first edition; however, I probably need to add a few words on the differences between the second edition and the first.

As you can see, I have changed the title. In 2011 when I finished *Graphene: Carbon in Two Dimensions*, there were no other books on graphene, and the accuracy of the title was probably not so important. I would also like to emphasize what is special about this book and in what respect it is different from the many others that have appeared in the market in the meantime. To my knowledge, this is the only book on graphene (yet) that focuses completely on fundamental issues of physics and completely ignores all aspects of fabrication, devices, applications, chemistry, etc. Hopefully, the new title, *The Physics of Graphene*, stresses this point clearly enough and helps potential readers to avoid any disappointment if they do not find something in the book which, in their mind, *should* be in a book on graphene. Of course, I do not mean that these aspects are not important; I just believe that I am not the proper person to write about them and that other people can do that much better.

In the field of graphene, eight years is a very long period of time, when many things have happened. To my surprise, when I started to work on the new edition, I did not find anything that should be eliminated from the book because it turned out to be fundamentally wrong or irrelevant for further development. Of course, there were some inaccuracies and mistakes, which hopefully have been fixed now, but even so, I think all old issues remain interesting and important. At the same time, many new concepts and facts have appeared that should be reflected in the new book. Therefore I have added three new chapters: Chapters 13 and 14 introduce the basic physics of an important new concept, van der Waals hetero-structures, and Chapter 15 gives a very brief summary of our progress in understanding many-body effects in graphene. Eight years ago we had the feeling that a single-particle picture of noninteracting Dirac fermions explained everything; this is no longer the case. Huge progress in the quality of graphene samples has opened a way to essentially observe many-body features of the electronic spectrum near the neutrality point.

My work on these subjects was essentially based on a collaboration with Nikita Astrakhantsev, Viktor Braguta, Annalisa Fasolino, Andre Geim, Yura Gornostyrev, Sasha Lichtenstein, Kostya Novoselov, Marco Polini, Burkhard Sachs, Guus Slotman, Misha Titov, Maksim Ulybyshev, Merel van Wijk, Tim Wehling, and Shengjun Yuan. Many thanks!

New material has also been added to the old chapters. The most important new points are:

(1) We now understand the physics and mathematics of chiral tunneling in single- and bilayer graphene much better, therefore Chapter 4 has been expanded. These new results were obtained in collaboration with many people, and I especially thank Koen Reijnders and Victor Kleptsyn.

(2) I have added a new section to Chapter 5 on a spectral flow of Dirac operator in multiconnected graphene flakes. Topological aspects of condensed matter physics have become really hot of late, and this provides a nice and fresh new example. This piece is based on our work with Vladimir Nazaikinskii, to whom I also give thanks.

(3) Chapter 9 was essentially rewritten. I have added new material on mechanical properties, which is based on our work with Jan Los and Annalisa Fasolino, and on thermal expansion of graphene. I thank Igor Burmistrov, Igor Gornyi, Paco Guinea, Valentin Kachorovskii, and Sasha Mirlin for collaboration and useful discussions of this subtle issue. I also thank Achille Mauri who found some inaccuracies in the old Chapter 9 and helped to fix them.

(4) In Chapter 11, I have added a discussion of edge scattering, which is based on our work with Vitaly Dugaev, to whom I am very thankful for his collaboration. Hydrodynamics of electronic liquid in graphene is a very fresh and popular subject now, and I cannot avoid it. When I wrote this part, discussions with Misha Titov and Marco Polini were very helpful.

(5) We now know much more about magnetism and spin-orbit effects in graphene and related two-dimensional materials, therefore Chapter 12 has also been updated. The common work with Andre Geim, Irina Grigorieva, Sasha Lichtenstein, Vladimir Mazurenko, and Sasha Rudenko provided essential insights on my new understanding of the subject.

I would like to repeat all of my acknowledgments from the preface to the first edition. Without all of these old and new collaborations and, of course, without full support from my wife Marina, this book would be impossible.

Preface to the first edition

I do not think that I need to explain, in the preface to a book that is all about graphene, what graphene is and why it is important. After the Nobel Prize for physics in 2010, everybody should have heard something about graphene. I do need, however, to explain why I wrote this book and what is special about it.

I hope it will not be considered a disclosure of insider information if I tell you that Andre Geim is a bit sarcastic (especially with theoreticians). Every time I mentioned that I was somewhat busy writing a book on graphene, he always replied "Go to Amazon.com and search for 'graphene'." Indeed, there are many books on graphene, many more reviews, and infinitely many collections of papers and conference proceedings (well, not really infinitely many ... in the main text I will use the mathematical terminology in a more rigorous way, I promise). Why, nevertheless, has this book been written and why may it be worthwhile for you to read it?

Of course, this is a personal view of the field. I do love it, and it has been my main scientific activity during the last seven years, from 2004 when graphene started to be the subject of intensive and systematic investigations. Luckily, I was involved in this development almost from the very beginning. It was a fantastic experience to watch a whole new world coming into being and to participate in the development of a new language for this new world. I would like to try to share this experience with the readers of this book.

The beauty of graphene is that it demonstrates in the most straightforward way many basic concepts of fundamental physics, from Berry's phase and topologically protected zero modes, to strongly interacting fluctuations and scaling laws for two-dimensional systems. It is also a real test bed for relativistic quantum phenomena such as Klein tunneling or vacuum reconstruction – "CERN on one's desk." I was not able to find a book that focused on these aspects of graphene, namely on its role in our general physical view of the world. I have tried to write such a book myself. The price is that I have sacrificed all practical aspects of graphene science and technology, so you will not find a single word here about the ways in which graphene is produced, and there is hardly anything about its potential applications. Well, there is a lot of literature on these subjects. Also, I have said very little about the chemistry of graphene, which is an extremely interesting subject in itself. It certainly deserves a separate book, and I am not chemist enough to write it.

The field is very young, and it is not easy to know what will not be out of date in just a couple of years. My choice is clear from the contents of this book. I do believe that it represents the core of graphene physics, which will not be essentially modified in the near future. I do not mean that this is the most interesting part; moreover, I am sure that there will be impressive progress, at least, in two more directions that are hardly mentioned in the book: in the many-body physics of graphene and in our understanding of electron transport near the neutrality point, where the semiclassical Boltzmann equation is obviously inapplicable. I think, however, that it is a bit too early to cover these subjects in a book, since too many things are not yet clear. Also, the mathematical tools required are not as easy as those used in this book, and I think it is unfair to force the reader to learn something technically quite complicated without a deep internal confidence that the results are relevant for the real graphene.

The way the book has been written is how I would teach a course with the title "Introduction to the Theory of Graphene." I have tried to make a presentation that is reasonably independent of other textbooks. I have therefore included some general issues such as Berry's phase, the statistical mechanics of fluctuating membranes, a quick overview of itinerant-electron magnetism, a brief discussion of basic nonequilibrium statistical mechanics, etc. The aims were, first, to show the physics of graphene in a more general context and, second, to make the reading easier.

It is very difficult to give an overview of a field that has developed so quickly as has that of graphene. So many papers appear, literally every day, that keeping permanently up to date would be an enterprise in the style of ancient myths, e.g., those of Sisyphus, the Danaïdes, and some of the labors of Hercules. I apologize therefore for the lack of many important references. I tried to do my best.

I cannot even list all of the scientific reviews on the basic physics of graphene that are available now (let alone reviews of applications and of popular literature). Let me mention at least several of them, in chronological order: Katsnelson (2007a), Geim and Novoselov (2007), Beenakker (2008), Castro Neto et al. (2009), Geim (2009), Abergel et al. (2010), Vozmediano, Katsnelson, and Guinea (2010), Peres (2010), Das Sarma et al. (2011), Goerbig (2011), and Kotov et al. (2012). There you can find different, complementary views on the field (with the possible exception of the first one). Of course, the Nobel lectures by Geim (2011) and Novoselov (2011) are especially strongly recommended. In particular, the lecture by Andre Geim contains a brilliant presentation of the prehistory and history of graphene research, so I do not need to discuss these unavoidably controversial issues in my book.

I am very grateful to Andre Geim and Kostya Novoselov, who involved me in this wonderful field before it became fashionable (otherwise I would probably never have dared to join such a brilliant company). I am especially grateful to Andre for regular and lengthy telephone conversations; when you have to discuss a theory using just words, without formulas and diagrams, and cannot even make faces, after several years it does improve your understanding of theoretical physics.

It is impossible to thank all my other collaborators in the field of graphene in a short preface, as well as other colleagues with whom I have had fruitful discussions. I have to thank, first of all, Annalisa Fasolino, Paco Guinea, Sasha Lichtenstein, and Tim Wehling for especially close and intensive collaboration. I am very grateful to the former and current members of our group in Nijmegen working on graphene: Misha Akhukov, Danil Boukhvalov, Jan Los, Koen Reijnders, Rafa Roldan, Timur Tudorovskiy, Shengjun Yuan, and Kostya Zakharchenko, and to my other collaborators and coauthors, especially Mark Auslender, Eduardo Castro, Hans De Raedt, Olle Eriksson, Misha Fogler, Jos Giesbers, Leonya Levitov, Tony Low, Jan Kees Maan, Hector Ochoa, Marco Polini, Sasha Rudenko, Mark van Schilfgaarde, Andrey Shytov, Alyosha Tsvelik, Maria Vozmediano, Oleg Yazyev, and Uli Zeitler.

I am grateful to the Faculty of Science of Radboud University and the Institute for Molecules and Materials for making available to me the time and resources for research and writing.

I am very grateful to Marina Katsnelson and Timur Tudorovskiy for their invaluable help with the preparation of the manuscript and for their critical reading. I am grateful to many colleagues for permission to reproduce figures from their papers and for providing some of the original figures used in the book. I am especially grateful to Annalisa Fasolino for the wonderful picture that is used on the cover.

Of course, the role of my wife Marina in this book amounts to much more than her help with the manuscript. You cannot succeed in such a long and demanding task without support from your family. I am very grateful for her understanding and full support.

The book is dedicated to the memory of two people who were very close to me, my teacher Serghey Vonsovsky (1910–1998) and my friend Sasha Trefilov (1951–2003). I worked with them for about 20 years, and they had a decisive influence on the formation of my scientific taste and my scientific style. I thought many times during these last seven years how sad it is that I cannot discuss some of the new and interesting physics about graphene with them. Also, in a more technical sense, I would not have been able to write this book without the experience of writing my previous books, Vonsovsky and Katsnelson (1989) and Katsnelson and Trefilov (2002).

1

The electronic structure of ideal graphene

1.1 The carbon atom

Carbon is the sixth element in the periodic table. It has two stable isotopes, ^{12}C (98.9% of natural carbon) with nuclear spin $I = 0$ and, thus, nuclear magnetic moment $\mu_n = 0$, and ^{13}C (1.1% of natural carbon) with $I = \frac{1}{2}$ and $\mu_n = 0.7024\mu_N$ (μ_N is the nuclear magneton); see Radzig and Smirnov (1985). Like most of the chemical elements, it originates from nucleosynthesis in stars (for a review, see the Nobel lecture by Fowler [1984]). Actually, it plays a crucial role in the chemical evolution of the universe.

The stars of the first generation produced energy only by proton–proton chain reaction, which results in the synthesis of one α-particle (nucleus ^{4}He) from four protons, p. Further nuclear fusion reactions might lead to the formation of either of the isotopes ^{5}He and ^{5}Li (p + α collisions) or of ^{8}Be (α + α collisions); however, all these nuclei are very unstable. As was first realized by F. Hoyle, the chemical evolution does not stop at helium only due to a lucky coincidence − the nucleus ^{12}C has an energy level close enough to the energy of three α-particles, thus, the *triple* fusion reaction $3\alpha \rightarrow {}^{12}$C, being resonant, has a high enough probability. This opens up a way to overcome the mass gap (the absence of stable isotopes with masses 5 and 8) and provides the prerequisites for nucleosynthesis up to the most stable nucleus, ^{56}Fe; heavier elements are synthesized in supernova explosions.

The reaction $3\alpha \rightarrow {}^{12}$C is the main source of energy for red giants. Carbon also plays an essential role in nuclear reactions in stars of the main sequence (heavier than the Sun) via the so-called CNO cycle.

The carbon atom has six electrons, two of them forming a closed $1s^2$ shell (helium shell) and four filling 2s and 2p states. The ground-state atomic configuration is $2s^2\ 2p^2$, with the total spin $S = 1$, total orbital moment $L = 1$ and total angular moment $J = 0$ (the ground-state multiplet $^3P_0$). The first excited state, with a $J = 1$, $^3P_1$ multiplet, has the energy 16.4 cm$^{-1} \approx 2$ meV (Radzig & Smirnov,

1985), which gives an estimate of the strength of the spin-orbit coupling in the carbon atom. The lowest-energy state with configuration $2s^1\, 2p^3$ has the energy 33,735.2 cm^{-1} ≈ 4.2 eV (Radzig & Smirnov, 1985), so this is the promotion energy for exciting a 2s electron into a 2p state. At first sight, this would mean that carbon should always be divalent, due to there being two 2p electrons while the 2s electrons are chemically quite inert. This conclusion is, however, wrong. Normally, carbon is tetravalent, due to a formation of hybridized sp electron states, according to the concept of "resonance" developed by L. Pauling (Pauling, 1960; see also Eyring, Walter, & Kimball, 1946).

When atoms form molecules or solids, the total energy decreases due to overlap of the electron wave functions at various sites and formation of molecular orbitals (in molecules) or energy bands (in solids); for a compact introduction to chemical bonding in solids, see section 1.7 in Vonsovsky and Katsnelson (1989). This energy gain can be sufficient to provide the energy that is necessary to promote a 2s electron into a 2p state in the carbon atom.

In order to maximize the energy gained during the formation of a covalent bond, the overlap of the wave functions with those at neighboring atoms should also be maximal. This is possible if the neighboring atoms are situated in such directions from the central atoms that the atomic wave functions take on maximum values. The larger these values are, the stronger the bond is. There are four basis functions corresponding to the spherical harmonics

$$Y_{0,0}(\vartheta, \varphi) = \frac{1}{\sqrt{4\pi}},$$

$$Y_{1,0}(\vartheta, \varphi) = i\sqrt{\frac{3}{4\pi}}\cos\vartheta, \tag{1.1}$$

$$Y_{1,\pm 1}(\vartheta, \varphi) = \mp i\sqrt{\frac{3}{8\pi}}\sin\vartheta\,\exp\left(\mp i\varphi\right),$$

where ϑ and φ are polar angles. Rather than take the functions $Y_{1,\,m}(\vartheta, \varphi)$ to be the basis functions, it is more convenient to choose their orthonormalized linear combinations of the form

$$\frac{i}{\sqrt{2}}[Y_{1,1}(\vartheta, \varphi) - Y_{1,-1}(\vartheta, \varphi)] = \sqrt{\frac{3}{4\pi}}\sin\vartheta\cos\varphi,$$

$$\frac{i}{\sqrt{2}}[Y_{1,1}(\vartheta, \varphi) + Y_{1,-1}(\vartheta, \varphi)] = \sqrt{\frac{3}{4\pi}}\sin\vartheta\sin\varphi, \tag{1.2}$$

$$-iY_{1,0}(\vartheta, \varphi) = \sqrt{\frac{3}{4\pi}}\cos\vartheta,$$

which are transformed under rotations as the Cartesian coordinates x, y, and z, respectively. The radial components of the s and p functions in the simplest approximation are supposed to be equal in magnitude (which is of course a very strong assumption) and may be omitted, together with the constant factor $1/\sqrt{4\pi}$ which is not important here. Then the angular dependence of the four basis functions that we will introduce in lieu of $Y_{1,m}(\vartheta, \varphi)$ can be represented as

$$|s\rangle = 1,$$
$$|x\rangle = \sqrt{3}\sin\vartheta\cos\varphi, \quad |y\rangle = \sqrt{3}\sin\vartheta\sin\varphi, \quad |z\rangle = \sqrt{3}\cos\vartheta. \tag{1.3}$$

We now seek linear combinations of the functions (1.3) that will ensure maximum overlap with the functions of the adjacent atoms. This requires that the value of $\alpha = \max_{\vartheta, \varphi} \psi$ be a maximum. With the normalization that we have chosen, $\alpha = 1$ for the s states and $\alpha = \sqrt{3}$ for the p functions of $|x\rangle$, $|y\rangle$, and $|z\rangle$. We then represent the function $|\psi\rangle$ as

$$|\psi\rangle = a|s\rangle + b_1|x\rangle + b_2|y\rangle + b_3|z\rangle, \tag{1.4}$$

where a and b_i are real-valued coefficients that satisfy the normalization condition

$$a^2 + b_1^2 + b_2^2 + b_3^2 = 1. \tag{1.5}$$

The function $|\psi\rangle$, then, is normalized in the same way as (1.3). This follows from their mutual orthogonality

$$\int do|\psi(\vartheta, \varphi)|^2 \equiv \langle\psi|\psi\rangle = a^2\langle s|s\rangle + b_1^2\langle x|x\rangle + b_2^2\langle y|y\rangle + b_3^2\langle z|z\rangle = 4\pi,$$

with do being an element of solid angle. For the time being, the orientation of the axes in our case is arbitrary.

Let us assume that in one of the functions ψ for which α is a maximum, this maximum value is reached in the direction along the diagonal of the cube $(1, 1, 1)$, with the carbon atom at its center and with the coordinate axes parallel to its edges (Fig. 1.1). Then $b_1 = b_2 = b_3 = b$. The $(1, 1, 1)$ direction is given by angles ϑ and φ such that

$$\sin\varphi = \cos\varphi = \frac{1}{\sqrt{2}}, \quad \cos\vartheta = \frac{1}{\sqrt{3}}, \quad \sin\vartheta = \sqrt{\frac{2}{3}},$$

so that

$$|x\rangle = |y\rangle = |z\rangle = 1.$$

In addition,

$$\alpha = a + 3b = a + \sqrt{3(1 - a^2)}, \tag{1.6}$$

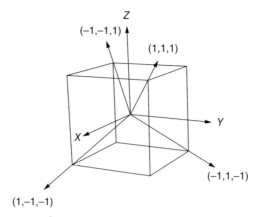

Fig. 1.1 Directions of sp^3 chemical bonds of the carbon atom.

where we have used the conditions (1.3). The maximum of α as a function of a is reached for $a = \frac{1}{2}$ and is equal to 2. The quantity b in this case is equal to $\frac{1}{2}$. Thus the first orbital with maximum values along the coordinate axes that we have chosen is of the form

$$|1\rangle = \frac{1}{2}\left(|s\rangle + |x\rangle + |y\rangle + |z\rangle\right). \qquad (1.7)$$

It can be readily shown that the functions

$$|2\rangle = \frac{1}{2}\left(|s\rangle + |x\rangle - |y\rangle - |z\rangle\right),$$

$$|3\rangle = \frac{1}{2}\left(|s\rangle - |x\rangle + |y\rangle - |z\rangle\right), \qquad (1.8)$$

$$|4\rangle = \frac{1}{2}\left(|s\rangle - |x\rangle - |y\rangle + |z\rangle\right)$$

correspond to the same value $\alpha = 2$. The functions $|i\rangle$ ($i = 1, 2, 3, 4$) are mutually orthogonal. They take on their maximum values along the $(1,1, 1)$, $(1, \bar{1}, \bar{1})$, $(\bar{1}, 1, \bar{1})$, and $(\bar{1}, \bar{1}, 1)$ axes, i.e., along the axes of the tetrahedron, and, therefore, the maximum gain in chemical-bonding energy corresponds to the tetrahedral environment of the carbon atom. In spite of being qualitative, the treatment that we have performed here nevertheless explains the character of the crystal structure of the periodic table group-IV elements (diamond-type lattice, Fig. 1.2) as well as the shape of the methane molecule, which is very close to being tetrahedral.

The wave functions (1.7) and (1.8) correspond to a so-called sp^3 state of the carbon atom, for which all chemical bonds are equivalent. Another option is that *three* sp electrons form hybrid covalent bonds, whereas one p electron has a special destiny, being distributed throughout the whole molecule (benzene) or the whole

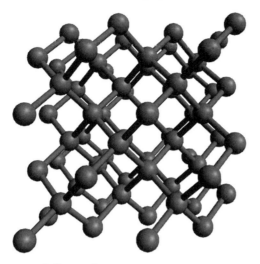

Fig. 1.2 The structure of diamond.

crystal (graphite or graphene). If one repeats the previous consideration for a smaller basis, including only functions, $|s\rangle$, $|x\rangle$ and, $|y\rangle$ one finds the following functions corresponding to the maximum overlap (Eyring, Walter, & Kimball, 1946):

$$|1\rangle = \frac{1}{\sqrt{3}}\left(|s\rangle + \sqrt{2}|x\rangle\right),$$

$$|2\rangle = \frac{1}{\sqrt{3}}|s\rangle - \frac{1}{\sqrt{6}}|x\rangle + \frac{1}{\sqrt{2}}|y\rangle, \tag{1.9}$$

$$|3\rangle = \frac{1}{\sqrt{3}}|s\rangle - \frac{1}{\sqrt{6}}|x\rangle - \frac{1}{\sqrt{2}}|y\rangle.$$

The corresponding orbits have maxima in the *xy*-plane separated by angles of $120°$. These are called σ *bonds*. The last electron with the p orbital perpendicular to the plane ($|z\rangle$ function) forms a π *bond*. This state (sp^2) is therefore characterized by threefold coordination of carbon atoms, in contrast with fourfold coordination for the sp^3 state. This is the case of *graphite* (Fig. 1.3).

1.2 π States in graphene

Graphene has a honeycomb crystal lattice as shown in Fig. 1.4(a). The Bravais lattice is triangular, with the lattice vectors

$$\vec{a}_1 = \frac{a}{2}\left(3, \sqrt{3}\right), \quad \vec{a}_2 = \frac{a}{2}\left(3, -\sqrt{3}\right), \tag{1.10}$$

Fig. 1.3 The structure of graphite.

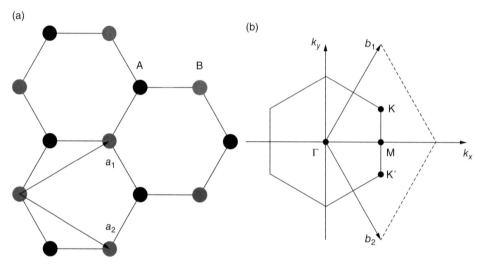

Fig. 1.4 (a) A honeycomb lattice: sublattices A and B are shown as black and gray. (b) Reciprocal lattice vectors and some special points in the Brillouin zone.

where $a \approx 1.42$ Å is the nearest-neighbor distance. It corresponds to a so-called conjugated carbon–carbon bond (like in benzene) intermediate between a single bond and a double bond, with lengths $r_1 \approx 1.54$ Å and $r_2 \approx 1.31$ Å, respectively.

The honeycomb lattice contains two atoms per elementary cell. They belong to two sublattices, A and B, each atom from sublattice A being surrounded by three atoms from sublattice B, and vice versa (a bipartite lattice). The nearest-neighbor vectors are

$$\vec{\delta}_1 = \frac{a}{2}\left(1, \sqrt{3}\right), \quad \vec{\delta}_2 = \frac{a}{2}\left(1, -\sqrt{3}\right), \quad \vec{\delta}_3 = a(-1, 0). \tag{1.11}$$

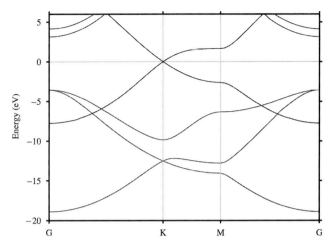

Fig. 1.5 The band structure of graphene.
(Reproduced with permission from Boukhvalov, Katsnelson, & Lichtenstein, 2008.)

The reciprocal lattice is also triangular, with the lattice vectors

$$\vec{b}_1 = \frac{2\pi}{3a}\left(1, \sqrt{3}\right), \quad \vec{b}_2 = \frac{2\pi}{3a}\left(1, -\sqrt{3}\right). \tag{1.12}$$

The Brillouin zone is presented in Fig. 1.4(b). Special high-symmetry points K, K′, and M are shown there, with the wave vectors

$$\vec{K}' = \left(\frac{2\pi}{3a}, \frac{2\pi}{3\sqrt{3}a}\right), \quad \vec{K} = \left(\frac{2\pi}{3a}, -\frac{2\pi}{3\sqrt{3}a}\right), \quad \vec{M} = \left(\frac{2\pi}{3a}, 0\right). \tag{1.13}$$

The electronic structures of graphene and graphite are discussed in detail in Bassani and Pastori Parravicini (1975). In Fig. 1.5 we show a recent computational result for graphene. The sp^2 hybridized states (σ states) form occupied and empty bands with a huge gap, whereas p_z (π) states form a single band, with a conical self-crossing point in K (the same point, by symmetry, exists also in K′). This conical point is a characteristic of the peculiar electronic structure of graphene and the origin of its unique electronic properties. It was first obtained by Wallace (1947) in the framework of a simple tight-binding model. Furthermore this model was developed by McClure (1957) and Slonczewski and Weiss (1958).

Let us start, following Wallace (1947), with the nearest-neighbor approximation for the π states only, with the hopping parameter t. The basis of electron states contains two π states belonging to the atoms from sublattices A and B. In the nearest-neighbor approximation, there are no hopping processes within the sub-lattices; hopping occurs only between them. The tight-binding Hamiltonian is therefore described by the 2×2 matrix

$$\hat{H}\left(\vec{k}\right) = \begin{pmatrix} 0 & tS\left(\vec{k}\right) \\ tS^*\left(\vec{k}\right) & 0 \end{pmatrix}, \tag{1.14}$$

where $\vec{k}$ is the wave vector and

$$S\left(\vec{k}\right) = \sum_{\vec{\delta}} e^{i\vec{k}\vec{\delta}} = 2\exp\left(\frac{ik_x a}{2}\right)\cos\left(\frac{k_y a\sqrt{3}}{2}\right) + \exp\left(-ik_x a\right). \tag{1.15}$$

The energy is, therefore,

$$E\left(\vec{k}\right) = \pm t\left|S\left(\vec{k}\right)\right| = \pm t\sqrt{3 + f\left(\vec{k}\right)}, \tag{1.16}$$

where

$$f\left(\vec{k}\right) = 2\cos\left(\sqrt{3}k_y a\right) + 4\cos\left(\frac{\sqrt{3}}{2}k_y a\right)\cos\left(\frac{3}{2}k_x a\right). \tag{1.17}$$

One can see immediately that $S(\vec{K}) = S(\vec{K}') = 0$, which means band crossing. On expanding the Hamiltonian near these points one finds

$$\hat{H}_{K'}(\vec{q}) \approx \frac{3at}{2}\begin{pmatrix} 0 & \alpha\left(q_x + iq_y\right) \\ \alpha^*\left(q_x - iq_y\right) & 0 \end{pmatrix}$$

$$\hat{H}_{K}(\vec{q}) \approx \frac{3at}{2}\begin{pmatrix} 0 & \alpha\left(q_x - iq_y\right) \\ \alpha^*\left(q_x + iq_y\right) & 0 \end{pmatrix} \tag{1.18}$$

where $\alpha = e^{5i\pi/6}$, with $\vec{q} = \vec{k} - \vec{K}$ and $\vec{k} - \vec{K}'$ respectively. The phase $5\pi/6$ can be excluded by a unitary transformation of the basis functions. Thus, the effective Hamiltonians near the points K and K$'$ take the form

$$\hat{H}_{K,K'}(\vec{q}) = \hbar v\begin{pmatrix} 0 & q_x \mp iq_y \\ q_x \pm iq_y & 0 \end{pmatrix}, \tag{1.19}$$

where

$$v = \frac{3a|t|}{2\hbar} \tag{1.20}$$

is the electron velocity at the conical points. The possible negative sign of t can be excluded by an additional phase shift by π.

On taking into account the next-nearest-neighbor hopping t', one finds, instead of Eq. (1.16),

$$E\left(\vec{k}\right) = \pm t \left|S\left(\vec{k}\right)\right| + t'f\left(\vec{k}\right) = \pm t\sqrt{3 + f\left(\vec{k}\right)} + t'f\left(\vec{k}\right). \qquad (1.21)$$

The second term breaks the electron–hole symmetry, shifting the conical point from $E = 0$ to $E = -3t'$, but it does not change the behavior of the Hamiltonian near the conical points. Actually, this behavior is symmetry-protected (and even topologically protected), as we will see in the next section.

Note that, contrary to the sign of t, the sign of t' describing the hopping within the same sublattice cannot be changed by unitary transformation.

The points K and $-$K$'$ differ by the reciprocal lattice vector $\vec{b} = \vec{b}_1 - \vec{b}_2$, so the point K$'$ is equivalent to $-$K. To show this explicitly, it is sometimes convenient to use a larger unit cell in the reciprocal space, with six conical points. The spectrum (1.16) in this representation is shown in Fig. 1.6.

The parameters of the effective tight-binding model can be found by fitting the results of first-principles electronic-structure calculations. According to Reich et al. (2002), the first three hopping parameters are $t = -2.97$ eV, $t' = -0.073$ eV and $t'' = -0.33$ eV. Experimental estimates (Kretinin et al., 2013) yield $t' \approx -0.3$ eV $\pm 15\%$. The smallness of t' in comparison with t means that the electron–hole symmetry of the spectrum is quite accurate not only in the vicinity of the conical points but also throughout the whole Brillouin zone.

There are saddle points of the electron energy spectrum at M (see Figs. 1.5 and 1.6), with Van Hove singularities in the electron density of states, $\delta N(E) \propto -\ln |E - E_M|$ (Bassani & Pastori Parravicini, 1975). The positions of these singularities with the parameters from Reich et al. (2002) are

$$E_{M-} = t + t' - 3t'' \approx -2.05\text{eV}$$

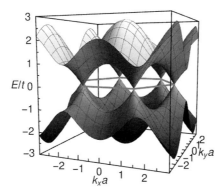

Fig. 1.6 The electron energy spectrum of graphene in the nearest-neighbor approximation.

and

$$E_{M+} = -t + t' + 3t'' \approx 1.91\text{eV}.$$

The Hamiltonian (1.14) in the representation (1.15) has an obvious trigonal symmetry (a symmetry with respect to rotation at $120°$). At the same time, it is not periodic in the reciprocal space, which may be inconvenient for some calculations (of course, its eigenvalues (1.16) are periodic). This can be fixed by the change of basis, e.g., by multiplying the A-component of the wave function by a factor $\exp(-i\vec{k}\vec{\delta}_3)$. Then, instead of Eq. (1.15) we will have the expression

$$S(\vec{k}) = 1 + e^{i\vec{k}(\vec{\delta}_1 - \vec{\delta}_3)} + e^{i\vec{k}(\vec{\delta}_2 - \vec{\delta}_3)},$$

which is obviously periodic but its trigonal symmetry is now hidden. The use of the representation is dictated by convenience for a specific problem.

1.3 Massless Dirac fermions in graphene

Undoped graphene has a Fermi energy coinciding with the energy at the conical points, with a completely filled valence band, an empty conduction band, and no band gap in between. This means that, from the point of view of a general band theory, graphene is an example of a *gapless semiconductor* (Tsidilkovskii, 1996). Three-dimensional crystals, such as HgTe and α-Sn (gray tin) are known to be gapless semiconductors. What makes graphene unique is not the gapless state itself but the very special, chiral nature of the electron states, as well as the high degree of electron–hole symmetry.

For any realistic doping, the Fermi energy is close to the energy at the conical point, $|E_F| \ll |t|$. To construct an effective model describing electron and hole states in this regime one needs to expand the effective Hamiltonian near one of the special points K and K$'$ and then make the replacements

$$q_x \to -i\frac{\partial}{\partial x}, \quad q_y \to -i\frac{\partial}{\partial y},$$

which corresponds to the effective mass approximation, or $\vec{k}\cdot\vec{p}$ perturbation theory (Tsidilkovskii, 1982; Vonsovsky & Katsnelson, 1989). From Eq. (1.19), one has

$$\hat{H}_K = -i\hbar v \vec{\sigma}\nabla, \tag{1.22}$$

$$\hat{H}_{K'} = \hat{H}_K^T, \tag{1.23}$$

where

$$\sigma_0 = \begin{pmatrix} 1 & 0 \\ 0 & 1 \end{pmatrix}, \quad \sigma_x = \begin{pmatrix} 0 & 1 \\ 1 & 0 \end{pmatrix}, \quad \sigma_y = \begin{pmatrix} 0 & -i \\ i & 0 \end{pmatrix}, \quad \sigma_z = \begin{pmatrix} 1 & 0 \\ 0 & -1 \end{pmatrix} \tag{1.24}$$

are Pauli matrices (only *x*- and *y*-components enter Eq. (1.22)) and T denotes a transposed matrix. A complete low-energy Hamiltonian consists of 4×4 matrices taking into account both two sublattices and two conical points (in terms of semiconductor physics, two valleys).

In the basis

$$\Psi = \begin{pmatrix} \psi_{KA} \\ \psi_{KB} \\ \psi_{K'A} \\ \psi_{K'B} \end{pmatrix}, \tag{1.25}$$

where ψ_{KA} means a component of the electron wave function corresponding to valley K and sublattice A, the Hamiltonian is a 2×2 block supermatrix

$$\hat{H} = \begin{pmatrix} \hat{H}_K & 0 \\ 0 & \hat{H}_{K'} \end{pmatrix}. \tag{1.26}$$

Sometimes it is more convenient to choose the basis as

$$\Psi = \begin{pmatrix} \psi_{KA} \\ \psi_{KB} \\ \psi_{K'B} \\ -\psi_{K'A} \end{pmatrix} \tag{1.27}$$

(Aleiner & Efetov, 2006; Akhmerov & Beenakker, 2008; Basko, 2008), then the Hamiltonian (1.26) takes the most symmetric form

$$\hat{H} = -i\hbar v \tau_0 \otimes \vec{\sigma} \nabla, \tag{1.28}$$

where τ_0 is the unit matrix in valley indices (we will use different notations for the same Pauli matrices acting on different indices, namely, $\vec{\sigma}$ in the sublattice space and $\vec{\tau}$ in the valley space).

For the case of an ideal graphene, the valleys are decoupled. If we add some inhomogeneities (external electric and magnetic fields, disorder, etc.) that are smooth at the atomic scale, the valleys remain independent, since the Fourier component of external potential with the *Umklapp* wave vector $\vec{b}$ is very small, and intervalley scattering is improbable. We will deal mainly with this case. However, one should keep in mind that any sharp (atomic-scale) inhomogeneities, e.g., boundaries or vacancies, will mix the states from different valleys, see Chapters 5 and 6.

The Hamiltonian (1.22) is a two-dimensional analog of the Dirac Hamiltonian for massless fermions (Bjorken & Drell, 1964; Berestetskii, Lifshitz, & Pitaevskii, 1971; Davydov, 1976). Instead of the velocity of light c, there is a parameter $v \approx 10^6 \, \mathrm{ms}^{-1} \approx c/300$ (we will discuss later, in Chapter 2, how this parameter has been found experimentally).

A formal similarity between ultrarelativistic particles (with energy much larger than the rest energy mc^2, such that one can consider the particles as massless) and electrons in graphene makes graphene a playground on which to study various quantum relativistic effects – "CERN on one's desk." These relationships between the physics of graphene and relativistic quantum mechanics will be considered in the next several chapters.

The internal degree of freedom, which is just spin for "true" Dirac fermions, is the sublattice index in the case of graphene. The Dirac "spinors" consist here of the components describing the distribution of electrons in sublattices A and B. We will call this quantum number *pseudospin*, so that pseudospin "up" means sublattice A and pseudospin "down" means sublattice B. Apart from the pseudospin, there are two more internal degrees of freedom, namely the valley label (sometimes called *isospin*) and *real spin*. So the most general low-energy Hamiltonian of electrons in graphene is an 8×8 matrix.

Spin-orbit coupling leads to a mixture of pseudospin and real spin and to the gap opening (Kane & Mele, 2005b). However, the value of the gap is supposed to be very small, of the order of 10^{-2} K for pristine graphene (Huertas-Hernando, Guinea, & Brataas, 2006). The reason is not only the lightness of carbon atoms but also the orientation of orbital moments for p_z states perpendicular to the graphene plane. In silicene and germanene, that is, Si and Ge analogs of graphene, the structure is buckled, which leads to a dramatic enhancement of the spin-orbit coupling (Acun et al., 2015). Defects can significantly enhance the spin-orbit coupling (Castro Neto & Guinea, 2009) and the corresponding effects are relevant, e.g., for spin relaxation in graphene (Huertas-Hernando, Guinea, & Brataas, 2009), but the influence of spin-orbit coupling on the electronic structure is negligible. Henceforth we will neglect these effects, until the end of the book (see Section 12.4).

For the case of "true" Dirac fermions in three-dimensional space, the Hamiltonian is a 4×4 matrix, due to two projections of spins and two values of a charge degree of freedom – particle versus antiparticle. For the two-dimensional case the latter is not independent of the former. Electrons and holes are just linear combinations of the states from the sublattices A and B. The 2×2 matrix $\hbar v \vec{\sigma} \vec{k}$ (the result of action of the Hamiltonian (1.22) on a plane wave with wave vector $\vec{k}$) is diagonalized by the unitary transformation

$$\hat{U}_{\vec{k}} = \frac{1}{\sqrt{2}} \left(1 + i \vec{m}_{\vec{k}} \vec{\sigma} \right), \tag{1.29}$$

where $\vec{m}_{\vec{k}} = \left(\sin \phi_{\vec{k}}, -\cos \phi_{\vec{k}} \right)$, and $\phi_{\vec{k}}$ is the polar angle of the vector $\vec{k}\left(\vec{m}_{\vec{k}} \perp \vec{k} \right)$. The eigenfunctions

$$\psi_{e,h}^{(K)}\left(\vec{k}\right) = \frac{1}{\sqrt{2}} \begin{pmatrix} \exp\left(-i\phi_{\vec{k}}/2\right) \\ \pm\exp\left(i\phi_{\vec{k}}/2\right) \end{pmatrix} \tag{1.30}$$

correspond to electron (e) and hole (h) states, with the energies

$$E_{e,h} = \pm\hbar vk. \tag{1.31}$$

For the valley K′ the corresponding states (in the basis (1.25)) are

$$\psi_{e,h}^{(K')}\left(\vec{k}\right) = \frac{1}{\sqrt{2}} \begin{pmatrix} \exp\left(i\phi_{\vec{k}}/2\right) \\ \pm\exp\left(-i\phi_{\vec{k}}/2\right) \end{pmatrix}. \tag{1.32}$$

Of course, this choice of the wave functions is not unique, they can be multiplied by an arbitrary phase factor; only the ratio of the components of the spinor corresponding to the sublattices A and B has a physical meaning.

For the electron (hole) states, by definition

$$\frac{\left(\vec{k}\vec{\sigma}\right)}{k}\psi_{e,h} = \pm\psi_{e,h}. \tag{1.33}$$

This means that the electrons (holes) in graphene have a definite pseudospin direction, namely parallel (antiparallel) to the direction of motion. Thus, these states are *chiral* (*helical*), as should be the case for massless Dirac fermions (Bjorken & Drell, 1964). This is of crucial importance for "relativistic" effects, such as Klein tunneling, which will be considered in Chapter 4.

The Dirac model for electrons in graphene results from the lowest-order expansion of the tight-binding Hamiltonian (1.14) near the conical points. If one takes into account the next, quadratic, term, one finds, instead of the Hamiltonian (1.28) (in the basis (1.27))

$$\hat{H} = \hbar v\tau_0\otimes\vec{\sigma}\vec{k} + \mu\tau_z\otimes\left[2\sigma_y k_x k_y - \sigma_x\left(k_x^2 - k_y^2\right)\right], \tag{1.34}$$

where $\mu = 3a^2t/8$. The additional term in Eq. (1.34) corresponds to a *trigonal warping* (Ando, Nakanishi, & Saito, 1998; McCann et al., 2006). Diagonalization of the Hamiltonian (1.34) gives the spectrum $E_{e,h}\left(\vec{k}\right) = \pm\varepsilon\left(\vec{k}\right)$, where

$$\varepsilon^2\left(\vec{k}\right) = \hbar^2 v^2 k^2 \mp 2\hbar v\mu k^3 \cos\left(3\phi_{\vec{k}}\right) + \mu^2 k^4, \tag{1.35}$$

with the signs $\mp$ corresponding to valleys K and K′. The dispersion law is no longer isotropic but has threefold (trigonal) symmetry. Importantly, $\varepsilon\left(\vec{k}\right) \neq \varepsilon\left(-\vec{k}\right)$, which means that the trigonal warping destroys an effective time-reversal symmetry for a given valley (the property $E\left(\vec{k}\right) = E\left(-\vec{k}\right)$ follows from the time-reversal symmetry [Vonsovsky & Katsnelson, 1989]). Of course, for the electron spectrum as a whole, taking into account the two valleys, the symmetry holds:

$$\varepsilon\left(\vec{k} + \vec{K}\right) = \varepsilon\left(-\vec{k} - \vec{K}\right). \tag{1.36}$$

At the end of this section we show, following Mañes, Guinea, and Vozmediano (2007), that the gapless state with the conical point is symmetry-protected. The proof is very simple and based on consideration of two symmetry operations: time reversal T and inversion I. We will use the basis (1.25) and the extended-Brillouin-zone representation of Fig. 1.6 assuming $\vec{K}' = -\vec{K}$. The time reversal changes the sign of the wave vector, or valley,

$$T\psi_{K(A,B)} = \psi_{K(A,B)}^* = \psi_{K'(A,B)}, \tag{1.37}$$

whereas the inversion also exchanges the sublattices:

$$I\psi_{KA} = \psi_{K'B}, \quad I\psi_{KB} = \psi_{K'A}. \tag{1.38}$$

Invariance under these symmetries imposes the following conditions for $\hat{H}_K$ and $\hat{H}_{K'}$:

$$T : H_K = H_{K'}^* = H_K, \tag{1.39}$$

$$I : H_K = \sigma_x H_{K'} \sigma_x = H_K. \tag{1.40}$$

Indeed,

$$\sigma_x \begin{pmatrix} a_{11} & a_{12} \\ a_{21} & a_{22} \end{pmatrix} \sigma_x = \begin{pmatrix} a_{22} & a_{21} \\ a_{12} & a_{11} \end{pmatrix}, \tag{1.41}$$

so the operation in (1.40) does exchange the A and B sublattices.

The conditions (1.39) and (1.40) establish relations between the Hamiltonians for the different valleys. If we use both these symmetry transformations we impose restrictions on H_K and $H_{K'}$ separately, e.g.,

$$TI : H_K = \sigma_x H_K^* \sigma_x = H_K. \tag{1.42}$$

If we write the Hamiltonian as

$$H_K = \sum_i \alpha_i \sigma_i$$

one can see immediately that $\alpha_z = 0$, which means the absence of the mass term. Thus, a perturbation that is invariant under T and I can, in principle, shift the conical point (we will see in Chapter 10 that it can indeed be done, by deformations), but cannot open the gap: $(H_K)_{11} = (H_K)_{22}$ and the bands split by $\pm|H_{12}|$.

If the sublattices are no longer equivalent, then there is no inversion symmetry, the mass term naturally appears and the gap opens. This is, for example, the case of graphene on top of hexagonal boron nitride, h-BN (Giovannetti et al., 2007; Sachs et al., 2011). This case will be considered in detail in Chapter 13.

1.4 The electronic structure of bilayer graphene

By exfoliation of graphene one can obtain *several* layers of carbon atoms. Bilayer graphene (Novoselov et al., 2006) is especially interesting. Its electronic structure can be understood in the framework of a tight-binding model (McCann & Fal'ko, 2006; McCann, Abergel, & Fal'ko, 2007).

The crystal structure of bilayer graphene is shown in Fig. 1.7. Like in graphite, the second carbon layer is rotated by $60°$ with respect to the first one. In graphite, such a configuration is repeated, which is called *Bernal stacking*. The sublattices A of the two layers lie exactly on top of one another, with a significant hopping parameter γ_1 between them, whereas there are no essential hopping processes between the sublattices B of the two layers. The parameter $\gamma_1 = t_\perp$ is usually taken as 0.4 eV, from data on the electronic structure of graphite (Brandt, Chudinov, & Ponomarev, 1988; Dresselhaus & Dresselhaus, 2002), which is an order of magnitude smaller than the nearest-neighbor in-plane hopping parameter $\gamma_0 = t$.

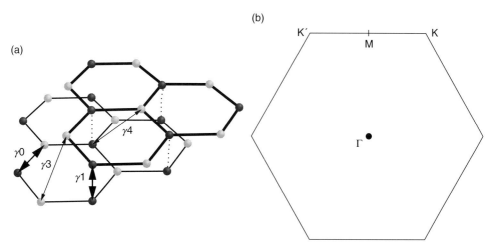

Fig. 1.7 (a) The crystal structure of bilayer graphene; hopping parameters are shown. (b) Special points in the Brillouin zone for the bilayer graphene.

The simplest model, which takes into account only these processes, is described by the Hamiltonian

$$
\hat{H}\left(\vec{k}\right) = \begin{pmatrix}
0 & tS\left(\vec{k}\right) & t_\perp & 0 \\
tS^*\left(\vec{k}\right) & 0 & 0 & 0 \\
t_\perp & 0 & 0 & tS^*\left(\vec{k}\right) \\
0 & 0 & tS\left(\vec{k}\right) & 0
\end{pmatrix}
\tag{1.43}
$$

with $S\left(\vec{k}\right)$ from Eq. (1.15). The basis states are ordered in the sequence first layer, sublattice A; first layer, sublattice B; second layer, sublattice A; second layer, sublattice B.

The matrix (1.43) can be easily diagonalized, with four eigenvalues

$$
E_i\left(\vec{k}\right) = \pm \frac{1}{2}t_\perp \pm \sqrt{\frac{1}{4}t_\perp^2 + t^2\left|S\left(\vec{k}\right)\right|^2}
\tag{1.44}
$$

with two independent $\pm$ signs. The spectrum is shown in Fig. 1.8(a). Two bands touch one another at the points K and K'. Near these points

$$
E_{1,2}\left(\vec{k}\right) \approx \pm \frac{t^2\left|S\left(\vec{k}\right)\right|^2}{t_\perp} \approx \pm \frac{\hbar^2 q^2}{2m^*},
\tag{1.45}
$$

where the effective mass is $m^* = \frac{|t_\perp|}{2v^2}$ (McCann, Abergel, & Fal'ko, 2007). The experimental data give a value $m^* \approx 0.028 m_e$, where m_e is the mass of a free electron (Mayorov et al., 2011). So, in contrast with the case of a single layer, bilayer graphene turns out to be a gapless semiconductor with *parabolic* band

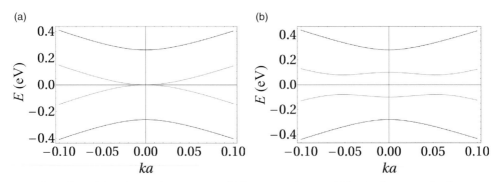

Fig. 1.8 (a) The electronic structure of bilayer graphene within the framework of the simplest model (nearest-neighbor hopping processes only). (b) The same, for the case of biased bilayer graphene (a voltage is applied perpendicular to the layers).

touching. Two other branches $E_{3,4}\left(\vec{k}\right)$ are separated by a gap $2|t_\perp|$ and are irrelevant for low-energy physics.

If one neglects intervalley scattering and replaces $\hbar q_x$ and $\hbar q_y$ by operators $\hat{p}_x = -i\hbar\partial/\partial x$ and $\hat{p}_y = -i\hbar\partial/\partial y$ as usual, one can construct the effective Hamiltonian; for single-layer graphene, this is the Dirac Hamiltonian (1.22). For the case of bilayer graphene, instead, we have (McCann & Fal'ko, 2006; Novoselov et al., 2006)

$$\hat{H}_{\mathrm{K}} = \frac{1}{2m^*}\begin{pmatrix} 0 & \left(\hat{p}_x - i\hat{p}_y\right)^2 \\ \left(\hat{p}_x + i\hat{p}_y\right)^2 & 0 \end{pmatrix}. \tag{1.46}$$

This is a new type of quantum-mechanical Hamiltonian that is different from both nonrelativistic (Schrödinger) and relativistic (Dirac) cases. The eigenstates of this Hamiltonian have special chiral properties (Novoselov et al., 2006), resulting in a special Landau quantization, special scattering, etc., as will be discussed later. Electron and hole states corresponding to the energies

$$E_{\mathrm{e,h}} = \pm\frac{\hbar^2 k^2}{2m^*} \tag{1.47}$$

(cf. Eq. (1.31)) have a form similar to Eq. (1.30), with the replacement $\phi_{\vec{k}} \to 2\phi_{\vec{k}}$:

$$\psi_{\mathrm{e,h}}^{(\mathrm{K})}\left(\vec{k}\right) = \frac{1}{\sqrt{2}}\begin{pmatrix} e^{-i\phi_{\vec{k}}} \\ \pm e^{i\phi_{\vec{k}}} \end{pmatrix}. \tag{1.48}$$

These are characterized by a helicity property similar to Eq. (1.33):

$$\frac{\left(k_x^2 - k_y^2\right)\sigma_x + 2k_x k_y\sigma_y}{k^2}\psi_{e,h} = \pm\psi_{e,h}. \tag{1.49}$$

By applying a voltage V perpendicular to the carbon planes one can open a gap in the energy spectrum (McCann & Fal'ko, 2006; Castro et al., 2007, 2010a). In this case, instead of the Hamiltonian (1.43), one has

$$H\left(\vec{k}\right) = \begin{pmatrix} V/2 & tS\left(\vec{k}\right) & t_\perp & 0 \\ tS^*\left(\vec{k}\right) & V/2 & 0 & 0 \\ t_\perp & 0 & -V/2 & tS^*\left(\vec{k}\right) \\ 0 & 0 & tS\left(\vec{k}\right) & -V/2 \end{pmatrix} \tag{1.50}$$

and, instead of the eigenvalues (1.44), we obtain

$$E_i^2\left(\vec{k}\right) = t^2 \left|S\left(\vec{k}\right)\right|^2 + \frac{t_\perp^2}{2} + \frac{V^2}{4} \pm \sqrt{\frac{t_\perp^4}{4} + \left(t_\perp^2 + V^2\right)t^2 \left|S\left(\vec{k}\right)\right|^2} \qquad (1.51)$$

For the two low-lying bands in the vicinity of the K (or K′) point the spectrum has the "Mexican hat" dispersion

$$E\left(\vec{k}\right) \approx \pm\left(\frac{V}{2} - \frac{V\hbar^2v^2}{t_\perp^2}k^2 + \frac{\hbar^4v^4}{t_\perp^2 V}k^4\right) \qquad (1.52)$$

where we assume, for simplicity, that $\hbar vk \ll V \ll |t_\perp|$. This expression has a maximum at $k = 0$ and a minimum at $k = V/(\sqrt{2}\hbar v)$ (see Fig. 1.8(b)). The opportunity to tune a gap in bilayer graphene is potentially interesting for applications. It was experimentally confirmed by Castro et al. (2007) and Oostinga et al. (2008).

Consider now the effect of larger-distance hopping processes, namely hopping between B sublattices ($\gamma_3 \approx 0.3$eV) (Brandt, Chudinov, & Ponomarev, 1988; Dresselhaus & Dresselhaus, 2002). Higher-order terms, such as $\gamma_4 \approx 0.04$ eV, are assumed to be negligible. These processes lead to a qualitative change of the spectrum near the K (K′) point. As was shown by McCann and Fal'ko (2006) and McCann, Abergel, and Fal'ko (2007), the effective Hamiltonian (1.46) is modified by γ_3 terms, giving

$$\hat{H}_K = \begin{pmatrix} 0 & \dfrac{\left(\hat{p}_x - i\hat{p}_y\right)^2}{2m^*} + \dfrac{3\gamma_3 a}{\hbar}\left(\hat{p}_x + i\hat{p}_y\right) \\ \dfrac{\left(\hat{p}_x + i\hat{p}_y\right)^2}{2m^*} + \dfrac{3\gamma_3 a}{\hbar}\left(\hat{p}_x - i\hat{p}_y\right) & 0 \end{pmatrix},$$

$$(1.53)$$

with the energy spectrum determined by the equation (assuming that $|\gamma_3| \ll |\gamma_0|$)

$$E^2\left(\vec{k}\right) \approx (3\gamma_3 a)^2 k^2 + \frac{3\gamma_3 a\hbar^2 k^3}{m^*}\cos\left(3\phi_{\vec{k}}\right) + \left(\frac{\hbar^2 k^2}{2m^*}\right)^2. \qquad (1.54)$$

This means that, at small enough wave vectors

$$ka \leq \left|\frac{\gamma_3\gamma_1}{\gamma_0^2}\right| \sim 10^{-2}, \qquad (1.55)$$

the parabolic dispersion law (1.47) is replaced by the linear one. The corresponding level of doping when the Fermi wave vector satisfies the conditions (1.55) is estimated as $n < 10^{11}$ cm^{-2} (McCann, Abergel, & Fal'ko, 2007).

The spectrum (1.54) is shown in Fig. 1.9. The term with $\cos\left(3\phi_{\vec{k}}\right)$ in Eq. (1.53) corresponds to the trigonal warping, which is more important for the bilayer than it

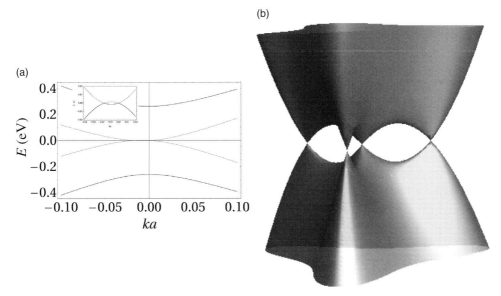

Fig. 1.9 The effect of trigonal warping on the electronic structure of bilayer graphene. (a) A cross-section of the dispersion surface at $\phi_{\vec{k}} = 0$; one can see the asymmetry of the spectrum (cf. Fig. 1.8(a)). (b) A general view of the dispersion surface.

is for the single layer: It leads to a reconstruction of isoenergetic lines when k grows. Instead of one point of parabolic touching of the bands at $k = 0$, there are now four conical points at $k = 0$ and $k = 6m^*\gamma_3 a/\hbar^2$, $\cos\left(3\phi_{\vec{k}}\right) = \pm 1$, where the signs $\pm$ correspond to K and K' valleys. The merging of four cones into one paraboloid with increasing energy is a particular case of the Lifshitz electronic topological transition associated with a Van Hove singularity of the electron density of states (Lifshitz, Azbel, & Kaganov, 1973; Abrikosov, 1988; Vonsovsky & Katsnelson, 1989).

1.5 Multilayer graphene

For the third layer of carbon atoms there are two options: It can be rotated with respect to the second layer by either $-60°$ or by $60°$. In the first case, the third layer lies exactly on top of the first layer, with the layer order aba. In the second case, we will denote the structure as abc. In bulk graphite, the most stable state corresponds to Bernal stacking, abab ... However, *rhombohedral* graphene with the stacking abcabc ... also exists, as does *turbostratic* graphite with an irregular stacking (on the energetics of these different stackings, see Savini et al., 2011).

Here we consider the evolution of the electronic structure of N-layer graphene with different stacking as N increases (Guinea, Castro Neto, & Peres, 2006;

Partoens & Peeters, 2006; Koshino & McCann, 2010). First we will discuss the case of Bernal stacking. We will restrict ourselves to considering only the simplest model with parameters $\gamma_0 = t$ and $\gamma_1 = t_\perp$, neglecting all other hopping parameters γ_i. For the case of bilayer graphene this corresponds to the Hamiltonian (1.43).

On introducing the basis functions $\psi_{n,A}\left(\vec{k}\right)$ and $\psi_{n,B}\left(\vec{k}\right)$ ($n = 1, 2, \ldots, N$ (N is the number of carbon layers, A and B label sublattices, and $\vec{k}$ is the two-dimensional wave vector in the layer) we can write the Schrödinger equation as

$$
\begin{aligned}
E\psi_{2n,A}\left(\vec{k}\right) &= tS\left(\vec{k}\right)\psi_{2n,B}\left(\vec{k}\right) + t_\perp\left[\psi_{2n-1,A}\left(\vec{k}\right) + \psi_{2n+1,A}\left(\vec{k}\right)\right], \\
E\psi_{2n,B}\left(\vec{k}\right) &= tS^*\left(\vec{k}\right)\psi_{2n,A}\left(\vec{k}\right), \\
E\psi_{2n+1,A}\left(\vec{k}\right) &= tS^*\left(\vec{k}\right)\psi_{2n+1,B}\left(\vec{k}\right) + t_\perp\left[\psi_{2n,A}\left(\vec{k}\right) + \psi_{2n+2,A}\left(\vec{k}\right)\right], \\
E\psi_{2n+1,B}\left(\vec{k}\right) &= tS\left(\vec{k}\right)\psi_{2n+1,A}\left(\vec{k}\right).
\end{aligned}
\tag{1.56}
$$

Excluding the components ψ_B from Eq. (1.56), one can write the equation

$$
\left(E - \frac{t^2\left|S\left(\vec{k}\right)\right|^2}{E}\right)\psi_{n,A}\left(\vec{k}\right) = t_\perp\left[\psi_{n+1,A}\left(\vec{k}\right) + \psi_{n-1,A}\left(\vec{k}\right)\right].
\tag{1.57}
$$

For an infinite sequence of layers (bulk graphite with Bernal stacking) one can try the solutions of Eq. (1.57) as

$$
\psi_{n,A}\left(\vec{k}\right) = \psi_A\left(\vec{k}\right)e^{in\xi},
\tag{1.58}
$$

which gives us the energies (Wallace, 1947)

$$
E\left(\vec{k}, \xi\right) = t_\perp\cos\xi \pm \sqrt{t^2\left|S\left(\vec{k}\right)\right|^2 + t_\perp^2\cos^2(\xi)}.
\tag{1.59}
$$

The parameter ξ can be written as $\xi = 2k_z c$, where k_z is the z-component of the wave vector, c is the interlayer distance and, thus, $2c$ is the lattice period in the z-direction. A more accurate tight-binding model of the electronic structure of graphite, taking into account more hoppings, γ_i, was proposed by McClure (1957) and Slonczewski and Weiss (1958); for reviews, see Dresselhaus and Dresselhaus (2002) and Castro Neto et al. (2009).

For the case of N-layer graphene ($n = 1, 2, \ldots, N$) one can still use Eq. (1.57), continuing it for $n = 0$ and $n = N + 1$, but with constraints

$$
\psi_{0,A} = \psi_{N+1,A} = 0
\tag{1.60}
$$

requiring the use of linear combinations of the solutions with ξ and $-\xi$; since $E(\xi) = E(-\xi)$ the expression for the energy (1.59) remains the same but ξ is now discrete. Owing to Eq. (1.60) we have

$$\psi_{n,A} \sim \sin(\xi_p n) \tag{1.61}$$

with

$$\xi_p = \frac{\pi p}{N+1}, \quad p = 1, 2, \ldots, N. \tag{1.62}$$

Eq. (1.59) and (1.62) formally solve the problem of the energy spectrum for N-layer graphene with Bernal stacking. For the case of bilayer graphene $\cos \xi_p = \pm\frac{1}{2}$, and we come back to Eq. (1.44). For $N = 3$, there are six solutions with $\cos \xi_p = 0, \pm 1/\sqrt{2}$:

$$E\left(\vec{k}\right) = \begin{cases} \pm t \left| S\left(\vec{k}\right) \right|, \\ \pm t_\perp \sqrt{2}/2 \pm \sqrt{t_\perp^2/2 + t^2 \left| S\left(\vec{k}\right) \right|^2}. \end{cases} \tag{1.63}$$

We have both conical (like in single-layer graphene) and parabolic (like in bilayer graphene) touching at K and K′ points where $S\left(\vec{k}\right) \to 0$.

For rhombohedral stacking (abc), instead of Eq. (1.56), we have the Schrödinger equation in the form

$$\begin{aligned}
E\psi_{1,A}\left(\vec{k}\right) &= tS\left(\vec{k}\right)\psi_{1,B}\left(\vec{k}\right) + t_\perp \psi_{2,A}\left(\vec{k}\right), \\
E\psi_{1,B}\left(\vec{k}\right) &= tS^*\left(\vec{k}\right)\psi_{1,A}\left(\vec{k}\right), \\
E\psi_{2,A}\left(\vec{k}\right) &= tS^*\left(\vec{k}\right)\psi_{2,B}\left(\vec{k}\right) + t_\perp \psi_{1,A}\left(\vec{k}\right), \\
E\psi_{2,B}\left(\vec{k}\right) &= tS\left(\vec{k}\right)\psi_{2,A}\left(\vec{k}\right) + t_\perp \psi_{3,A}\left(\vec{k}\right), \\
E\psi_{3,A}\left(\vec{k}\right) &= tS\left(\vec{k}\right)\psi_{3,B}\left(\vec{k}\right) + t_\perp \psi_{2,B}\left(\vec{k}\right), \\
E\psi_{3,B}\left(\vec{k}\right) &= tS^*\left(\vec{k}\right)\psi_{3,A}\left(\vec{k}\right).
\end{aligned} \tag{1.64}$$

On excluding from Eq. (1.64) $\psi_{1,B}$ and $\psi_{3,B}$ one obtains

$$\left(E - \frac{t^2 \left| S\left(\vec{k}\right) \right|^2}{E} \right) \psi_{1,A}\left(\vec{k}\right) = t_\perp \psi_{2,A}\left(\vec{k}\right),$$

$$\left(E - \frac{t^2 \left| S\left(\vec{k}\right) \right|^2}{E} \right) \psi_{3,A}\left(\vec{k}\right) = t_\perp \psi_{2,B}\left(\vec{k}\right), \tag{1.65}$$

so we have just two equations for $\psi_{2,A}$, a and $\psi_{2,B}$

$$E\left(1 - \frac{t_\perp^2}{E^2 - t^2\left|S(\vec{k})\right|^2}\right)\psi_{2,A}(\vec{k}) = tS^*(\vec{k})\psi_{2,B}(\vec{k}),$$

$$E\left(1 - \frac{t_\perp^2}{E^2 - t^2\left|S(\vec{k})\right|^2}\right)\psi_{2,B}(\vec{k}) = tS(\vec{k})\psi_{2,A}(\vec{k}),$$

(1.66)

and, finally, the equation for the energy

$$E^2\left(1 + \frac{t_\perp^2}{t^2\left|S(\vec{k})\right|^2 - E^2}\right)^2 = t^2\left|S(\vec{k})\right|^2.$$

(1.67)

Near the K and K' points when $S(\vec{k}) \to 0$ there is a solution of Eq. (1.67) that behaves as

$$E(\vec{k}) \approx \pm\frac{t^3\left|S(\vec{k})\right|^3}{t_\perp^2} \propto \pm q^3,$$

(1.68)

where $\vec{q} = \vec{k} - \vec{K}$ or $\vec{k} - \vec{K}'$. So, in trilayer graphene with rhombohedral stacking we have a gapless semiconducting state with *cubic* touching of the conduction and valence bands.

If we have a rhombohedral stacking of N layers (each layer is rotated with respect to the previous one by $+60°$), the low-lying part of the spectrum behaves, similarly to Eq. (1.68), according to

$$E(\vec{q}) \propto \pm\frac{t^N}{t_\perp^{N-1}}q^N$$

(1.69)

(Mañes, Guinea, & Vozmediano, 2007).

Effects of γ_j beyond the simplest model were discussed by Koshino and McCann (2010).

To finish this chapter, we calculate the density of states

$$N(E) = 2\int\frac{d^2k}{(2\pi)^2}\delta\left(E - E(\vec{k})\right),$$

(1.70)

where integration is over the Brillouin zone of the honeycomb lattice and the factor 2 takes into account spin degeneracy. For small energies $E \to 0$ the contribution to (1.70) comes only from the vicinity of the K and K' points and $E = E(|\vec{q}|)$

depends, to a first approximation (neglecting trigonal warping), only on the modulus of the wave vector. Thus, one gets

$$N(E) = 2 \cdot 2 \int_0^\infty \frac{dqq}{2\pi} \delta(E - E(\vec{q})) = \frac{2}{\pi} \frac{q(E)}{\left| \frac{dE}{dq} \right|}. \tag{1.71}$$

For the case of single-layer graphene, according to Eq. (1.31)

$$N(E) = \frac{2}{\pi} \frac{|E|}{\hbar^2 v^2}, \tag{1.72}$$

and the density of states vanishes linearly as $E \to 0$. For bilayer graphene, due to Eq. (1.47)

$$N(E) = \frac{2m^*}{\pi \hbar^2}, \tag{1.73}$$

and the density of states is constant. Finally, for the spectrum (1.69) the density of states is divergent at $E \to 0$, $N > 2$:

$$N(E) \propto \frac{1}{|E|^{1-2/N}}. \tag{1.74}$$

At large enough energies the density of states has Van Hove singularities (related to the M point) that are relevant for the optical properties and will be discussed in Chapter 7.

2

Electron states in a magnetic field

2.1 The effective Hamiltonian

The reality of massless Dirac fermions in graphene has been demonstrated by Novoselov et al. (2005a) and Zhang et al. (2005) using quantized magnetic fields. The discovery of the anomalous (half-integer) quantum Hall effect in these works was the real beginning of the "graphene boom." Discussion of the related issues allows us to clarify in the most straightforward way possible the basic properties of charge-carrier states in graphene, such as chirality, Berry's phase, etc. So, it seems natural, both historically and conceptually, to start our consideration of the electronic properties of graphene with a discussion of the effects of the magnetic field.

We proceed with the derivation of the effective Hamiltonian of band electrons in a magnetic field (Peierls, 1933); our presentation will mainly follow Vonsovsky and Katsnelson (1989). It is assumed that the magnetic length

$$l_B = \sqrt{\frac{\hbar c}{|e| B}} \tag{2.1}$$

(B is the magnetic induction) is much larger than the interatomic distance:

$$l_B \gg a \tag{2.2}$$

which is definitely the case for any experimentally available fields; it would be violated only for $B \gg 10^4 T$.

Another approximation is that we will take into account only π electrons and neglect transitions to other electron bands (e.g., σ bands). Since the distance between π and σ bands is of the order of the π bandwidth (see Fig. 1.5) one can prove that the approximation is justified under the same condition (2.2) (see the discussion of magnetic breakdown at the end of this section). A rigorous theory of the effect of magnetic fields on Bloch states has been developed by Kohn (1959) and Blount (1962). It is rather cumbersome, and its use for the case of graphene, with its simple band structure, would obviously be overkill.

The original Hamiltonian is

$$H = \frac{\hat{\vec{\pi}}^2}{2m} + V(\vec{r}), \tag{2.3}$$

where

$$\hat{\vec{\pi}} = \hat{\vec{p}} - \frac{e}{c}\vec{A}, \quad \vec{p} = -i\hbar\vec{\nabla}, \tag{2.4}$$

$\vec{A}$ is the vector potential

$$\vec{B} = \vec{\nabla} \times \vec{A}, \tag{2.5}$$

m is the mass of a free electron, and $V(\vec{r})$ is a periodic crystal potential. The operators $\hat{\pi}_\alpha$ satisfy the commutation relations

$$\left[\hat{\pi}_x, \hat{\pi}_y\right] = -\left[\hat{\pi}_y, \hat{\pi}_x\right] = \frac{ie}{\hbar c}B, \tag{2.6}$$

other commutators being zero (we assume that the magnetic induction is along the z-axis).

We can try a general solution of the Schrödinger equation

$$H\psi = E\psi, \tag{2.7}$$

as an expansion in the Wannier basis $\varphi_i(\vec{r})$ (we will omit the band label since we will consider only π states):

$$\psi = \sum_i c_i\varphi_i(\vec{r}). \tag{2.8}$$

The Wannier function on state i can be represented as

$$\varphi_i(\vec{r}) = \varphi_0(\vec{r} - \vec{R}_i) = \exp\left\{-\frac{i}{\hbar}\vec{R}_i\hat{\vec{p}}\right\}\varphi_0(\vec{r}) \tag{2.9}$$

where φ_0 is the function corresponding to the zero site.

For future use, we have to specify the gauge. Here we will use a radial gauge

$$\vec{A} = \frac{1}{2}\vec{B} \times \vec{r} = \left(-\frac{By}{2}, \frac{Bx}{2}, 0\right). \tag{2.10}$$

Then, instead of the expansions (2.8) and (2.9), it is convenient to choose another basis, namely

$$\psi = \sum_i a_i\tilde{\varphi}_i(\vec{r}),$$

$$\tilde{\varphi}_i(\vec{r}) = \exp\left\{-\frac{i}{\hbar}\vec{R}_i\hat{\vec{\Pi}}\right\}\varphi_0(\vec{r}) \tag{2.11}$$

where

$$\hat{\Pi} = \hat{p} + \frac{e}{c}\vec{A}. \tag{2.12}$$

The point is that the operators $\hat{\Pi}_\alpha$ commute with $\hat{\pi}_\beta$ and, thus, with the kinetic energy term in Eq. (2.3):

$$[\hat{\pi}_\alpha, \hat{\Pi}_\beta] = -\frac{ie\hbar}{c}\left(\frac{\partial A_\beta}{\partial x_\alpha} + \frac{\partial A_\alpha}{\partial x_\beta}\right) = 0 \tag{2.13}$$

due to Eq. (2.10). Using the identity

$$\exp\left(\hat{A} + \hat{B}\right) = \exp\left(\hat{A}\right)\exp\left(\hat{B}\right)\exp\left(-\frac{1}{2}[\hat{A}, \hat{B}]\right) \tag{2.14}$$

(assuming $[\hat{A}, [\hat{A}, \hat{B}]] = [\hat{B}, [\hat{A}, \hat{B}]] = 0$), see Vonsovsky and Katsnelson (1989), one can prove that the operator

$$\exp\left(\frac{i}{\hbar}\vec{R}_i\hat{\Pi}\right) = \exp\left\{\frac{ie}{2\hbar c}(\vec{R}_i \times \vec{B})\vec{r}\right\}\exp\left\{\frac{i}{\hbar}\vec{R}_i\hat{p}\right\} \tag{2.15}$$

commutes also with the potential energy $V(\vec{r})$ due to translational invariance of the crystal:

$$\exp\left\{\frac{i}{\hbar}\vec{R}_i\hat{p}\right\}V(\vec{r})\ldots = V(\vec{r} + \vec{R}_i)\exp\left\{\frac{i}{\hbar}\vec{R}_i\hat{p}\right\}\ldots = V(\vec{r})\exp\left\{\frac{i}{\hbar}\vec{R}_i\hat{p}\right\}\ldots \tag{2.16}$$

and, thus, the Hamiltonian matrix in the basis (2.12) has the form

$$H_{ij} = \int d\vec{r}\varphi_0^*(\vec{r})\hat{H}\exp\left(\frac{i}{\hbar}\vec{R}_i\hat{\Pi}\right)\exp\left(-\frac{i}{\hbar}\vec{R}_j\hat{\Pi}\right)\varphi_0(\vec{r}). \tag{2.17}$$

Using, again, Eq. (2.14) one finds

$$\exp\left(\frac{i}{\hbar}\vec{R}_i\hat{\Pi}\right)\exp\left(-\frac{i}{\hbar}\vec{R}_j\hat{\Pi}\right) = \exp\left[\frac{i}{\hbar}\hat{\Pi}(\vec{R}_i - \vec{R}_j)\right]\exp\left\{-\frac{ie}{2\hbar c}(\vec{R}_i \times \vec{R}_j)\vec{B}\right\}$$

$$= \exp\left\{\frac{ie}{2\hbar c}[(\vec{R}_i - \vec{R}_j) \times \vec{B}]\vec{r}\right\}$$

$$\times \exp\left\{-\frac{ie}{2\hbar c}(\vec{R}_i \times \vec{R}_j)\vec{B}\right\}\exp\left\{\frac{i}{\hbar}\hat{p}(\vec{R}_i - \vec{R}_j)\right\}. \tag{2.18}$$

The Wannier functions are localized within a region of extent of a few interatomic distances, so, to estimate the various terms in (2.18), one has to assume $r \approx a$ and $|\vec{R}_i - \vec{R}_j| \approx a$ and take into account Eq. (2.2).

Thus,

$$H_{ij} \approx \exp\left\{\frac{ie}{2\hbar c}(\vec{R}_l \times \vec{R}_j)\vec{B}\right\} t_{ij}, \tag{2.19}$$

where $t_{ij} = H_{ij}(\vec{B} = 0)$ is the hopping parameter without a magnetic field. With the same accuracy, one can prove that the basis (2.11) is orthonormal.

Further straightforward transformations (Vonsovsky & Katsnelson, 1989) show that the change of the hopping parameters (2.11) corresponds to a change of the band Hamiltonian $t(\vec{p})$ (where $\vec{p} = \hbar\vec{k}$) by

$$\hat{H}_{\text{eff}} = t\left(\hat{\vec{\pi}}\right) \tag{2.20}$$

and, thus, the Schrödinger equation (2.7) takes the form

$$t\left(\hat{\vec{\pi}}\right)\psi = E\psi. \tag{2.21}$$

Instead of the operators $\hat{\pi}_x$ and $\hat{\pi}_y$ satisfying the commutation relations (2.6), it is convenient to introduce the standard Bose operators $\hat{b}$ and $\hat{b}^+$ by writing

$$\hat{\pi}_- = \hat{\pi}_x - i\hat{\pi}_y = \sqrt{\frac{2|e|\hbar B}{c}}\hat{b},$$

$$\hat{\pi}_+ = \hat{\pi}_x + i\hat{\pi}_y = \sqrt{\frac{2|e|\hbar B}{c}}\hat{b}^+ \tag{2.22}$$

in such a way that

$$\left[\hat{b}, \hat{b}^+\right] = 1. \tag{2.23}$$

We will see later that this representation is very convenient for the cases of both single-layer and, especially, bilayer graphene.

To finish this section, we should discuss the question of neglected transitions to other bands (magnetic breakdown). If the distance between the bands is of the order of their bandwidth (which is the case for σ and π bands in graphene), the condition (2.2) still suffices to allow us to neglect the transitions. If the gap between the states $\Delta \ll |t|$, the magnetic breakdown can be neglected if

$$\frac{|e|B}{\hbar c} = \frac{1}{l_B^2} \ll \left(\frac{\Delta}{t}\right)^2 \frac{1}{a^2},$$

where we assume that $t \approx \frac{\hbar^2}{ma^2}$ (Vonsovsky & Katsnelson, 1989).

Similarly to the derivation of equations for the electron spectrum of a semiconductor with impurities in the effective-mass approximation (Tsidilkovskii, 1982), one can prove that, if the magnetic induction $\vec{B}(x, y)$ is inhomogeneous but the spatial scale of this inhomogeneity is much larger than a, the Hamiltonian (2.20) still works.

2.2 Landau quantization for massless Dirac fermions

Let us apply the general theory to electrons in graphene in the vicinity of the point K. It follows from Eq. (1.22), (2.20), and (2.22) that the effective Hamiltonian is

$$\hat{H} = v \begin{pmatrix} 0 & \hat{\pi}_- \\ \hat{\pi}_+ & 0 \end{pmatrix} = \sqrt{\frac{2\,|e|\,\hbar B v^2}{c}} \begin{pmatrix} 0 & \hat{b} \\ \hat{b}^+ & 0 \end{pmatrix} \qquad (2.24)$$

and the Schrödinger equation (2.21) for the two-component spinor reads

$$\hat{b}\,\psi_2 = \varepsilon\psi_1,$$
$$\hat{b}^+\psi_1 = \varepsilon\psi_2, \qquad (2.25)$$

where we have introduced a dimensionless quantity ε, such that

$$E = \sqrt{\frac{2\,|e|\,\hbar B v^2}{c}}\,\varepsilon \equiv \frac{\sqrt{2}\hbar v}{l_B}\,\varepsilon. \qquad (2.26)$$

We assume here that $B > 0$ (magnetic field up). For the second valley K′, ψ_1 and ψ_2 exchange their places in Eq. (2.25).

First, one can see immediately from (2.25) that a zero-energy solution exists with $\psi_1 = 0$, and $\psi_2 \equiv |0\rangle$ is the ground state of a harmonic oscillator:

$$\hat{b}|0\rangle = 0. \qquad (2.27)$$

This solution is 100% polarized in pseudospin; that is, for a given direction of the magnetic field for the valleys K and K′, electrons in this state belong completely to sublattices A and B, respectively, or conversely if the direction of the magnetic field is reversed.

To find the complete energy spectrum, one has to act with the operator $\hat{b}^+$ on the first equation of (2.25), which gives us immediately

$$\hat{b}^+\hat{b}\psi_2 = \varepsilon^2\psi_2, \qquad (2.28)$$

with the well-known eigenvalues

$$\varepsilon_n^2 = n = 0, 1, 2, \ldots. \qquad (2.29)$$

Thus, the eigenenergies of massless Dirac electrons in a uniform magnetic field are given by

$$E_n^{(\pm)} = \pm\hbar\omega_c\sqrt{n}, \qquad (2.30)$$

where the quantity

$$\hbar\omega_c = \frac{\sqrt{2}\hbar v}{l_B} = \sqrt{\frac{2\hbar\,|e|\,Bv^2}{c}} \qquad (2.31)$$

will be called the "cyclotron quantum." In the context of condensed-matter physics, this spectrum was first derived by McClure (1956) in his theory of the diamagnetism of graphite. This spectrum is drastically different from that for nonrelativistic electrons with $t\left(\hat{\vec{\pi}}\right) = \hat{\vec{\pi}}^2/(2m)$, where (Landau, 1930)

$$\varepsilon_n = \hbar\tilde{\omega}_c\left(n + \frac{1}{2}\right), \qquad \tilde{\omega}_c = \frac{|e|\,B}{mc}. \qquad (2.32)$$

Discrete energy levels of two-dimensional electrons in magnetic fields are called *Landau levels*.

First, the spectrum (2.31), in contrast with (2.32), is not equidistant. Second, and more importantly, the zero Landau level ($n = 0$) has zero energy and, due to the electron–hole symmetry of the problem, is equally shared by electrons and holes. The states at this level are chiral; that is, they belong to only one sublattice, as was explained previously. The existence of the zero-energy Landau level has deep topological reasons and leads to dramatic consequences for the observable properties of graphene, as will be discussed later in this chapter.

To better understand the relations between relativistic and nonrelativistic Landau spectra, let us calculate the Hamiltonian (2.24) squared, taking into account the commutation relations (2.6):

$$\hat{H}^2 = v^2\left(\hat{\vec{\sigma}}\hat{\vec{\pi}}\right)^2 = v^2\hat{\vec{\pi}}^2 + iv^2\hat{\vec{\sigma}}\left(\hat{\vec{\pi}} \times \hat{\vec{\pi}}\right) = v^2\hat{\vec{\pi}}^2 - \frac{v^2\hbar|e|B}{c}\sigma^z. \qquad (2.33)$$

The spectrum of the operator (2.33) can be immediately found from the solution of the nonrelativistic problem if one puts $m = 1/(2v^2)$. Then,

$$E_n^2 = \frac{2\hbar\,|e|\,Bv^2}{c}\left(n + \frac{1}{2}\right) \mp \frac{v^2\hbar\,|e|\,B}{c} = \frac{2\hbar\,|e|\,Bv^2}{c}\left(n + \frac{1}{2} \mp \frac{1}{2}\right), \qquad (2.34)$$

where ∓ 1 are eigenstates of the operator $\hat{\sigma}^z$. The last term in Eq. (2.33) looks like Zeeman splitting, and the existence of the zero Landau level in these terms results from an exact cancellation of the cyclotron energy and the Zeeman energy. Actually, for free electrons, for which the same mass is responsible for both the orbital motion and the internal magnetic moment, the situation is exactly the same:

$$E_{n,\sigma} = \frac{\hbar\,|e|\,B}{mc}\left(n + \frac{1}{2}\right) \mp \frac{\hbar\,|e|\,B}{2mc}. \qquad (2.35)$$

In semiconductors, the effective electron mass is usually much smaller than the effective electron mass, and the Zeeman term just gives small corrections to Landau quantization. For the case of graphene, the pseudo-Zeeman term also originates from the orbital motion, namely from hopping processes between neighboring sites.

To find the eigenfunctions corresponding to the eigenenergies (2.30), one needs to specify a gauge for the vector potential. The choice (2.10) gives us solutions with radial symmetry. It is more convenient, however, to use the Landau gauge

$$\vec{A} = (0, Bx, 0). \tag{2.36}$$

Then Eq. (2.25) takes the form

$$
\begin{aligned}
\left(\frac{\partial}{\partial x} - i\frac{\partial}{\partial y} - \frac{x}{l_B^2}\right)\psi_2 &= \frac{iE}{\hbar v}\psi_1, \\
\left(\frac{\partial}{\partial x} + i\frac{\partial}{\partial y} + \frac{x}{l_B^2}\right)\psi_1 &= \frac{iE}{\hbar v}\psi_2.
\end{aligned}
\tag{2.37}
$$

In the gauge (2.36), y is the cyclic coordinate, and the solutions of Eq. (2.37) can be tried in the form

$$\psi_{1,2}(x, y) = \psi_{1,2}(x)\exp(ik_y y), \tag{2.38}$$

which transforms Eq. (2.37) into

$$
\begin{aligned}
\left(\frac{\partial}{\partial x} - \frac{x - x_0}{l_B^2}\right)\psi_2 &= \frac{iE}{\hbar v}\psi_1, \\
\left(\frac{\partial}{\partial x} + \frac{x - x_0}{l_B^2}\right)\psi_1 &= \frac{iE}{\hbar v}\psi_2.
\end{aligned}
\tag{2.39}
$$

where

$$x_0 = l_B^2 k_y \tag{2.40}$$

is the coordinate of the center of the electron orbit (Landau, 1930). On introducing a dimensionless coordinate

$$X = \frac{\sqrt{2}}{l_B}(x - x_0) \tag{2.41}$$

and a dimensionless energy (2.26), one can transform Eq. (2.37) to

$$\left(\frac{d^2}{dX^2} + \varepsilon^2 + \frac{1}{2} - \frac{X^2}{2}\right)\psi_1(X) = 0. \tag{2.42}$$

$$\psi_2(X) = -\frac{i}{\varepsilon}\left(\frac{d}{dX} + \frac{X}{2}\right)\psi_1(X). \tag{2.43}$$

We assume in the second equation that $\varepsilon \neq 0$, otherwise

$$\psi_1(X) \sim \exp\left(-\frac{X^2}{4}\right),$$

$$\psi_2(X) = 0.$$

(2.44)

The only solution of Eq. (2.42) vanishing at $X \to -\infty$ (the second one is exponentially growing) is, with an accuracy to within a constant multiplier

$$\psi_1(X) = D_{\varepsilon^2}(-X),$$ (2.45)

where $D_\nu(X)$ is the Weber function (Whittaker & Watson, 1927) and

$$\psi_2(X) = i\varepsilon D_{\varepsilon^2-1}(-X).$$ (2.46)

If the sample is not restricted for both $X \to -\infty$ and $X \to \infty$, the solutions (2.45) and (2.46) are normalizable only for integer ε^2, which again gives us the quantization condition (2.29). For an integer index n, the Weber functions

$$D_n(X) = (-1)^n \exp\left(\frac{X^2}{4}\right) \frac{d^n}{dX^n} \exp\left(-\frac{X^2}{2}\right)$$ (2.47)

decay as $\exp(-X^2/4)$ for $X \to \pm\infty$.

The energy is not dependent on the quantum number k_y or, equivalently, on the position of the center of the orbit x_0. This means that the Landau levels (2.30) have a macroscopically large degeneracy g. To calculate it, it is convenient to use a periodic (Born–von Kármán) boundary condition in the y-direction

$$\psi_{1,2}(x, y) = \psi_{1,2}(x, y + L_y)$$ (2.48)

(for large enough samples the density of states does not depend on boundary conditions [Vonsovsky & Katsnelson, 1989]). Thus,

$$k_y = \frac{2\pi}{L_y} n,$$ (2.49)

where $n = 0, \pm 1, \ldots$ the maximum value of n is determined by the condition that the center of the orbit should be within the sample: $0 < x_0 < L_x$ (L_x is the width of the sample in the x-direction), or

$$|k_y| < \frac{L_x}{l_B^2} = \frac{|e|B}{\hbar c} L_x.$$ (2.50)

Thus, the total number of solutions is

$$g = \frac{|e|\,B}{\hbar c}\frac{L_x L_y}{2\pi} = \frac{|e|\,B}{\hbar c}\frac{A}{2\pi} = \frac{\Phi}{\Phi_0}, \tag{2.51}$$

where $A = L_x L_y$ is the sample area, Φ is the total magnetic flux though the sample, and

$$\Phi_0 = \frac{hc}{|e|} \tag{2.52}$$

is the flux quantum. Keeping in mind further applications to graphene, one should multiply the degeneracy (2.51) by a factor of 4, namely a factor of 2 for the two valleys K and K' and a further factor of 2 for the two spin projections. The latter is possible since the ratio of the Zeeman energy $E_Z = |e|\,\hbar B/(2mc)$ to the cyclotron quantum $\hbar\omega_c$ is always very small (about 0.01 in fields $B \approx 10\text{--}30$ T).

2.3 Topological protection of the zero-energy states

The existence of the zero-energy Landau level is the consequence of one of the most important theorems of modern mathematical physics, the Atiyah–Singer index theorem (Atiyah & Singer, 1968, 1984). This theorem has important applications in quantum field and superstring theories (Kaku, 1988; Nakahara, 1990). In its simplest version, being applied to the operator

$$\hat{H} = v\vec{\hat{\sigma}}\left(-i\hbar\vec{\nabla} - \frac{e}{c}\vec{A}(x,y)\right) \tag{2.53}$$

acting on a torus (that is, with periodic boundary conditions both in the y- and in the x-direction), the theorem states that the *index* of this operator is proportional to the total flux, namely

$$\text{index}\left(\hat{H}\right) = N_+ - N_- = \frac{\Phi}{\Phi_0}, \tag{2.54}$$

for an inhomogeneous magnetic field as well as for a homogeneous one. Here N_+ is the number of solutions with zero energy and positive chirality

$$\hat{H}\psi_1 = 0, \quad \psi_2 = 0, \tag{2.55}$$

and N_- is the number of solutions with zero energy and negative chirality

$$\psi_1 = 0, \quad \hat{H}\psi_2 = 0. \tag{2.56}$$

For the case of a homogeneous magnetic field, $N_+ = g$ is given by Eq. (2.51) and $N_- = 0$. Strictly speaking, we did not consider the case of a torus; instead, we considered periodic boundary conditions in the y-direction only; the case of a torus is analyzed by Tenjinbayashi, Igarashi, and Fujiwara (2007), and the result for the number of zero modes is the same. A simplified (in comparison with the general case) formal discussion of the Atiyah−Singer theorem for the Hamiltonian (2.53) can be found in Katsnelson and Prokhorova (2008).

The index theorem tells us that the zero-energy Landau level is topologically protected; that is, it is robust with respect to possible inhomogeneities of the magnetic field (Novoselov et al., 2005a; Katsnelson, 2007a). This statement is important for real graphene since the effective magnetic field there *should* be inhomogeneous due to the effect of so-called *ripples*, as will be discussed in Chapter 10.

The simplest way (at least for physicists) to understand the robustness of zero-energy modes is to explicitly construct the solutions for zero-energy states in an inhomogeneous magnetic field. This was done by Aharonov and Casher (1979) for the case of an infinite sample with the magnetic flux Φ localized in a restricted region.

Let us assume, first, that the vector potential satisfies the condition

$$\vec{\nabla}\vec{A} = 0; \tag{2.57}$$

otherwise, one can always use the gauge transformation

$$\vec{A} \rightarrow \vec{A} + \vec{\nabla}\chi, \qquad \psi \rightarrow \psi \exp\left(\frac{ie}{\hbar c}\chi\right), \tag{2.58}$$

choosing χ to provide Eq. (2.57). Thus, one can introduce a scalar "potential" $\varphi(x, y)$ such that

$$A_x = -\frac{\partial\varphi}{\partial y}, \quad A_y = \frac{\partial\varphi}{\partial x} \tag{2.59}$$

and, due to Eq. (2.5),

$$B = \nabla^2\varphi. \tag{2.60}$$

Then, Eq. (2.55) and (2.56) can be written in the form

$$\left(\frac{\partial}{\partial x} + i\sigma\frac{\partial}{\partial y} + \frac{ie}{\hbar c}\frac{\partial\varphi}{\partial x} + \frac{\sigma e}{\hbar c}\frac{\partial\varphi}{\partial y}\right)\psi_{1,2} = 0, \tag{2.61}$$

where $\sigma = 1$ and -1 for ψ_1 and ψ_2, respectively. The potential φ can be excluded by the substitution

$$\psi_{1,2} = \exp\left(-\frac{\sigma e}{\hbar c}\varphi\right)f_{1,2}, \tag{2.62}$$

which transforms Eq. (2.61) into the equation

$$\left(\frac{\partial}{\partial x} + i\sigma\frac{\partial}{\partial y}\right)f_{1,2} = 0. \tag{2.63}$$

This means that f_1 and f_2 are analytic and complex-conjugated analytic entire functions of $z = x + iy$, respectively.

Eq. (2.60) has a solution

$$\varphi(\vec{r}) = \int d\vec{r}' G(\vec{r}, \vec{r}') B(\vec{r}'), \tag{2.64}$$

where

$$G(\vec{r}, \vec{r}') = \frac{1}{2\pi} \ln \left(\frac{|\vec{r} - \vec{r}'|}{r_0} \right) \tag{2.65}$$

is the Green function of the Laplace operator in two dimensions (Jackson, 1962), where r_0 is an arbitrary constant. At $r \to \infty$

$$\varphi(r) \approx \frac{\Phi}{2\pi} \ln \left(\frac{r}{r_0} \right) \tag{2.66}$$

and

$$\psi_{1,2}(\vec{r}) = \left(\frac{r_0}{r} \right)^{\frac{\sigma e \Phi}{2\pi \hbar c}} f_{1,2}(\vec{r}), \tag{2.67}$$

where

$$\Phi = \int d\vec{r} \vec{B}(\vec{r}) \tag{2.68}$$

is the total magnetic flux. Since the entire function $f(z)$ cannot go to zero in all directions at infinity, ψ_i can be normalizable only assuming that $\sigma e \Phi > 0$; that is, zero-energy solutions can only exist for one (pseudo)spin direction, depending on the sign of the total flux.

Let us now count how many independent solutions of Eq. (2.63) we have. As a basis, we can choose just polynomials searching the solutions of the form

$$f_1(z) = z^j \tag{2.69}$$

(to be specific, we consider the case $e\Phi > 0$), where $j = 0, 1, 2, \ldots$ One can see from Eq. (2.67) that the solution is integrable with the square, only assuming that $j < N$, where N is the integer part of

$$\frac{e\Phi}{2\pi \hbar c} = \frac{\Phi}{\Phi_0}.$$

Thus, the number of the states with zero energy for one (pseudo)spin projection is equal to N, and there are no such solutions for another spin projection. This agrees with Eq. (2.54).

2.4 Semiclassical quantization conditions and Berry's phase

The exact spectrum (2.30) of Dirac electrons in a uniform magnetic field B seems to be in a contradiction with the Lifshitz–Onsager semiclassical quantization condition (Lifshitz, Azbel, & Kaganov, 1973; Abrikosov, 1988; Vonsovsky & Katsnelson, 1989)

$$S(E_n) = \frac{2\pi \mid e \mid B}{\hbar c}\left(n + \frac{1}{2}\right),$$ (2.70)

where $S(E_n)$ is the area of k-space inside the line determined by the equation

$$E(k_x, k_y) = E_n.$$ (2.71)

For massless Dirac electrons this is just a circle of radius $k(E) = E/(\hbar v)$ and

$$S(E) = \pi \frac{E^2}{(\hbar v)^2},$$ (2.72)

so the term with $\frac{1}{2}$ in Eq. (2.70) should not exist. Strictly speaking, the semiclassical condition (2.70) is only valid for highly excited states, $n \gg 1$; however, for these states it should give us not only the leading, but also the subleading, term correctly, which is not the case now.

The replacement $n \to n + \frac{1}{2}$ follows from the existence of two turning points for a classical periodic orbit; in a more general case, it is related to the so-called Keller–Maslov index. The simplest way to derive it is probably by using the saddle-point approximation in the path-integral formulation of quantum mechanics (Schulman, 1981). It turns out that the case of electrons in single-layer (as well as in bilayer, see later) graphene is special, and, for Dirac fermions, the correct semiclassical condition is

$$S(E_n) = \frac{2\pi \mid e \mid B}{\hbar c}n,$$ (2.73)

which gives us, together with Eq. (2.72), the *exact* spectrum (2.30), including the existence of a zero mode at $n = 0$. Of course, in general, we are not always so lucky, and for the case of bilayer graphene (Section 2.5) the situation is different.

The mystery of the missing term $\frac{1}{2}$ is a good way to introduce one of the deepest concepts of modern quantum mechanics, namely Berry's (or the geometrical) phase (Berry, 1984; Schapere & Wilczek, 1989).

Let us start with the following simple observation. If we rotate the $\vec{k}$ vector by the angle 2π, the wave functions (1.30) change sign:

$$\psi_{e,h}\left(\phi_{\vec{k}} = 2\pi\right) = -\psi_{e,h}\left(\phi_{\vec{k}} = 0\right).$$ (2.74)

This is not surprising when rotating spin $\frac{1}{2}$ in spin space, but we are talking about rotations in real physical space, and our "spin" is just a label for sublattices! This property (2.74) has a deep geometrical and topological meaning.

Berry (1984) considered a general adiabatic evolution of a quantum system. To be specific, we will apply these ideas to the evolution of electron states in $\vec{k}$ space (Zak, 1989; Chang & Niu, 2008; Xiao, Chang, & Niu, 2010).

The Bloch states

$$\psi_{n\vec{k}}(\vec{r}) = u_{n\vec{k}}(\vec{r}) \exp\left(i\vec{k}\vec{r}\right), \tag{2.75}$$

where $u_{n\vec{k}}(\vec{r})$ is the Bloch amplitude periodic in the real space, evolve under the action of external electric and magnetic fields. If they are time-independent, or their time dependence is slow in comparison with typical electron times of the order of $\hbar/W$ (W is the bandwidth), this evolution is mainly within the same band n, with an exponentially small probability of interband transitions (electric or magnetic breakdown; Vonsovsky & Katsnelson, 1989).

By substituting Eq. (2.75) into the Schrödinger equation one can derive the equation for the Bloch amplitude with a slowly varying wave vector $\vec{k}(t)$

$$i\hbar \frac{\partial |u(t)\rangle}{\partial t} = \hat{H}_{\text{eff}}\left(\vec{k}(t)\right)|u(t)\rangle \tag{2.76}$$

(an explicit form of the Hamiltonian H_{eff} is not essential here). The time-dependent band states $\left|n\vec{k}\right\rangle$ satisfy a *stationary* Schrödinger equation

$$\hat{H}_{\text{eff}}\left(\vec{k}\right)\left|n,\vec{k}\right\rangle = E_n\left(\vec{k}\right)\left|n,\vec{k}\right\rangle, \tag{2.77}$$

where $\left|n\vec{k}\right\rangle = u_{n\vec{k}}(\vec{r})$. Neglecting interband transitions, one can try the solution of Eq. (2.76) with an initial condition

$$|u(0)\rangle = \left|n,\vec{k}(0)\right\rangle \tag{2.78}$$

$$|u(t)\rangle = |u(0)\rangle \exp\left\{-\frac{i}{\hbar}\int_0^t dt' E_n\left(\vec{k}(t')\right)\right\} \exp\left\{i\gamma_n(t)\right\}\left|n,\vec{k}(t)\right\rangle. \tag{2.79}$$

On substituting Eq. (2.79) into Eq. (2.76), one finds

$$\frac{\partial \gamma_n(t)}{\partial t} = i\left\langle n,\vec{k}(t)\left|\vec{\nabla}_{\vec{k}}\right|n,\vec{k}(t)\right\rangle \frac{d\vec{k}(t)}{dt}. \tag{2.80}$$

If we consider a periodic motion $\vec{k}(\tau) = \vec{k}(0)$, then, on integrating Eq. (2.80) over the period of motion τ, one finds for the *Berry phase*

$$\gamma_n = i \oint_C d\vec{k} \left\langle n,\vec{k} \middle| \vec{\nabla}_{\vec{k}} \middle| n,\vec{k} \right\rangle = -\mathrm{Im} \oint_C d\vec{k} \left\langle n,\vec{k} \middle| \vec{\nabla}_{\vec{k}} \middle| n,\vec{k} \right\rangle, \qquad (2.81)$$

where C is a line drawn by the end of the vector $\vec{k}(t)$ (the real part of the integral vanishes identically: $2\mathrm{Re} \oint_C d\vec{k} \left\langle n,\vec{k} \middle| \vec{\nabla}_{\vec{k}} \middle| n,\vec{k} \right\rangle = \int d\vec{r} \oint_C d\vec{k} \vec{\nabla}_{\vec{k}} |u_{n\vec{k}}|^2 = 0$). For non-degenerate bands, it is obvious that $\gamma_n = 0$. However, this is not the case for a degenerate spectrum and, in particular, for the case in which conical points exist, like in graphene.

Using Stokes' theorem, Eq. (2.81) can be written in terms of the surface integral over the area, restricted by the contour C:

$$\gamma_n(C) = -\mathrm{Im} \int d\vec{S} \cdot \vec{\nabla}_{\vec{k}} \times \left\langle n,\vec{k} \middle| \vec{\nabla}_{\vec{k}} \middle| n,\vec{k} \right\rangle = -\mathrm{Im} \int d\vec{S} \left\langle \vec{\nabla}_{\vec{k}} n \middle| \times \middle| \vec{\nabla}_{\vec{k}} n \right\rangle \qquad (2.82)$$

with obvious notations, e.g., $\left| \vec{\nabla}_{\vec{k}} n \right\rangle = \vec{\nabla}_{\vec{k}} \left| n,\vec{k} \right\rangle$.

To explicitly demonstrate the role of crossing points of the energy spectrum (such as the conical points in graphene), we introduce, following Berry (1984), the summation over a complete set of eigenstates $|m\rangle$:

$$\left\langle \vec{\nabla}_{\vec{k}} n \middle| \times \middle| \vec{\nabla}_{\vec{k}} n \right\rangle = \sum_m \left\langle \vec{\nabla}_{\vec{k}} n \middle| m \right\rangle \times \left\langle m \middle| \vec{\nabla}_{\vec{k}} n \right\rangle. \qquad (2.83)$$

The term with $m = n$ in Eq. (2.83) is obviously zero and can be omitted since, due to the normalization condition $\langle n|n \rangle = 1$, $\left\langle \vec{\nabla}_{\vec{k}} n \middle| n \right\rangle = -\left\langle n \middle| \vec{\nabla}_{\vec{k}} n \right\rangle$.

On differentiating Eq. (2.77) with respect to $\vec{k}$ one has

$$\vec{\nabla}_{\vec{k}} \hat{H}_{\mathrm{eff}} |n\rangle + \left(\hat{H}_{\mathrm{eff}} - E_n \right) \left| \vec{\nabla}_{\vec{k}} n \right\rangle = \vec{\nabla}_k E_n |n\rangle. \qquad (2.84)$$

On multiplying Eq. (2.84) by $\langle m|$ from the left and taking into account that $\langle m|\hat{H}_{\mathit{eff}} = \langle m|E_m$ and $\langle m| n \rangle = 0$ at $m \neq n$, one finds

$$\left\langle m \middle| \vec{\nabla}_{\vec{k}} n \right\rangle = \frac{\langle m|\vec{\nabla}_{\vec{k}} \hat{H}_{\mathrm{eff}} |n\rangle}{E_n - E_m}. \qquad (2.85)$$

Finally, by substituting Eq. (2.85) into Eq. (2.83) we derive the following expression for the Berry phase:

$$\gamma_n(C) = -\int d\vec{S}\vec{V}_n\left(\vec{k}\right),$$

where

$$\vec{V}_n = \mathrm{Im}\sum_{m\neq n}\frac{\langle n|\vec{\nabla}_{\vec{k}}\hat{H}_{\mathrm{eff}}|m\rangle \times \langle m|\vec{\nabla}_{\vec{k}}\hat{H}_{\mathrm{eff}}|n\rangle}{\left(E_m - E_n\right)^2}. \tag{2.86}$$

This vector is called *Berry curvature*.

Suppose we have two neighboring bands described by the effective Hamiltonian

$$\hat{H}_{\mathrm{eff}} = \frac{1}{2}\vec{R}\left(\vec{k}\right)\vec{\sigma} \tag{2.87}$$

with the eigenenergies

$$E_{\pm}\left(\vec{k}\right) = \pm\frac{1}{2}\left|R\left(\vec{k}\right)\right| \tag{2.88}$$

and the corresponding eigenstates $|\psi_{\pm}\rangle = \frac{1}{\sqrt{2R(R\mp R_z)}}\begin{pmatrix}\pm R_-\\ R\mp R_z\end{pmatrix}$, where $R = |\vec{R}|, R_{\pm} = R_x \pm iR_y$. Assuming $\vec{k} = (k_x, k_y)$, after long but straightforward calculations one finds:

$$\vec{V}(\vec{R}) = \pm\frac{\partial(R_x, R_y)}{\partial(k_x, k_y)}\frac{\vec{R}}{2R^3}, \tag{2.89}$$

where $\frac{\partial(R_x, R_y)}{\partial(k_x, k_y)}$ is the corresponding Jacobian. At last, we make the replacement of variables in the integral (2.86): $\vec{k} \to \vec{R}$. Then, $\frac{\partial(R_x, R_y)}{\partial(k_x, k_y)}d\vec{S} = d\vec{S}_{\vec{R}}$ and the expression for the Berry phase is dramatically simplified:

$$\gamma_{\pm}(C) = \mp\int d\vec{S}_{\vec{R}}\frac{\vec{R}}{2R^3}, \tag{2.90}$$

which is nothing other than the electric flux through the contour C created by the charge $\frac{1}{2}$ at the point $\vec{R} = 0$. The answer is obvious:

$$\gamma_{\pm}(C) = \mp\frac{1}{2}\Omega(C), \tag{2.91}$$

where $\Omega(C)$ is the solid angle of the contour (Fig. 2.1).

For the case of massless Dirac fermions $\vec{R}\left(\vec{k}\right) \sim \vec{k}$ is the two-dimensional vector (k_x, k_y), and the solid angle is 2π, so the Berry phase is $\gamma_{\pm} = \mp\pi$, in agreement with Eq. (2.74).

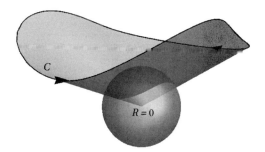

Fig. 2.1 The derivation of Berry's phase (Eq. (2.90)).

As was demonstrated by Kuratsuji and Iida (1985), the Berry phase enters the semiclassical quantization condition. Their approach was based on the path-integral formalism (Schulman, 1981). Here we will present in the simplest way just a general idea of the derivation. Instead of $\vec{k}(t)$ we will consider a general set of slowly varying with time (adiabatic) variables $\vec{x}(t)$.

Let us consider a periodic process with $x_i(\tau) = x_i(0)$. We are interested in calculating the evolution operator

$$\hat{K}(\tau) = \hat{T} \exp\left\{ -\frac{i}{\hbar} \int_0^\tau dt \hat{H}[x_i(t)] \right\},\qquad (2.92)$$

where $\hat{H}$ is the Hamiltonian dependent on $\vec{x}(t)$ and $\hat{T}$ is the time-ordering operator. To calculate the expression (2.92) via a path integral, one has to discretize the time interval, $t_n = n\varepsilon$, where $n = 0, 1, \ldots, N - 1$ and $\varepsilon = \tau/N$ ($N \to \infty$):

$$\hat{K}(\tau) = \mathrm{Tr}\left\{ \exp\left[-\frac{i\varepsilon}{h}\hat{H}(t_0) \right] \exp\left[-\frac{i\varepsilon}{h}\hat{H}(t_1) \right] \ldots \exp\left[-\frac{i\varepsilon}{h}\hat{H}(t_{N-1}) \right] \right\}. \quad (2.93)$$

In the adiabatic approximation, the evolution involves only the transitions between the same states of the Hamiltonian:

$$\hat{K}(\tau) = \sum_n \langle n(t_0)| \exp\left[-\frac{i\varepsilon}{\hbar}\hat{H}(t_0) \right] |n(t_1)\rangle \langle n(t_1)| \exp\left[-\frac{i\varepsilon}{\hbar}\hat{H}(t_1) \right] |n(t_2)\rangle$$

$$\ldots \langle n(t_{N-1})| \exp\left[-\frac{i\varepsilon}{\hbar}\hat{H}(t_{N-1}) \right] |n(t)\rangle.$$

$$(2.94)$$

At $\varepsilon \to 0$, the overlap integral

$$\langle n(t)|n(t+\varepsilon)\rangle \approx \langle n(t)|n(t)\rangle + \varepsilon\frac{d\vec{x}}{dt}\left\langle n(t)|\vec{\nabla}_{\vec{x}}n(t)\right\rangle$$

$$= 1 + \varepsilon\frac{d\vec{x}}{dt}\left\langle n(t)|\vec{\nabla}_{\vec{x}}n(t)\right\rangle \tag{2.95}$$

$$\approx \exp\left[\frac{\varepsilon d\vec{x}}{dt}\left\langle n(t)|\vec{\nabla}_{\vec{x}}n(t)\right\rangle\right]$$

and each term in $\langle n|\ldots|n\rangle$ in Eq. (2.94), apart from the standard dynamical contribution, has an additional phase factor

$$\prod_{n=0}^{N-1}\langle n(t_n)|n(t_{n+1})\rangle = \exp\left[\int_0^\tau dt\frac{d\vec{x}}{dt}\left\langle n|\vec{\nabla}_{\vec{x}}n\right\rangle\right] = \exp\left[i\gamma_n(C)\right] \tag{2.96}$$

(cf. Eq. (2.81)), which leads to the change of the effective action of the system $S \to S + \hbar\gamma$. On repeating a standard derivation of the semiclassical quantization condition, one can see that $n + \frac{1}{2}$ is replaced by $n + \frac{1}{2} - \gamma/(2\pi)$. In particular, for Bloch electrons in a magnetic field, instead of Eq. (2.70), one has

$$S(E_n) = \frac{2\pi|e|B}{\hbar c}\left(n + \frac{1}{2} - \frac{\gamma}{2\pi}\right) \tag{2.97}$$

(Mikitik & Sharlai, 1999). For $\gamma = \pi$ one has the quantization condition (2.73).

Again, we see that anomalous quantization of Landau levels for the case of graphene is related to the nontrivial topological properties of a system with a conical point in its energy spectrum.

This derivation is, however, too schematic; whereas it gives the correct result for the case of massless Dirac fermions, under the condition (2.91), for the massive case, one needs to be more careful. The detailed analysis (Fuchs et al., 2010) shows that in this case, what enters the semiclassical condition is not the full Berry phase (2.90) but only its "topological part" related to the "winding number" (number of rotations of the pseudospin vector under the cycle), and one should still put $\gamma = \pi$ into Eq. (2.97).

2.5 Landau levels in bilayer graphene

Consider now the case of bilayer graphene (McCann & Fal'ko, 2006; Novoselov et al., 2006; McCann, Abergel, & Fal'ko, 2007; Fal'ko, 2008).

Let us start with the simplest Hamiltonian (1.46), which means intermediate energies

$$|t_\perp|\frac{\gamma_3^2}{t^2} \ll |E| \ll |t_\perp|. \tag{2.98}$$

At lower energies (cf. Eq. (1.55)) trigonal warping terms in the Hamiltonian (1.53) become important, and at higher energies all four bands (1.44) become relevant. For realistic parameters, this means energies of the order of tens of meV. Later we will consider a more general case.

On combining Eq. (1.46) with Eq. (2.20) and (2.22) we find the Hamiltonian for the case of a uniform magnetic field:

$$\hat{H} = \hbar\omega_c^* \begin{pmatrix} 0 & \hat{b}^2 \\ \left(\hat{b}^+\right)^2 & 0 \end{pmatrix}, \tag{2.99}$$

where

$$\omega_c^* = \frac{|e|B}{m^*c} \tag{2.100}$$

is the cyclotron frequency for nonrelativistic electrons with effective mass m^*. Then, instead of Eq. (2.25) for single-layer graphene, one has the Schrödinger equation

$$\hat{b}^2 \psi_2 = \varepsilon \psi_1,$$
$$\left(\hat{b}^+\right)^2 \psi_1 = \varepsilon \psi_2, \tag{2.101}$$

where the dimensionless energy ε is introduced now by writing

$$E = \hbar\omega_c^* \varepsilon. \tag{2.102}$$

Again, for the case of valley K' one has to exchange ψ_1 and ψ_2.

First, one can see immediately from Eq. (2.102) that there are zero modes with $\varepsilon = 0$ and $\psi_2 = 0$, and their number is twice as great as for the case of a single layer. Indeed, both the states of the harmonic oscillator with $n = 0$ and those with $n = 1$ satisfy the equation $\hat{b}^2 |\psi\rangle = 0$:

$$\hat{b}|0\rangle = 0, \quad \hat{b}^2|1\rangle = \hat{b}\left(\hat{b}|1\rangle\right) = \hat{b}|0\rangle = 0. \tag{2.103}$$

On multiplying the first of the Eq. (2.101) by $(b^+)^2$ from the left, one finds

$$\left(\hat{b}^+\right)^2 \hat{b}^2 \psi_1 = \varepsilon^2 \psi_1. \tag{2.104}$$

Since

$$\left(\hat{b}^+\right)^2 \hat{b}^2 = \left(\hat{b}^+\hat{b}\right)\left(\hat{b}^+\hat{b} - 1\right) \tag{2.105}$$

we have immediately the spectrum

$$E_n = \pm \hbar \omega_c^* \sqrt{n(n-1)} \tag{2.106}$$

with $n = 0, 1, 2, \ldots$

The counting of the degeneracy of Landau levels (2.106) can be done in exactly the same way as in Section 2.2, and one finds, instead of Eq. (2.51),

$$g_n = \frac{\Phi}{\Phi_0}, \quad n \geq 2, \tag{2.107}$$

and

$$g_0 = \frac{2\Phi}{\Phi_0} \tag{2.108}$$

(the latter follows from the fact that the zero and first levels are degenerate, Eq. (2.103)).

One can prove that Eq. (2.108) follows from the Atiyah–Singer index theorem and remains correct if the magnetic field is inhomogeneous (Katsnelson & Prokhorova, 2008). This fact is quite simple and follows from the property that the index of a product of operators equals the sum of their indices. An explicit construction of zero modes for the Hamiltonian (2.99) that is similar to the Aharonov–Casher construction for the case of the Dirac equation (see Section 2.3) was done by Kailasvuori (2009).

For $n \gg 1$, the spectrum (2.106) is described by the expression

$$|E_n| \approx \hbar \omega_c^* \left(n - \frac{1}{2} \right), \tag{2.109}$$

in agreement with the semiclassical quantization condition

$$S(E_n) = \frac{2\pi |e| B}{\hbar c} \left(n - \frac{1}{2} \right). \tag{2.110}$$

It follows from the general quantization law (2.97) assuming that the Berry phase

$$\gamma = 2\pi. \tag{2.111}$$

This is indeed the case (McCann & Fal'ko, 2006; Novoselov et al., 2006), although the description in terms of the winding number seems to be more accurate (Mañes, Guinea, & Vozmediano, 2007; Katsnelson & Prokhorova, 2008; Park & Marzari, 2011). The Hamiltonian (1.46) has the form (2.87) with

$$\left(R_x, R_y \right) \sim \left(k_x^2 - k_y^2, 2k_x k_y \right)$$

or

$$(R_x + iR_y) \sim (k_x + ik_y)^2. \tag{2.112}$$

It is clear, therefore, that when the vector $\vec{k}$ runs over the closed loop the vector $\vec{R}$ runs over the same loop twice, and the Berry phase should be twice as large as for the case of a single layer. Actually, the Berry phase and the index are proportional; they are both related to the winding number of the vector $\vec{R}$ in the Hamiltonian (Katsnelson & Prokhorova, 2008). For the case of a rhombohedral N-layer system (1.69), the number of zero modes is equal to $N\Phi/\Phi_0$ and the Berry phase is $\gamma = N\pi$.

2.6 The case of bilayer graphene: trigonal warping effects

Consider now the case of small energies

$$|E| \sim \gamma_3^2 \frac{|t_\perp|}{t^2}. \tag{2.113}$$

Thus, the effects of trigonal warping should be taken into account, and one has to proceed with the Hamiltonian (1.53). Instead of the Hamiltonian (2.99) we have for the case of a uniform magnetic field

$$\hat{H} = \hbar\omega_c^* \begin{pmatrix} 0 & \hat{b}^2 + \alpha\hat{b}^+ \\ \left(\hat{b}^+\right)^2 + \alpha\hat{b} & 0 \end{pmatrix}, \tag{2.114}$$

where

$$\alpha = \frac{3\gamma_3 am^*}{\hbar^2}\sqrt{\frac{2\hbar c}{|e|B}} \tag{2.115}$$

is a dimensionless parameter characterizing the role of trigonal warping. The Schrödinger equation (2.101) is modified to the form

$$\left(\hat{b}^2 + \alpha\hat{b}^+\right)\psi_2 = \varepsilon\psi_1,$$
$$\left(\left(\hat{b}^+\right)^2 + \alpha\hat{b}\right)\psi_1 = \varepsilon\psi_2. \tag{2.116}$$

First, let us consider zero modes with $\varepsilon = 0$ and $\psi_1 = 0$. Taking into account that in dimensionless coordinates, (2.40) and (2.41),

$$\hat{b} = -i\left(\frac{\partial}{\partial X} + \frac{X}{2}\right),$$

$$\hat{b}^+ = -i\left(\frac{\partial}{\partial X} - \frac{X}{2}\right), \tag{2.117}$$

the first of the Eq. (2.116) for $\varepsilon = 0$ reads

$$\frac{d^2\psi_2}{dX^2} + (X + i\alpha)\frac{d\psi_2}{dX} + \left(\frac{1}{2} + \frac{X^2}{4} - \frac{iX\alpha}{2}\right)\psi_2 = 0. \tag{2.118}$$

The substitution

$$\psi_2(X) = \exp\left(-\frac{X^2}{4} - \frac{i\alpha}{2}X\right)\varphi(X) \tag{2.119}$$

eliminates the first derivative $\partial/\partial X$ in Eq. (2.118), so

$$\frac{\partial^2}{\partial X^2}\varphi + \left(\frac{\alpha^2}{4} - iX\alpha\right)\varphi = 0. \tag{2.120}$$

At $\alpha = 0$ there are two independent solutions of Eq. (2.120), $\varphi_0 = 1$ and $\varphi_1 = X$. For finite α there are still two solutions, and they can be expressed in terms of Bessel functions of order $\pm\frac{1}{3}$ (Whittaker & Watson, 1927). Anyway, both of the solutions (2.119) vanish at $X \to \pm\infty$ due to the factor $\exp(-X^2/4)$ and, therefore, the number of zero modes remains the same at $\alpha \neq 0$. Obviously, the second of the Eq. (2.116) has no normalizable solutions at $\varepsilon = 0$. These results are not surprising; they are related to a general statement that index(H) is determined solely by the terms with the highest order of derivatives (Katsnelson & Prokhorova, 2008).

To consider the effects of the trigonal warping on other Landau levels, one has to square the Hamiltonian (2.114) or just act by the operator $\left(\left(\hat{b}^+\right)^2 + \alpha\hat{b}\right)$ from the left on the first equation of Eq. (2.116). The result is

$$\hat{L}\psi_2 = \varepsilon^2\psi_2, \tag{2.121}$$

where

$$\hat{L} = \left(\hat{b}^+\hat{b}\right)^2 - (1 - \alpha^2)\hat{b}^+\hat{b} + \alpha\left(\hat{b}^3 + \left(\hat{b}^+\right)^3\right).$$

Using a standard perturbation theory in α one can find a strange result: only the level with $n = 2$ has corrections of the order of α^2

$$\varepsilon_2^2 = 2 - \frac{\alpha^2}{3}, \tag{2.122}$$

whereas the leading corrections to the levels with $n > 2$ are proportional to α^4 and positive.

To qualitatively understand the opposite case of a very large α (or very weak magnetic fields), it is convenient to use the semiclassical approximation (Dresselhaus, 1974). In this regime, one can consider energy levels belonging independently to each of four cones of the spectrum (see Fig. 1.9). The energy

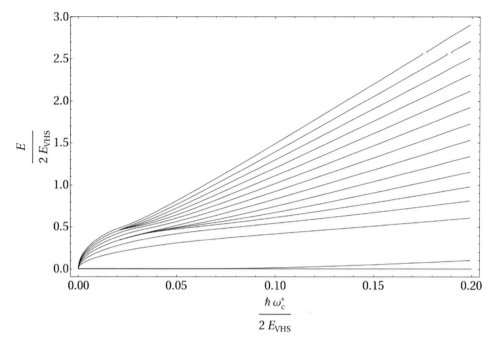

Fig. 2.2 The energy spectrum for bilayer graphene in a magnetic field, with the trigonal warping effects taken into account. Here $\hbar\omega_c^*$ is the cyclotron quantum and E_{VHS} is the energy of the Van Hove singularity at the merging of four conical legs.

level with $n = 2$ tends to zero at $\alpha \to \infty$, since one more zero mode should appear for three independent (in this limit) side cones: the zero mode corresponding to the central cone is associated (for a given direction of the magnetic field) with another valley.

For intermediate α, Eq. (2.116) can be solved numerically (McCann & Fal'ko, 2006; Mayorov et al., 2011a). The results are shown in Fig. 2.2.

Finally, we analyze the effects of trigonal warping on the Berry phase. One can demonstrate by a straightforward calculation (Mikitik & Sharlai, 2008) that each of the three side conical points contributes π to the Berry phase and the central one contributes $-\pi$, so the total Berry phase is $3\pi - \pi = 2\pi$, in agreement with Eq. (2.111). One can also straightforwardly see that the winding number of the transformation

$$(R_x + iR_y) \sim (k_x + ik_y)^2 + \alpha(k_x - ik_y) \tag{2.123}$$

is the same (two) as for Eq. (2.112).

The distribution of the Berry "vector potential" $\vec{\Omega}(\vec{k}) = -i\langle n|\vec{\nabla}_{\vec{k}}|n\rangle$, demonstrating singularities at four conical points is shown in Fig. 2.3.

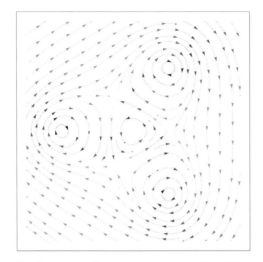

Fig. 2.3 The distribution of the Berry vector potential in bilayer graphene, with the trigonal warping effects taken into account.

2.7 A unified description of single-layer and bilayer graphene

Consider now the case of magnetic fields large enough that

$$|E| \geq |t_\perp|. \tag{2.124}$$

At these energies, a parabolic dispersion transforms to a conical one. Neglecting the trigonal warping and using Eq. (2.20) and (1.43), one has the 4×4 Hamiltonian

$$\hat{H} = \begin{pmatrix} 0 & v\hat{\pi}_+ & t_\perp & 0 \\ v\hat{\pi}_- & 0 & 0 & 0 \\ t_\perp & 0 & 0 & v\hat{\pi}_- \\ 0 & 0 & v\hat{\pi}_+ & 0 \end{pmatrix}. \tag{2.125}$$

Using the operator (2.22) and dimensionless units (2.26) and introducing the notation

$$t_\perp = \Gamma \sqrt{\frac{2|e|hBv^2}{c}}, \tag{2.126}$$

one can represent the Schrödinger equation with the Hamiltonian (2.125) as

$$\hat{b}\,\psi_2 + \Gamma\psi_3 = \varepsilon\psi_1,$$
$$\hat{b}^+\psi_1 = \varepsilon\psi_2,$$
$$\Gamma\psi_1 + \hat{b}^+\psi_4 = \varepsilon\psi_3, \tag{2.127}$$
$$\hat{b}\psi_3 = \varepsilon\psi_4.$$

On excluding ψ_4 and ψ_2 from Eq. (2.127), one obtains

$$\frac{1}{\varepsilon}\hat{b}\,\hat{b}^{+}\psi_1 + \Gamma\psi_3 = \varepsilon\psi_1,$$

$$\Gamma\psi_1 + \frac{1}{\varepsilon}\hat{b}^{+}\hat{b}\psi_3 = \varepsilon\psi_3. \tag{2.128}$$

One can see that ψ_i are eigenfunctions of the operator $\hat{n} = \hat{b}^{+}\hat{b}$ whose eigenvalues are $n = 0, 1, 2, \ldots$ On replacing $\hat{b}^{+}\hat{b}$ by n and $\hat{b}\hat{b}^{+}$ by $n + 1$ in Eq. (2.128) we find the eigenenergies ε_n as

$$\varepsilon_n^2 = \frac{\Gamma^2 + 2n + 1}{2} \pm \sqrt{\left(\frac{\Gamma^2 + 2n + 1}{2}\right)^2 - n(n+1)}. \tag{2.129}$$

This formula (Pereira, Peeters, & Vasilopoulos, 2007) gives a unified description of Landau levels for the cases of both single-layer and bilayer graphene (without trigonal warping effects). On putting $\Gamma = 0$ we come to the case of two independent layers, with

$$\varepsilon_n^2 = n + \frac{1}{2} \pm \frac{1}{2}, \tag{2.130}$$

which exactly coincides with Eq. (2.34). For large Γ (the case of relatively low energies, Eq. (2.98)) we have

$$\varepsilon_{n1}^2 = \frac{n(n+1)}{\Gamma^2} \tag{2.131}$$

and

$$\varepsilon_{n2}^2 = \Gamma^2 + 2n + 1. \tag{2.132}$$

Eq. (2.131) gives the Landau levels for low-lying bands in the parabolic approximation (1.46). The energies

$$\varepsilon_{n2} \approx \pm\left[\Gamma + \frac{1}{\Gamma}\left(n + \frac{1}{2}\right)\right] \tag{2.133}$$

following from Eq. (2.132) are nothing other than the Landau levels for two-gapped bands in the parabolic approximation.

The condition $\Gamma \approx 1$ for which nonparabolic band effects in the Landau-level spectrum of bilayer graphene become very important, corresponds to magnetic fields of the order of

$$B_c \approx \frac{2}{9}\left(\frac{t_\perp}{t}\right)^2 \frac{\hbar c}{|e|a^2} \approx 70\,\text{T},$$

which is too high to be attained in present-day experiments. However, even in fields of $20-30$ T the effects of nonparabolicity should be quite noticeable.

2.8 Magnetic oscillations in single-layer graphene

Magneto-oscillation effects in quantized magnetic fields make possible one of the most efficient ways to probe the electron-energy spectra of metals and semiconductors (Schoenberg, 1984). The basic idea of the oscillations is quite simple: since most of the properties are dependent on what happens in the close vicinity of the Fermi level, whenever, on changing the magnetic induction or chemical potential μ, one of the Landau levels coincides with the Fermi energy, the properties should have some anomalies that repeat periodically as a function of the inverse magnetic field (the latter follows from the semiclassical quantization condition (2.97) $\Delta n \sim (1/B)\Delta E$). These anomalies are smeared by temperature and disorder; so, to observe the oscillations, one needs, generally speaking, low temperatures and clean enough samples. It was the observation of magneto-oscillation effects (Novoselov et al., 2005a; Zhang et al., 2005) that demonstrated the massless Dirac behavior of charge carriers in graphene. Experimentally, oscillations of the conductivity (the Shubnikov–de Haas effect) were studied first; it is more difficult (but quite possible, see later in this section) to observe the oscillations of thermodynamic properties, e.g., magnetization (the de Haas–van Alphen effect) in a single layer of atoms. However, physics of these two effects is just the same, but theoretical treatment of thermodynamic properties can be done in a more clear and rigorous way. Here we will consider, following Sharapov, Gusynin, and Beck (2004), de Haas–van Alphen magnetic oscillations for two-dimensional Dirac fermions, i.e., for single-layer graphene.

The standard expression for the thermodynamic potential of the grand canonical ensemble for noninteracting fermions with energies E_λ is (Landau & Lifshitz, 1980)

$$\Omega = -T \sum_\lambda \left[1 + \exp\left(\frac{\mu - E_\lambda}{T} \right) \right] = -T \int\limits_{-\infty}^{\infty} d\varepsilon N(\varepsilon) \ln \left[1 + \exp\left(\frac{\mu - \varepsilon}{T} \right) \right],$$

(2.134)

where

$$N(\varepsilon) = \sum_\lambda \delta(\varepsilon - E_\lambda)$$

(2.135)

is the density of states. However, one should be careful at this point, since statistical mechanics assumes that the energy spectrum is bounded from below, which is not the case for the Dirac equation. One can either use a complete tight-binding Hamiltonian, where the spectrum is bound, and carefully analyze the limit of the continuum model, or just write the answer from considerations of relativistic invariance (Cangemi & Dunne, 1996). The correct relativistic answer is

$$\Omega = -T \int_{-\infty}^{\infty} d\varepsilon N(\varepsilon) \ln \left[2 \cosh \left(\frac{\varepsilon - \mu}{2T} \right) \right], \tag{2.136}$$

which differs from Eq. (2.134) by the term

$$\Delta\Omega = \frac{1}{2} \int_{-\infty}^{\infty} d\varepsilon N(\varepsilon)(\varepsilon - \mu). \tag{2.137}$$

This term is, in general, infinite and temperature-independent. If the spectrum is symmetric, namely $N(-\varepsilon) = N(\varepsilon)$ (which is necessary for relativistic invariant theories), and the chemical potential is chosen in such a way that $\mu = 0$ for the half-filled case (all hole states are occupied and all electron states are empty), then the correction (2.137) vanishes in that situation.

The expression (2.136) is still not well defined, but its derivatives with respect to μ, temperature, and magnetic field are convergent. For example, the compressibility is proportional to the "thermodynamic density of states"

$$D(\mu) = \frac{\partial n}{\partial \mu} = -\frac{\partial^2 \Omega}{\partial \mu^2} = \int_{-\infty}^{\infty} d\varepsilon N(\varepsilon) \left(-\frac{\partial f(\varepsilon)}{\partial \varepsilon} \right), \tag{2.138}$$

where $f(\varepsilon)$ is the Fermi function

$$-\frac{\partial f(\varepsilon)}{\partial \varepsilon} = \frac{1}{4T \cosh^2 \left(\frac{\varepsilon - \mu}{2T} \right)}, \tag{2.139}$$

and this expression is certainly well defined, with the difference between Eq. (2.134) and (2.136) becoming irrelevant. The quantity (2.138) is directly measurable as the *quantum capacitance* (John, Castro, & Pulfrey, 2004); for the case of graphene, see Ponomarenko et al. (2010), Yu et al. (2013).

At zero temperature, the expression (2.138) is just a sum of delta-functional contributions:

$$D_{T=0}(\mu) = 4 \frac{\Phi}{\Phi_0} \left[\delta(E) + \sum_{v=1}^{\infty} \delta(E - \hbar\omega_c \sqrt{v}) + \delta(E + \hbar\omega_c \sqrt{v}) \right] \tag{2.140}$$

(see Eq. (2.30), (2.31), and (2.51); we have taken into account a factor of 4 due to the valley and spin degeneracy). Using the identities

$$\delta(E - x) + \delta(E + x) = 2|E|\delta(E^2 - x^2), \tag{2.141}$$

$$\delta(E) = \frac{d\Theta(E)}{dE} \tag{2.142}$$

($\Theta(x > 0) = 1$, $\Theta(x < 0) = 0$ is the step function) and

$$\sum_{n=1}^{\infty} \Theta(a - xn) = \Theta(a)\left[-\frac{1}{2} + \frac{a}{x} + \sum_{k=1}^{\infty} \frac{\sin\left(2\pi k\frac{a}{x}\right)}{\pi k}\right], \qquad (2.143)$$

one can find the closed expression

$$D_{T=0}(\mu) = 4\frac{\Phi}{\Phi_0}\,\mathrm{sgn}(\mu)\frac{d}{d\mu}\left\{\frac{\mu^2}{\varepsilon_c^2} + \frac{1}{\pi}\tan^{-1}\left[\cot\left(\frac{2\pi\mu^2}{\varepsilon_c^2}\right)\right]\right\}, \qquad (2.144)$$

where $\varepsilon_c = \hbar\omega_c$ (Sharapov, Gusynin, & Beck, 2004). Eq. (2.143) is the partial case of the Poisson summation formula

$$\sum_{n=-\infty}^{\infty} \delta(x - n) = \sum_{n=-\infty}^{\infty} \exp(2\pi i kx) \qquad (2.145)$$

and, thus,

$$\sum_{n=1}^{\infty} f(n) = \sum_{k=-\infty}^{\infty} \int_{\alpha}^{\infty} dx\, f(x)\exp(2\pi i kx) \qquad (2.146)$$

($0 < a < 1$) for any $f(x)$, and the identity

$$\sum_{n=1}^{\infty} \frac{\sin(\pi nx)}{n} = \tan^{-1}\left(\frac{\sin(\pi x)}{1 - \cos(\pi x)}\right) \qquad (2.147)$$

is used when deriving (2.144).

To consider the case of finite temperatures, it is convenient to use the expansion of $-\partial f(E)/\partial E$ into the Fourier integral:

$$-\frac{\partial f(E)}{\partial E} = \int_{-\infty}^{\infty} \frac{dt}{2\pi}\exp[i(\mu - E)t]R(t), \qquad (2.148)$$

where

$$R(t) = \frac{\pi Tt}{\sinh[(\pi Tt)]}. \qquad (2.149)$$

On substituting Eq. (2.148), together with Eq. (2.141) and (2.142), into the definition (2.138) one finds

$$D(\mu) = 4\frac{\Phi}{\pi\Phi_0}\int\int dE\,dt\,R(t)\exp[i(\mu - E)t]|E|\left[\frac{1}{\varepsilon_c^2} + \frac{2}{\varepsilon_c^2}\sum_{k=1}^{\infty}\cos\left(2\pi k\frac{E^2}{\varepsilon_c^2}\right)\right].$$

$$(2.150)$$

The sum over k describes oscillations of the thermodynamic density of states. To proceed further, one can use the saddle-point method (or "the method of steepest descent") for integrals of strongly oscillating functions (Fedoryuk, 1977). The procedure is as follows. If we have a multidimensional integral

$$I(\lambda) = \int d^n x\, f(x) \exp\left(i\lambda\Phi(x)\right) \tag{2.151}$$

with a large parameter λ, then the main contribution follows from the stationary point x_0 of the phase $\Phi(x)$, where

$$\frac{\partial\Phi}{\partial x_k} = 0, \tag{2.152}$$

since the oscillations are weakest in the vicinity of these points. On expanding $\Phi(x)$ near x_0,

$$\Phi(x) \approx \Phi(x_0) + \frac{1}{2}\sum_{kl}\left(\frac{\partial^2\Phi}{\partial x_k \partial x_l}\right)_0 (x_k - x_{k_0})(x_l - x_{l_0}), \tag{2.153}$$

one finds

$$I(\lambda) \approx f(x_0)\frac{(2\pi)^{n/2}}{\prod_k(-i\mu_k)^{1/2}} \exp\left[i\lambda\Phi(x_0)\right], \tag{2.154}$$

where μ_k are eigenvalues of the matrix

$$\left(\frac{\partial^2\Phi}{\partial x_k \partial x_l}\right).$$

If there is more than one stationary point, their contributions are just summed.

The oscillating part of the expression (2.150) can be estimated by this method, choosing

$$\Phi(E,t) = (\mu - E)t \pm \frac{2\pi k E^2}{\varepsilon_c^2}, \tag{2.155}$$

which gives us immediately

$$E_0 = \mu,$$

$$t_0 = \mp\frac{4\pi k\mu}{\varepsilon_c^2}. \tag{2.156}$$

Finally, one obtains

$$D_{\text{osc}}(\mu) \approx \frac{8A|\mu|}{\pi\hbar^2 v^2}\sum_{k=1}^{\infty}\frac{zk}{\sinh(zk)}\cos\left(\frac{\pi kc\mu^2}{\hbar|e|Bv^2}\right), \tag{2.157}$$

where

$$z = \frac{2\pi^2 Tc|\mu|}{\hbar|e|Bv^2} \qquad (2.158)$$

and A is the sample area. A formal condition of applicability of the saddle-point method is that the resulting oscillations are fast enough; that is, the argument of the cosine in Eq. (2.157) is much larger than 1.

Disorder will broaden Landau levels and smear the delta-functional peaks in the density of states, suppressing the oscillations. This effect, too, can be taken into account (Sharapov, Gusynin, & Beck, 2004; Ponomarenko et al., 2010).

A general semiclassical consideration for an arbitrary energy dispersion law (the Lifshitz–Kosevich theory; see Lifshitz, Azbel, & Kaganov [1973] and Abrikosov [1988]) leads to a similar temperature dependence of the oscillations, with

$$z = \frac{2\pi^2 Tcm^*}{\hbar|e|B}, \qquad (2.159)$$

where

$$m^* = \frac{1}{2\pi} \frac{\partial S(E)}{\partial E}\bigg|_{E=\mu} \qquad (2.160)$$

is the effective cyclotron mass. For the massless Dirac fermions

$$m^* = \frac{|m|}{v^2}, \qquad (2.161)$$

which is nothing other than the famous Einstein relation $E = mc^2$ with a replacement of c by v. For two-dimensional systems $S = \pi k_F^2 \propto n$, where n is the charge-carrier concentration, and, thus, for massless Dirac fermions one can expect

$$m^* \sim \sqrt{n}. \qquad (2.162)$$

The experimental observation of this dependence (Novoselov et al., 2005a; Zhang et al., 2005) was the first demonstration of the reality of massless Dirac fermions in graphene (see Fig. 2.4). This also gives us a value $v \approx 10^6 \text{ms}^{-1} \approx c/300$. Note that what was measured experimentally in these works was the conductivity, not $D(\mu)$, but the temperature dependence should be the same. Oscillations of $D(\mu)$ were measured later via quantum capacitance (Ponomarenko et al., 2010). They are well pronounced even at room temperature (see Fig. 2.5); their broadening is determined by disorder effects.

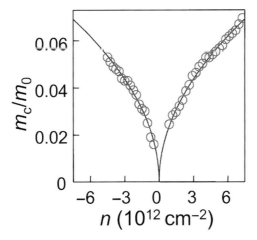

Fig. 2.4 The concentration dependence of the cyclotron mass for charge carriers in single-layer graphene; m_0 is the free-electron mass.
(Reproduced with permission from Novoselov et al., 2005a.)

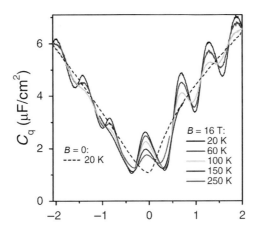

Fig. 2.5 Magnetic oscillations of the quantum capacitance (thermodynamic density of states) as a function of the gate voltage (which is proportional to the charge carrier concentration), for the magnetic field $B = 16$ T and various temperatures.
(Reproduced with permission from Ponomarenko et al., 2010.)

2.9 The anomalous quantum Hall effect in single-layer and bilayer graphene

The anomalous character of the quantum Hall effect in single-layer (Novoselov et al., 2005a; Zhang et al., 2005) and bilayer (Novoselov et al., 2006) graphene is probably the most striking demonstration of the unusual nature of the charge carriers therein. We do not need to present a real introduction to the theory of the quantum Hall effect in general (see Prange & Girvin, 1987). However, it would seem useful to provide some basic information, to emphasise the relation to the Berry phase and the existence of topologically protected zero modes.

If we consider the motion of electrons in the crossed magnetic $(\vec{B})$ and electric $(\vec{E})$ fields, the Lorentz force acting on an electron moving with a velocity $\vec{v}$ is

$$\vec{F} = e\left(\vec{E} + \frac{1}{c}\vec{v} \times \vec{B}\right). \qquad (2.163)$$

In the crossed fields $\vec{B} \parallel Oz$ and $\vec{E} \parallel Oy$, this will result in a steady drift of the electrons along the x-axis with a velocity of

$$v_x = c\frac{E}{B}. \qquad (2.164)$$

This effect results in the appearance of an off-diagonal (Hall) conductivity proportional to the total electron concentration and inversely proportional to the magnetic field:

$$\sigma_{xy} = \frac{nec}{B}. \qquad (2.165)$$

The standard theory of the quantum Hall effect assumes that all the states between Landau levels are localized due to disorder (Anderson localization), see Fig. 2.6. This means that, if the Fermi energy lies between the Landau levels, then only the states belonging to the occupied Landau levels contribute to transport and the Hall conductivity is merely proportional to the number of occupied levels N:

$$\sigma_{xy} = Ng_s g_v \frac{\Phi}{\Phi_0} \frac{1}{A} \frac{ec}{B} = g_s g_v N \frac{e^2}{h}, \qquad (2.166)$$

where g_s and g_v are the spin and valley degeneracy factors (for graphene $g_s = g_v = 2$) and we take into account Eq. (2.51) for the number of states per Landau level. Thus, the Hall conductivity should have plateaux as a function of the electron concentration: it remains constant and integer (in the units of e^2/h per valley per spin) when we pass from one occupied Landau level to the next one.

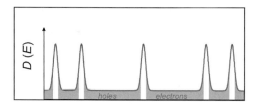

Fig. 2.6 A sketch of the density of states under quantum Hall effect conditions in graphene. The zero-energy Landau level separates electron and hole states and is equally shared by electrons and holes. Regions of localized and extended states are shown in gray and white, respectively.

However, in the case of graphene the zero-energy Landau level is equally shared by electrons and holes. This means that when counting only electrons ($\mu > 0$) or only holes ($\mu < 0$) it contains half as many states as do all other Landau levels. Thus, instead of Eq. (2.166), one has (Schakel, 1991; Gusynin & Sharapov, 2005; Novoselov et al., 2005a; Zhang et al., 2005; Castro Neto, Guinea, & Peres, 2006)

$$\sigma_{xy} = g_s g_v \left(N + \frac{1}{2} \right) \frac{e^2}{h}. \qquad (2.167)$$

This is exactly the behavior observed experimentally (the *half-integer quantum Hall effect*). For the case of bilayer graphene, the zero-energy level contains twice as many states as for single-layer graphene, and the quantum Hall effect is integer, but, in contrast with the case of a conventional electron gas, there is no plateau at zero Fermi energy (Novoselov et al., 2006). These two cases are shown in Fig. 2.7.

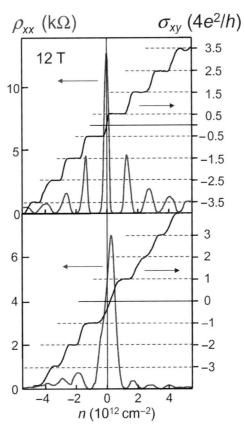

Fig. 2.7 The resistivity and Hall conductivity as functions of the charge-carrier concentration in single-layer (top) and bilayer (bottom) graphene.
(Reproduced with permission from Novoselov et al., 2005a [top] and Novoselov, 2006 [bottom].)

Thus, the anomalous quantum Hall effect in graphene is related to the existence of zero-energy modes and, thus, to the Atiyah–Singer theorem.

Further understanding of geometrical and topological aspects of the anomalies can be attained within an approach developed by Thouless et al. (1982); see also Kohmoto (1985, 1989), Hatsugai (1997). The main observation is that the Hall conductivity can be represented in a form very similar to that for the Berry phase. Actually, the work by Thouless et al. (1982) was done before that by Berry (1984); the relation under discussion has been emphasized by Simon (1983).

Let us consider, again, a general two-dimensional electron system in a periodic potential plus uniform magnetic field (Section 2.1). We will prove later (Section 13.6) that, if the flux per elementary cell is rational (in units of the flux quantum), the eigenstates of this problem can be rigorously characterized by the wave vector $\vec{k}$ and considered as Bloch states in some supercell (for a formal discussion, see Kohmoto, 1985). We will label them as $|\lambda\rangle = |n\vec{k}\rangle$, where n is the band index.

We will use a linear response theory leading to a so-called Kubo formula (Kubo, 1957). The Hall effect was first considered in this way by Kubo, Hasegawa, and Hashitsume (1959); for a detailed derivation and discussions, see Ishihara (1971) and Zubarev (1974). For the single-electron case it can be essentially simplified.

Let A be a one-electron operator that can be represented in a secondary quantized form as

$$\hat{A} = \sum_{12} A_{12} \hat{c}_1^+ \hat{c}_2 \tag{2.168}$$

(here the numerical indices will label electron states in some basis; $\hat{c}_i^+$ and $\hat{c}_i$ are fermionic creation and annihilation operators). Thus, its average over an arbitrary state is

$$\langle \hat{A} \rangle = \sum_{12} A_{12} \langle \hat{c}_1^+ \hat{c}_2 \rangle = Tr(\hat{A}\hat{\rho}), \tag{2.169}$$

where

$$\rho_{21} = \langle \hat{c}_1^+ \hat{c}_2 \rangle \tag{2.170}$$

is the single-electron density matrix. For noninteracting electrons, the Hamiltonian of the system has the same form:

$$\hat{H} = \sum_{12} H_{12} \hat{c}_1^+ \hat{c}_2, \tag{2.171}$$

and, using the commutation relation

$$\left[\hat{c}_1^+ \hat{c}_2, \hat{c}_3^+ \hat{c}_4 \right] = \delta_{23} \hat{c}_1^+ \hat{c}_4 - \delta_{14} \hat{c}_3^+ \hat{c}_2, \tag{2.172}$$

one can prove that the density matrix $\hat{\rho}$ satisfies the communication relations

$$i\hbar \frac{\partial \hat{\rho}}{\partial t} = \left[\hat{H}, \hat{\rho} \right], \tag{2.173}$$

where the matrix multiplication is performed in the single-particle space, e.g.,

$$\left(\hat{H}\hat{\rho} \right)_{12} = \sum_{3} H_{13}\rho_{32}. \tag{2.174}$$

Let $\hat{H}(t) = \hat{H}_0 + \hat{V}(t)$, where $\hat{H}_0$ is diagonal (E_i are its eigenenergies) and $\hat{V}(t)$ is a small perturbation depending on time as $\exp(-i\omega t + \delta t)|_{\delta \to +0}$. Then, the correction to the density matrix, $\hat{\rho}' \exp(-i\omega t + \delta t)$ is given by the expression (see Vonsovsky & Katsnelson, 1989)

$$\rho'_{12} = \frac{f_1 - f_2}{E_2 - E_1 + \hbar(\omega + i\delta)} V_{12}, \tag{2.175}$$

where $f_i = f(E_i)$ is the Fermi function and the perturbation of an observable A is $\delta A \exp(-i\omega t + \delta t)$, where

$$\delta A = \text{Tr}\left(\hat{A}\hat{\rho}' \right) = \sum_{12} \frac{f_1 - f_2}{E_2 - E_1 + \hbar(\omega + i\delta)} V_{12}A_{21}. \tag{2.176}$$

To calculate the Hall conductivity one has to consider a perturbation

$$V = -e\vec{r}\vec{E}, \tag{2.177}$$

where $\vec{E}$ is the electric field, the coordinate operator is

$$\vec{r} = i\vec{\nabla}_{\vec{k}} \tag{2.178}$$

(see Vonsovsky & Katsnelson, 1989), and the current operator is

$$\hat{\vec{j}} = e\frac{d\hat{\vec{r}}}{dt} = \frac{ie}{\hbar}\left[\hat{H}, \hat{\vec{r}} \right]. \tag{2.179}$$

Using the identity (2.84) and restricting ourselves to the static case only ($\omega = \eta = 0$), one finds, for the case $T = 0$

$$\sigma_{\text{H}} = -\frac{2e^2}{A\hbar} \text{Im} \int \frac{d\vec{k}}{(2\pi)^2} \sum_{E_m < \mu} \sum_{E_n > \mu} \frac{\langle m|\partial H/\partial k_x|n\rangle \langle n|\partial H/\partial k_y|m\rangle}{(E_n - E_m)^2}, \tag{2.180}$$

where the integral is taken over the Brillouin zone of the magnetic supercell; we remind that A is the sample area. This is exactly the same expression as in

Eq. (2.86), and, thus, as in Eq. (2.82). Using Stokes' theorem one can represent Eq. (2.180) as a contour integral over the boundary of the Brillouin zone:

$$\sigma_H = -\frac{e^2}{2\pi h}\operatorname{Im}\sum_n^{occ}\oint d\vec{k}\langle n|\vec{\nabla}_{\vec{k}}|n\rangle, \qquad (2.181)$$

where the sum is taken over all occupied bands. The contour integral gives us the change of the phase of the state $|n\rangle$ when rotating by 2π in $\vec{k}$-space. If all the states are topologically trivial (i.e., there is no Berry phase), all these changes should be integer (in the units of 2π), and, thus, Eq. (2.181) gives us the quantization of the Hall conductivity (2.166). In the case of graphene, the Berry phase π should be added, which changes the quantization condition to Eq. (2.167). Of course, this is just an explanation and not derivation: One also needs to prove that Berry phase π enters the integral in Eq.(2.181) an odd number of times; this fact was confirmed by straightforward calculations by Watanabe, Hatsugai, and Aoki (2010).

The real situation is more complicated since the consideration by Thouless et al. (1982) does not take into account disorder effects, in particular, Anderson localization, which are actually crucial for a proper understanding of the quantum Hall effect. A more complete mathematical theory requires the use of noncommutative geometry (Bellissard, van Elst, & Schulz-Baldes, 1994) and is too complicated to review here. Just to make this statement not completely esoteric one has to refer to the properties of the operators (2.15) describing translations in the presence of the magnetic field. Generally speaking, they do not commute (see, e.g., Eq. (2.18)), and noncommutative translations generate noncommutative geometry.

Keeping in mind the case of graphene, it was demonstrated by Ostrovsky, Gornyi, and Mirlin (2008) that, actually, the quantum Hall effect can be either anomalous (half-integer) or normal (integer) depending on the type of disorder. Short-range scatterers induce a strong mixture of the states from different valleys and restore the ordinary (integer) quantum Hall effect. Of course, this is beyond the "Dirac" physics, which is valid assuming that the valleys are essentially independent.

The cyclotron quantum (2.31) in graphene is much higher than in most semiconductors. The energy gap between the Landau levels with $n = 0$ and $n = 1$ is $\Delta E \approx 2{,}800$ K for the largest currently available permanent magnetic fields, $B = 45$ T ($\Delta E \approx 1{,}800$ K for $B = 20$ T). This makes graphene a unique system exhibiting the quantum Hall effect at room temperature (Novoselov et al., 2007).

Here we discuss only the background to quantum Hall physics in graphene. The real situation is much more complicated, both theoretically (involving the role of disorder and electron–electron interactions) and experimentally (Zhang et al., 2006; Giesbers et al., 2007; Jiang et al., 2007b; Checkelsky, Li, & Ong, 2008;

Giesbers et al., 2009). In particular, at high enough magnetic fields the spin and, probably, valley degeneracies are destroyed and additional plateaux appear, in addition to the fact that the gap opens at $n = 0$. The nature of these phenomena is still controversial. Last, but not least, the fractional quantum Hall effect has been observed for freely suspended graphene samples (Bolotin et al., 2009; Du et al., 2009). This is an essentially many-body phenomenon (Prange & Girvin, 1987). We will come back to the physics of the quantum Hall effect in graphene many times in this book.

2.10 Effects of smooth disorder and an external electric field on the Landau levels

In reality, all Landau levels are broadened due to disorder. If the latter can be described by a scalar potential $V(x, y)$ that is smooth and weak enough, the result will just be a modulation of the Landau levels by this potential (Prange & Girvin, 1987)

$$E_\nu(x, y) \approx E_\nu + V(x, y). \tag{2.182}$$

The weakness means that

$$|V(x, y)| \ll \hbar\omega_c, \tag{2.183}$$

and the smoothness means that a typical spatial scale of $V(x, y)$ is large in comparison with the magnetic length (2.1). The calculations for the case of graphene are especially simple and transparent if one assumes a one-dimensional modulation, such that V is dependent only on the y-coordinate (Katsnelson & Novoselov, 2007). Thus, instead of Eq. (2.37) one has

$$\left(\frac{\partial}{\partial x} - \frac{x}{l_B^2} - i\frac{\partial}{\partial y}\right)\psi_2 = \frac{iE}{\hbar v}\psi_1 - \frac{iV(y)}{\hbar v}\psi_1,$$

$$\left(\frac{\partial}{\partial x} + \frac{x}{l_B^2} + i\frac{\partial}{\partial y}\right)\psi_1 = \frac{iE}{\hbar v}\psi_2 - \frac{iV(y)}{\hbar v}\psi_2. \tag{2.184}$$

We can try the solutions of Eq. (2.184) as an expansion in the basis of the solutions (2.45) of the unperturbed problem ($V = 0$):

$$\psi_i(x, y) = \sum_{n=0}^{\infty} \int_{-\infty}^{\infty} \frac{dk_y}{2\pi} c_n^{(i)}(k_y) \exp\left(ik_y y\right) A_n D_n\left(\frac{\sqrt{2}\left(x - l_B^2 k_y\right)}{l_B}\right), \tag{2.185}$$

where A_n is the normalization factor (the basis functions are supposed to be normalized with respect to unity).

After straightforward calculations, one obtains a set of equations for the expansion coefficients $c_n^{(i)}(k_y)$:

$$-\frac{\sqrt{2}}{l_B}(1-\delta_{n,0})c_n^{(2)}(k_y)=\frac{iE}{\hbar v}c_n^{(1)}(k_y)-\frac{i}{\hbar v}\sum_{n'=0}^{\infty}\int_{-\infty}^{\infty}\frac{dq_y}{2\pi}v(k_y-q_y)c_{n'}^{(1)}(q_y)\left\langle n,k_y|n',q_y\right\rangle$$

$$\frac{\sqrt{2}}{l_B}(1+n)c_n^{(1)}(k_y)=\frac{iE}{\hbar v}c_n^{(2)}(k_y)-\frac{i}{\hbar v}\sum_{n'=0}^{\infty}\int_{-\infty}^{\infty}\frac{dq_y}{2\pi}v(k_y-q_y)c_{n'}^{(2)}(q_y)\left\langle n,k_y|n',q_y\right\rangle$$

$$(2.186)$$

where $v(q)$ is a Fourier component of $V(y)$

$$\left\langle n,k_y|n',q_y\right\rangle = A_nA_{n'}\int_{-\infty}^{\infty}dxD_n\left(\frac{\sqrt{2}(x-l_B{}^2k_y)}{l_B}\right)D_{n'}\left(\frac{\sqrt{2}(x-l_B{}^2q_y)}{l_B}\right).$$

$$(2.187)$$

If the potential is smooth and weak enough, one can use the adiabatic approximation and neglect the terms with $n'\neq n$ in Eq. (2.186) describing transitions between the Landau levels. Then,

$$-(1-\delta_{n,0})\tilde{c}_n^{(2)}(k_y) = i\varepsilon c_n^{(1)}(k_y) - i\int_{-\infty}^{\infty}\frac{dq_y}{2\pi}v(k_y-q_y)\left\langle n,k_y|n,q_y\right\rangle c_n^{(1)}(q_y)$$

$$nc_n^{(1)}(k_y) = i\varepsilon\tilde{c}_n^{(2)}(k_y) - i\int_{-\infty}^{\infty}\frac{dq_y}{2\pi}v(k_y-q_y)\left\langle n,k_y|n,q_y\right\rangle\tilde{c}_n^{(2)}(q_y),\quad (2.188)$$

where $\tilde{c}_n^{(2)} = c_{n-1}^{(2)}$, and we use a dimensionless energy (2.26). For $n=0$, the components 1 and 2 are decoupled and we have

$$\varepsilon c(k_y) = \int_{-\infty}^{\infty}\frac{dq_y}{2\pi}v(k_y-q_y)\exp\left[-\frac{l_B^2}{4}(k_y-q_y)^2\right]c(q_y),\quad (2.189)$$

where c is either $c_0^{(1)}$ or $\tilde{c}_0^{(2)}$ and we calculate explicitly $\langle 0,k_y|0,q_y\rangle$.
   Coming back to real space,

$$c(k_y) = \int_{-\infty}^{\infty}dy\exp\left(-ik_yy\right)c(k_y),\quad (2.190)$$

one can transform Eq. (2.189) to the form

$$\left(\varepsilon - \tilde{V}(y)\right)c(y) = 0, \tag{2.191}$$

where

$$\tilde{V}(y) = \frac{1}{\hbar\omega_c}\int\limits_{-\infty}^{\infty}\frac{dq_y}{2\pi}v\left(q_y\right)\exp\left[-\frac{l_B^2 q_y^2}{4}+iq_y y\right] = \frac{1}{\hbar\omega_c}\int\limits_{-\infty}^{\infty}dy'V(y')\frac{1}{\sqrt{\pi}l_B}\exp\left[-\frac{(y-y')^2}{l_B^2}\right] \tag{2.192}$$

is a convolution of the potential $V(y)$ with the ground-state probability density of a harmonic oscillator. If the potential is smooth in comparison with l_B, then $\tilde{V}(y) \approx V(y)$.

Eq. (2.191) has solutions

$$c(y) = \delta(y - Y), \\ \varepsilon = V(Y), \tag{2.193}$$

which means that the zero-energy Landau level broadens via just a modulation by the scalar potential. However, a random *vector* potential does not broaden the zero-energy level, due to the index theorem (Section 2.3). All other Landau levels are broadened both by scalar and by vector potentials. For a scalar potential only, one has in general

$$E_n(Y) \pm \frac{\hbar v_F}{l_B}\sqrt{2n} = V(Y). \tag{2.194}$$

There is some experimental evidence that the zero-energy Landau levels in graphene are narrower than the other ones (Giesbers et al., 2007). The most natural explanation is that there exist random pseudomagnetic fields in graphene due to ripples (corrugations; Morozov et al., 2006). The origin of these pseudomagnetic fields will be discussed later, in Chapter 10.

For the case of a constant electric field E

$$V(x) = -eEx, \tag{2.195}$$

the problem has a beautiful, exact solution that is based on relativistic invariance of the Dirac equation (Lukose, Shankar, & Baskaran, 2007). The Lorentz transformation

$$y' = \frac{y - \beta vt}{\sqrt{1 - \beta^2}}, \quad t' = \frac{t - \beta y/v}{\sqrt{1 - \beta^2}}, \tag{2.196}$$

corresponding to the coordinate system moving with the velocity βv, with $\beta < 1$ (we remind the reader that for our Dirac equation v plays the role of the velocity of light), changes the electric field $\vec{E}\|Oy$ and magnetic field $\vec{B}\|Oz$ according to (Jackson, 1962)

$$E' = \frac{E - \beta\frac{vB}{c}}{\sqrt{1 - \beta^2}}, \qquad \frac{vB'}{c} = \frac{\frac{vB}{c} - \beta E}{\sqrt{1 - \beta^2}}. \tag{2.197}$$

This means that, if the electric field is weak enough

$$E < \frac{v}{c}B, \tag{2.198}$$

it can be excluded by the Lorentz transformation with

$$\beta^* = \frac{cE}{vB}. \tag{2.199}$$

In the opposite case

$$E > \frac{v}{c}B,$$

one can, vice versa, exclude the magnetic field, see Shytov et al. (2009).

Thus, the effective magnetic field is

$$B_{\mathrm{eff}} = B\sqrt{1 - \beta^{*2}}. \tag{2.200}$$

As a result, the energy spectrum of the problem is (Lukose, Shankar, & Baskaran, 2007)

$$E_n(k_y) = \pm\hbar\omega_c\sqrt{n}\left(1 - \beta^{*2}\right)^{3/4} - \hbar v\beta^* k_y. \tag{2.201}$$

The distances between Landau levels are decreased by the factor $(1 - \beta^{*2})^{3/4}$. The last term in Eq. (2.201) (as well as the additional factor $\sqrt{1 - \beta^{*2}}$ in the first term) is nothing other than the result of Lorentz transformation of energy and momentum. It transforms the Landau *levels* into Landau *bands*, in qualitative agreement with Eq. (2.194).

3

Quantum transport via evanescent waves

3.1 *Zitterbewegung* as an intrinsic disorder

The Berry phase, the existence of a topologically protected zero-energy level and the anomalous quantum Hall effect are striking manifestations of the peculiar, "ultrarelativistic" character of charge carriers in graphene.

Another amazing property of graphene is the finite minimal conductivity, which is of the order of the conductance quantum e^2/h per valley per spin (Novoselov et al., 2005a; Zhang et al., 2005). Numerous considerations of the conductivity of a two-dimensional massless Dirac fermion gas do give us this value of the minimal conductivity with an accuracy of some factor of the order of one (Fradkin, 1986; Lee, 1993; Ludwig *et al.,* 1994; Nersesyan, Tsvelik, & Wenger, 1994; Shon & Ando, 1998; Ziegler, 1998; Gorbar et al., 2002; Yang & Nayak, 2002; Katsnelson, 2006a; Tworzydlo et al., 2006; Ryu et al., 2007).

It is really surprising that in the case of massless two-dimensional Dirac fermions there is a finite conductivity for an *ideal* crystal, that is, in the absence of any scattering processes (Ludwig et al., 1994; Katsnelson, 2006a; Tworzydlo et al., 2006; Ryu et al., 2007). This was first noticed by Ludwig et al. (1994) using a quite complicated formalism of conformal field theory (see also a more detailed and complete discussion in Ryu et al., 2007). After the discovery of the minimal conductivity in graphene (Novoselov et al., 2005a; Zhang et al., 2005), I was pushed by my experimentalist colleagues to give a more transparent physical explanation of this fact, which has been done in Katsnelson (2006a) on the basis of the concept of *Zitterbewegung* (Schrödinger, 1930) and the Landauer formula (Beenakker & van Houten, 1991; Blanter & Büttiker, 2000). The latter approach was immediately developed further and used to calculate the shot noise (Tworzydlo et al., 2006), which turns out to be similar to that in *strongly disordered* metals (a "pseudodiffusive transport"). There are now more theoretical (Prada et al., 2007; Katsnelson & Guinea, 2008; Rycerz, Recher, & Wimmer, 2009; Schuessler et al.,

2009; Katsnelson, 2010a) and experimental (Miao et al., 2007; Danneau et al., 2008; Mayorov et al., 2011a) works studying this regime in the context of graphene. This situation is very special. For a conventional electron gas in semiconductors, in the absence of disorder, the states with definite energy (eigenstates of the Hamiltonian) can simultaneously be the states with definite current (eigenstates of the current operator), and it is the disorder that results in the nonconservation of the current and finite conductivity. In contrast, for the Dirac fermions the current operator does not commute with the Hamiltonian (*Zitterbewegung*), which can be considered as a kind of intrinsic disorder (Katsnelson, 2006a; Auslender & Katsnelson, 2007). Therefore, a more detailed understanding of the pseudodiffusive transport in graphene is not only important for physics of graphene devices but also has a great general interest for quantum statistical physics and physical kinetics.

The *Zitterbewegung* is a quantum relativistic phenomenon that was first discussed by Schrödinger as early as in 1930 (Schrödinger, 1930). Only very recently was it observed experimentally for trapped ions (Gerritsma et al., 2010). This phenomenon seems to be important if one wishes to qualitatively understand the peculiarities of electron transport in graphene at its small doping (Katsnelson, 2006a; Auslender & Katsnelson, 2007). Other aspects of the *Zitterbewegung* in graphene physics, in particular, possibilities for its direct experimental observation, are discussed by Cserti and Dávid (2006) and Rusin and Zawadzki (2008, 2009). Here we will explain this basic concept for the case of two-dimensional massless Dirac fermions. In a secondary quantized form, the Dirac Hamiltonian reads

$$\hat{H} = v \sum_{\vec{p}} \widehat{\Psi}_{\vec{p}}^{+} \vec{\sigma} \vec{p} \hat{\Psi}_{\vec{p}} \equiv \sum_{\vec{p}} \widehat{\Psi}_{\vec{p}}^{+} \hat{h}_{\vec{p}} \hat{\Psi}_{\vec{p}}, \tag{3.1}$$

and the corresponding expression for the current operator is

$$\hat{\vec{j}} = ev \sum_{\vec{p}} \widehat{\Psi}_{\vec{p}}^{+} \vec{\sigma} \hat{\Psi}_{\vec{p}} \equiv \sum_{\vec{p}} \hat{\vec{j}}_{\vec{p}}, \tag{3.2}$$

where $\vec{p}$ is the momentum and $\widehat{\Psi}_{\vec{p}}^{+} = \left(\widehat{\Psi}_{\vec{p}1}^{+}, \widehat{\Psi}_{\vec{p}2}^{+} \right)$ are pseudospinor electron-creation operators. The expression (3.2) follows from Eq. (3.1) and the gauge invariance, which requires (Abrikosov, 1998)

$$\hat{\vec{j}}_{\vec{p}} = e \frac{\delta \hat{h}_{\vec{p}}}{\delta \vec{p}}. \tag{3.3}$$

Here we omit spin and valley indices (so, keeping in mind applications to graphene, the results for the conductivity should be multiplied by 4, due to there being two spin projections and two conical points per Brillouin zone).

Straightforward calculations for the time evolution of the electron operators give $\Psi(t) = \exp\left(i\hat{H}t\right)\Psi\exp\left(-i\hat{H}t\right)$ (here we will put $\hbar = 1$)

$$\hat{\Psi}_{\vec{p}}(t) = \frac{1}{2}\left\{\left[\exp\left(-i\varepsilon_{\vec{p}}t\right)\right]\left(\frac{1 + \vec{p}\sigma}{p}\right) + \left[\exp\left(i\varepsilon_{\vec{p}}t\right)\right]\left(\frac{1 - \vec{p}\sigma}{p}\right)\right\}\hat{\Psi}_{\vec{p}} \qquad (3.4)$$

and for the current operator

$$\hat{\vec{j}}(t) = \hat{\vec{j}}_0(t) + \hat{\vec{j}}_1(t) + \hat{\vec{j}}_1^+(t)$$

$$\hat{\vec{j}}_0(t) = ev\sum_{\vec{p}}\hat{\Psi}_{\vec{p}}^+\frac{\vec{p}(\vec{p}\sigma)}{p^2}\hat{\Psi}_{\vec{p}}$$

$$\hat{\vec{j}}_1(t) = \frac{ev}{2}\sum_{\vec{p}}\hat{\Psi}_{\vec{p}}^+\left[\frac{\sigma - \vec{p}(\vec{p}\sigma)}{p^2} + \frac{i}{p}\sigma\times\vec{p}\right]\hat{\Psi}_{\vec{p}}\exp\left(2i\varepsilon_{\vec{p}}t\right), \qquad (3.5)$$

where $\varepsilon_{\vec{p}} = vp$ is the particle energy. The last term in Eq. (3.5) corresponds to the *Zitterbewegung*.

Its physical interpretation is usually given in terms of the Landau–Peierls generalization of the Heisenberg uncertainty principle (Landau & Peierls, 1931; Berestetskii, Lifshitz, & Pitaevskii, 1971; Davydov, 1976). Attempts to measure the coordinate of a relativistic particle with a very high accuracy require an amount of energy that is sufficient to create particle–antiparticle pairs and, thus, we will inevitably lose our initial particle, being unable to distinguish it from one of the created particles (according to quantum statistics, all the particles are equivalent). This pair creation corresponds to the oscillating terms with frequency $2\varepsilon_{\vec{p}}$ in Eq. (3.5).

In terms of condensed-matter physics, the *Zitterbewegung* is nothing other than a special kind of interband transition with the creation of virtual electron–hole pairs. The unitary transformation generated by the operator (1.29) diagonalizes the Hamiltonian and thus introduces electron and hole states with the energies $\pm vp$; after this transformation the oscillating term in Eq. (3.5) obviously corresponds to the interband transitions, e.g.,

$$U_{\vec{p}}^+ j_{\vec{p}}^x U_{\vec{p}} = ev\begin{pmatrix} -\cos\phi_{\vec{p}} & -i\sin\phi_{\vec{p}}\exp\left(-i\phi_{\vec{p}} + 2i\varepsilon_{\vec{p}}t\right) \\ i\sin\phi_{\vec{p}}\exp\left(i\phi_{\vec{p}} - 2i\varepsilon_{\vec{p}}t\right) & \cos\phi_{\vec{p}} \end{pmatrix}. \qquad (3.6)$$

To calculate the conductivity $\sigma(\omega)$ one can first try to use the Kubo formula (Kubo, 1957), which reads, for the two-dimensional isotropic case

$$\sigma(\omega) = \frac{1}{A} \int\limits_0^\infty dt \exp{(i\omega t)} \int\limits_0^\beta d\lambda \left\langle \hat{\vec{j}}(t - i\lambda)\hat{\vec{j}} \right\rangle, \tag{3.7}$$

where $\beta = T^{-1}$ is the inverse temperature and A is the sample area. In the static limit $\omega = 0$, taking into account the Onsager relations and the analyticity of the correlators $\left\langle \hat{\vec{j}}(z)\hat{\vec{j}} \right\rangle$ for $-\beta < \mathrm{Im} \ z \leq 0$ (Zubarev, 1974), one has

$$\sigma = \frac{\beta}{2A} \int\limits_{-\infty}^\infty dt \left\langle \hat{\vec{j}}(t)\hat{\vec{j}} \right\rangle. \tag{3.8}$$

Usually, for ideal crystals, the current operator commutes with the Hamiltonian and thus $\vec{j}(t)$ does not depend on time. In that case, due to Eq. (3.7), the frequency-dependent conductivity in the ground state contains only the Drude peak

$$\sigma_D(\omega) = \frac{\pi}{A} \lim_{T \to 0} \frac{\left\langle \hat{\vec{j}}^2 \right\rangle}{T} \delta(\omega). \tag{3.9}$$

Either the spectral weight of the Drude peak is finite and, thus, the static conductivity is infinite, or it is equal to zero. It is easy to check that for the system under consideration, the spectral weight of the Drude peak is proportional to the modulus of the chemical potential $|\mu|$ and thus vanishes at zero doping ($\mu = 0$). It is the *Zitterbewegung*, i.e., the oscillating term $\hat{\vec{j}}_1(t)$, which is responsible for the non-trivial behavior of the conductivity for zero temperature and zero chemical potential. A straightforward calculation gives the formal result

$$\sigma = \frac{\pi e^2}{h} \int\limits_0^\infty d\varepsilon\varepsilon\delta^2(\varepsilon), \tag{3.10}$$

where one delta-function originates from the integration over t in Eq. (3.8) and the second one from the derivative of the Fermi distribution function appearing in the calculation of the average over the product of Fermi operators. Of course, the square of the delta-function is not a well-defined object, and thus Eq. (3.10) is meaningless before specification of how one should regularize the delta-functions. After regularization, the integral in Eq. (3.10) is finite, but its value depends on the regularization procedure (for a detailed discussion of this uncertainty, see Ryu et al., 2007). Although this derivation cannot give us a correct numerical factor, it opens a new path to qualitative understanding of more complicated situations. For example, the minimal conductivity of the order of e^2/h per channel has been observed experimentally also for bilayer graphene (Novoselov et al., 2006), with an energy spectrum drastically different from that for the single-layer case. Bilayer graphene is a zero-gap semiconductor with *parabolic* touching of the electron and hole bands described by the single-particle Hamiltonian (1.46). The Hamiltonian can be diagonalized by the unitary transformation $U_{\vec{p}}$ with the replacement $\phi_{\vec{p}} \rightarrow 2\phi_{\vec{p}}$. Thus, the current operator after the transformation takes the form (3.6) with the replacements $v \rightarrow p/m$ and $\exp\left(-i\phi_{\vec{p}}\right) \rightarrow \exp\left(-2i\phi_{\vec{p}}\right)$. In contrast with the single-layer case, the density of electron states for the Hamiltonian (1.46) is finite at zero energy but the square of the current is, vice versa, linear in energy. As a result, we have the same estimate as Eq. (3.10).

3.2 The Landauer-formula approach

A deeper understanding of the origin of finite conductivity without charge carriers can be reached using the Landauer-formula approach (Beenakker & van Houten, 1991; Blanter & Büttiker, 2000). Following Katsnelson (2006a) we consider the simplest possible geometry, choosing the sample as a ring of length L_y in the y-direction; we will use the Landauer formula to calculate the conductance in the x-direction (see Fig. 3.1). As we will see, the conductivity turns out to be dependent on the shape of the sample. To have a final transparency we should keep L_x finite. On the other hand, periodic boundary conditions in the y-direction are nonphysical, and we have to choose L_y as large as possible in order to weaken their effects. Thus, for the two-dimensional situation one should choose $L_x \ll L_y$.

In the coordinate representation the Dirac equation at zero energy takes the form

$$\left(\frac{\partial}{\partial x} + i\frac{\partial}{\partial y}\right)\psi_1 = 0,$$

$$\left(\frac{\partial}{\partial x} - i\frac{\partial}{\partial y}\right)\psi_2 = 0. \tag{3.11}$$

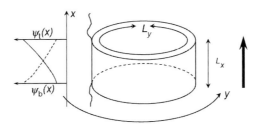

Fig. 3.1 The geometry of the sample. The thick arrow shows the direction of the current. Solid and dashed lines represent wave functions of the edge states localized near the top ($\psi_t(x)$) and bottom ($\psi_b(x)$) of the sample, respectively.

General solutions of these equations are just arbitrary analytic (or complex-conjugated analytic) functions:

$$\psi_1 = \psi_1(x + iy),$$
$$\psi_2 = \psi_2(x - iy). \tag{3.12}$$

Owing to periodicity in the y-direction, both wave functions should be proportional to $\exp(ik_y y)$, where $k_y = 2\pi n/L_y$, $n = 0, \pm 1, \pm 2, \ldots$ This means that the dependence on x is also fixed: The wave functions are proportional to $\exp(\pm 2\pi n x/L_y)$. They correspond to the states localized near the bottom and top of the sample (see Fig. 3.1).

To use the Landauer formula, we should introduce boundary conditions at the sample edges ($x = 0$ and $x = L_x$). To be specific, let us assume that the leads are made of doped graphene with the potential $V_0 < 0$ and the Fermi energy $E_F = vk_F = -V_0$. The wave functions in the leads are supposed to have the same y-dependence, namely $\psi_{1,2}(x, y) = \psi_{1,2}(x) \exp(ik_y y)$. Thus, one can try the solution of the Dirac equation in the following form that is consistent with Eq. (1.30):

$$\psi_1(x) = \begin{cases} \exp(ik_x x) + r \, \exp(-ik_x x), & x < 0, \\ a \, \exp(k_y x), & 0 < x < L_x, \\ t \, \exp(ik_x x), & x > L_x, \end{cases}$$

$$\psi_2(x) = \begin{cases} \exp(ik_x x + i\phi) + r \, \exp(-ik_x x - i\phi), & x < 0, \\ b \, \exp(-k_y x), & 0 < x < L_x, \\ t \, \exp(ik_x x + i\phi), & x > L_x, \end{cases} \tag{3.13}$$

where $\sin\phi = k_y/k_F$ and $k_x = \sqrt{k_F^2 - k_y^2}$. From the conditions of continuity of the wave functions, one can find the transmission coefficient

$$T_n = |t(k_y)|^2 = \frac{\cos^2\phi}{\cosh^2(k_y L_x) - \sin^2\phi}. \tag{3.14}$$

Further, one should assume that $k_F L_x \gg 1$ and put $\phi \cong 0$ in Eq. (3.14), so

$$T_n = \frac{1}{\cosh^2(k_y L_x)}. \tag{3.15}$$

The conductance G (per spin per valley) and Fano factor F of the shot noise (Blanter & Büttiker, 2000) are expressed via the transmission coefficients (3.15):

$$G = \frac{e^2}{h} \sum_{n=-\infty}^{\infty} T_n \tag{3.16}$$

and

$$F = 1 - \frac{\sum\limits_{n=-\infty}^{\infty} T_n^2}{\sum\limits_{n=-\infty}^{\infty} T_n}. \tag{3.17}$$

Note that in the ballistic regime, where the transmission probability for a given channel is either one or zero, $F = 0$ (the current is noiseless), whereas if all $T_n \ll 1$ (e.g., current through tunnel junctions) $F \approx 1$.

Thus, the trace of the transparency, which is just the conductance (in units of e^2/h), is

$$\operatorname{Tr} T = \sum_{n=-\infty}^{\infty} \frac{1}{\cosh^2\left(k_y L_x\right)} \cong \frac{L_y}{\pi L_x}. \tag{3.18}$$

Assuming that the conductance is equal to $\sigma L_y/L_x$ one finds a contribution to the conductivity per spin per valley equal to $e^2/(\pi h)$ (Katsnelson, 2006a; Tworzydlo et al., 2006). This result seems to be confirmed experimentally (Miao et al., 2007; Mayorov et al., 2011a). Also note that for the case of nanotubes ($L_x \gg L_y$) one has a conductance e^2/h per channel, in accordance with known results (Tian & Datta, 1994; Chico et al., 1996). For the Fano factor one has

$$F = \frac{1}{3} \tag{3.19}$$

(Tworzydlo et al., 2006). This result is very far from the ballistic regime and coincides with that for strongly disordered metals (Beenakker & Büttiker, 1992; Nagaev, 1992). This means that, in a sense, the *Zitterbewegung* works as an intrinsic disorder.

Instead of periodic boundary conditions in the y-direction, one can consider closed boundaries with zigzag-type or infinite-mass boundary conditions (we will discuss these later). The result (Tworzydlo et al., 2006) is just a replacement of the allowed values of the wave vectors in Eq. (3.15). One can write, in general (Rycerz, Recher, & Wimmer, 2009)

$$k_y(n) = \frac{g\pi(n + \gamma)}{L_y}, \tag{3.20}$$

where $g = 1$ and $\gamma = \frac{1}{2}$ for closed boundary conditions and $g = 2$ and $\gamma = 0$ for periodic boundary conditions. The results (3.18) and (3.19) for the case $L_x \gg L_y$ remain the same.

The case of bilayer graphene (Katsnelson, 2006b; Cserti, Csordás, & Dávid, 2007; Snyman & Beenakker, 2007) is more subtle. Even if we neglect the trigonal warping and use the Hamiltonian (1.46), an additional spatial scale

$$l_\perp = \frac{\hbar v}{t_\perp} \approx 10a \tag{3.21}$$

arises in the problem (Snyman & Beenakker, 2007), and the results for the conductance and the Fano factor depend on the sequence of the limits $L_x/l_\perp \to \infty$ and $E_F \to 0$. Moreover, when we cross the energy of trigonal warping and k_F satisfies the inequality (1.55), all four conical points work and the results are changed again (Cserti, Csordás, & Dávid, 2007).

3.3 Conformal mapping and Corbino geometry

Thus, electron transport in undoped graphene is due to zero modes of the Dirac operator, which are represented by analytic functions of $z = x + iy$ determined by boundary conditions. For the geometry shown in Fig. 3.1, these functions are just exponents:

$$\psi_{1n}(z) = \exp\left(\frac{2\pi n z}{L_y}\right), \tag{3.22}$$

so a generic wave function inside a graphene flake can be written as

$$\Psi(x, y) \equiv \sum_{n=-\infty}^{\infty} \left[a_n \left(\begin{array}{c} \exp\left(\frac{2\pi n z}{L_y}\right) \\ 0 \end{array} \right) + b_n \left(\begin{array}{c} 0 \\ \exp\left(\frac{2\pi n \bar{z}}{L_y}\right) \end{array} \right) \right], \tag{3.23}$$

where the coefficients a_n and b_n are determined by the boundary conditions. Let the Fermi wavelength in the leads be much smaller than the geometrical lengths of the flake. Then, for most of the modes one can write the boundary conditions assuming normal incidence $\phi = 0$:

$$\psi_{in} \equiv \left(\begin{array}{c} 1 + r \\ 1 - r \end{array} \right)$$

$$\psi_{out} \equiv \left(\begin{array}{c} t \\ t \end{array} \right), \tag{3.24}$$

where subscripts "in" and "out" label the values of the wave functions at the boundaries between the leads and the sample. In this approximation it is very easy to solve the problem of electron transport through a graphene quantum dot of arbitrary shape using a conformal mapping of this shape to the strip (Katsnelson & Guinea, 2008; Rycerz, Recher, & Wimmer, 2009). For example, the mapping

$$w(z) = R_1 \exp\left(\frac{2\pi z}{L_y}\right) \tag{3.25}$$

with

$$\exp\left(\frac{2\pi L_x}{L_y}\right) = \frac{R_2}{R_1}$$

transforms the rectangular strip $L_x \times L_y$ into a circular ring with inner and outer radii R_1 and R_2, respectively. Indeed, for $z = x + iy$, with $0 < x < L_x$ and $0 < y < L_y$, the transformation (3.25) leads to $0 \le \arg w < 2\pi$ and $R_1 \le |w| \le R_2$. Instead of Eq. (3.23) one can try in this case

$$\Psi(x, y) \equiv \sum_{n=-\infty}^{\infty} \left[a_n \begin{pmatrix} \frac{z^n}{0} \end{pmatrix} + b_n \begin{pmatrix} 0 \\ \bar{z}^n \end{pmatrix} \right]. \tag{3.26}$$

The conformal mapping allows us to find immediately the solution for Corbino geometry where "in" and "out" leads are attached to the inner and outer edges of the ring, respectively (see Fig. 3.2); in this case periodic boundary conditions in the y-direction should naturally be used. Moreover, the solution of the problem for any shape of the flake that is topologically equivalent to the ring can be written automatically in terms of the corresponding conformal mapping (Rycerz, Recher, & Wimmer, 2009). Earlier (Katsnelson & Guinea, 2008), this method was applied to the case of graphene quantum dots with thin leads attached.

If we just repeat the derivation of Eq. (3.15) using the boundary conditions (3.24), one can see that

$$\cosh\left(k_y L_x\right) = \frac{1}{2} \left[\exp\left(k_y L_x\right) + \exp\left(-k_y L_x\right) \right]$$
$$= \frac{1}{2} \left[\frac{\psi_1(x = L_x)}{\psi_1(x = 0)} + \frac{\psi_1(x = 0)}{\psi_1(x = L_x)} \right] \tag{3.27}$$

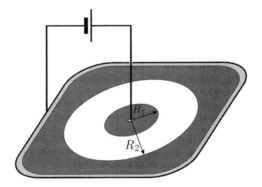

Fig. 3.2 The Corbino geometry: radial electric current in the ring.

and

$$\frac{\psi_1(x = L_x)}{\psi_1(x = 0)} = \frac{\psi_2(x = 0)}{\psi_2(x = L_x)}. \tag{3.28}$$

Under the conformal mapping (3.25)

$$\frac{\psi_1(x = L_x)}{\psi_1(x = 0)} = \exp\left(\frac{2\pi L_x}{L_y}\right) \rightarrow \frac{\psi_1(r = R_2)}{\psi_1(r = R_1)} = \frac{R_2}{R_1}, \tag{3.29}$$

and the result for the transmission coefficient reads

$$T_n = \frac{4}{\left(\frac{R_2}{R_1}\right)^n + \left(\frac{R_1}{R_2}\right)^n}. \tag{3.30}$$

We should be careful, however, since up to now we have not taken into account the Berry phase π for massless Dirac fermions. When we pass along the circle within the disc we have not periodic but *anti*periodic boundary conditions:

$$\psi_1(|w|, \arg w) = -\psi_1(|w|, \arg w + 2\pi), \tag{3.31}$$

which means that n in (3.30) should be replaced by $n + \frac{1}{2}$. Finally, one has (Rycerz, Recher, & Wimmer, 2009)

$$T_j = \frac{1}{\cosh^2\left[j\ln\left(\frac{R_2}{R_1}\right)\right]}, \quad j = \pm\frac{1}{2}, \pm\frac{3}{2}, \pm\frac{5}{2}, \dots, \tag{3.32}$$

and the summation over integer n in Eq. (3.16) and (3.17) should be replaced by a summation over half-integer j. For a ring that is thin enough, $|R_2 - R_1| \ll R_1$, the result is

$$G \approx \frac{2e^2}{h}\frac{1}{\ln\left(\frac{R_2}{R_1}\right)}, \quad F \approx \frac{1}{3}. \tag{3.33}$$

This agrees with the result (3.18) if we take into account that the thin ring is equivalent to the rectangular strip with $L_x = R_2 - R_1$ and $L_y = 2\pi R_1$. In the opposite limit $R_1 \ll R_2$ one has

$$G \approx \frac{8e^2}{h}\frac{R_1}{R_2}, \quad F \approx 1 - G\frac{h}{8e^2}. \tag{3.34}$$

Thus, for zero doping, the conductance of a graphene flake of arbitrary shape can be found without explicit solution of the Dirac equation, by a conformal mapping to a rectangle.

3.4 The Aharonov–Bohm effect in undoped graphene

The Aharonov–Bohm effect (Aharonov & Bohm, 1959; Olariu & Popescu, 1985) is the shift of interference patterns from different electron trajectories by the magnetic flux through the area between the trajectories. This leads to oscillations of observable quantities, such as conductance as a function of the magnetic flux. The Aharonov–Bohm effect in graphene has been studied both theoretically (Recher et al., 2007; Jackiw et al., 2009; Katsnelson, 2010a; Rycerz, 2010; Wurm et al., 2010) and experimentally (Russo et al., 2008; Huefner et al., 2009) for the case of a finite doping. It is not clear a priori whether this effect is observable or not in undoped graphene, where the transport is determined by evanescent waves. The analysis of Katsnelson (2010a) and Rycerz (2010) shows that, whereas for the case of very thin rings the Aharonov–Bohm oscillations are exponentially small, for a reasonable ratio of radii, such as, e.g., $R_2/R_1 = 5$, the effect is quite observable.

By combining the conformal-mapping technique with a general consideration of zero-energy states for massless Dirac fermions one can derive simple and general rigorous formulas for any graphene flakes topologically equivalent to the ring, avoiding both numerical simulations and explicit solutions of the Schrödinger equation for some particular cases (Katsnelson, 2010a). Note that for the case of a circular ring and a constant magnetic field the problem can be solved exactly for any doping (Rycerz, 2010), but, of course, the mathematics required is much more cumbersome. In the corresponding limits, the results are the same.

The effect of magnetic fields on the states with zero energy can be considered by employing the method of Aharonov and Casher (1979) (see Section 2.3). The general solutions have the form (2.62), where f_1 and $\bar{f}_1$ are analytic and complex-conjugated analytic functions. The boundary conditions following from Eq. (3.24) are

$$
\begin{aligned}
1 + r &= \psi_+^{(1)}, \\
1 - r &= \psi_-^{(1)}, \\
t &= \psi_+^{(2)}, \\
t &= \psi_-^{(2)},
\end{aligned}
\tag{3.35}
$$

where superscripts 1 and 2 label the boundaries attached to the corresponding leads.

If the boundary of the sample is simply connected, one can always choose $\varphi = 0$ at the boundary and, thus, the magnetic fields disappear from Eq. (3.35); this fact was used by Schuessler et al. (2009) as an elegant way to prove that a random vector potential has no effect on the value of the minimal conductivity. Further, we will consider a ring where the scalar potential φ is still constant at each boundary but these constants, φ_1 and φ_2, are different. Also, by symmetry (cf. Eq. (3.28)),

$$\frac{f_+^{(2)}}{f_+^{(1)}} = \frac{f_-^{(1)}}{f_-^{(2)}}. \tag{3.36}$$

The answer for the transmission coefficient $T = |t|^2$ for the case of a ring has the form

$$T_j = \frac{1}{\cosh^2\left[(j+a)\ln\left(\dfrac{R_2}{R_1}\right)\right]}, \tag{3.37}$$

the only difference from Eq. (3.32) being the shift of j by

$$a = \frac{e}{\hbar c}\frac{\varphi_2 - \varphi_1}{\ln\left(\dfrac{R_2}{R_1}\right)}, \tag{3.38}$$

which generalized the corresponding result of Rycerz, Recher, and Wimmer (2009) on the case of finite magnetic fields. The conductance G (per spin per valley) and Fano factor of the shot noise F are expressed via the transmission coefficients (3.37) by Eq. (3.16) and (3.17). To calculate the sums one can use the Poisson summation formula (2.145). On substituting Eq. (3.37) into (3.16) and (3.17) one finds a compact and general answer for the effect of a magnetic field on the transport characteristics:

$$G = \frac{2e^2}{h\ln(R_2/R_1)}\left[1 + 2\sum_{k=1}^{\infty}(-1)^k\cos(2\pi ka)\alpha_k\right], \tag{3.39}$$

$$F = 1 - \frac{2}{3}\left[\frac{1 + 2\sum_{k=1}^{\infty}(-1)^k\cos(2\pi ka)\alpha_k\left(1 + \pi^2k^2/\ln^2\left(\dfrac{R_2}{R_1}\right)\right)}{1 + 2\sum_{k=1}^{\infty}(-1)^k\cos(2\pi ka)\alpha_k}\right], \tag{3.40}$$

where

$$\alpha_k = \frac{\pi^2 k/\ln\left(\dfrac{R_2}{R_1}\right)}{\sinh\left(\pi^2 k/\ln\left(\dfrac{R_2}{R_1}\right)\right)}. \tag{3.41}$$

Eq. (2.60) can be solved explicitly for radially symmetric distributions of the magnetic field $B(r)$:

$$\varphi_2 - \varphi_1 = \frac{\Phi}{\Phi_0} \ln \left(\frac{R_2}{R_1} \right) + \int_{R_1}^{R_2} \frac{dr}{r} \int_{R_1}^{r} dr' r' B(r'), \tag{3.42}$$

where Φ is the magnetic flux though the inner ring. In the case of the Aharonov–Bohm effect where the whole magnetic flux is concentrated within the inner ring one has

$$a = \frac{\Phi}{\Phi_0}. \tag{3.43}$$

Owing to the large factor π^2 in the argument of sinh in Eq. (3.41), only the terms with $k = 1$ should be kept in Eq. (3.36) and (3.37) for all realistic shapes, thus

$$G = G_0 \left[1 - \frac{4\pi^2}{\ln \left(\frac{R_2}{R_1} \right)} \exp \left(-\frac{\pi^2}{\ln \left(\frac{R_2}{R_1} \right)} \right) \cos \left(\frac{e\Phi}{\hbar c} \right) \right], \tag{3.44}$$

$$F = \frac{1}{3} + \frac{8\pi^4}{3 \ln^3 \left(\frac{R_2}{R_1} \right)} \exp \left(-\frac{\pi^2}{\ln \left(\frac{R_2}{R_1} \right)} \right) \cos \left(\frac{e\Phi}{\hbar c} \right), \tag{3.45}$$

where G_0 is the conductance of the ring without magnetic field (3.33).

Oscillating contributions to G and F are exponentially small for very thin rings but are certainly measurable if the ring is thick enough. For $R_2/R_1 = 5$ their amplitudes are 5.3% and 40%, respectively.

Consider now a generic case with the magnetic field $B = 0$ within the flake. Then, the solution of Eq. (2.60) is a harmonic function, that is, the real or imaginary part of an analytic function. It can be obtained from the solution for the disc by the same conformal transformation as that which we use to solve the Dirac equation. One can see immediately that Eq. (3.35) remains the same. The expressions (3.44) and (3.45) can be rewritten in terms of an experimentally measurable quantity G_0

$$G = G_0 \left[1 - \frac{4\pi^2}{\beta} \exp \left(-\frac{\pi^2}{\beta} \right) \cos \left(\frac{e\Phi}{\hbar c} \right) \right], \tag{3.46}$$

$$F = \frac{1}{3} + \frac{8\pi^4}{3\beta^3} \exp \left(-\frac{\pi^2}{\beta} \right) \cos \left(\frac{e\Phi}{\hbar c} \right), \tag{3.47}$$

where $\beta = 2e^2/(hG_0)$ and we assume that $\beta \ll \pi^2$.

Thus, conformal transformation (Katsnelson & Guinea, 2008; Rycerz, Recher, & Wimmer, 2009) is a powerful tool with which to consider pseudodiffusive transport in undoped graphene flakes of arbitrary shape, not only in the absence of a magnetic field but also in the presence of magnetic fluxes in the system. An experimental study of the Aharonov–Bohm oscillations and comparison with the simple expressions (3.46) and (3.47) derived here would be a suitable way to check whether the ballistic (pseudodiffusive) regime is reached or not in a given experimental situation.

To conclude this chapter, we note that undoped graphene is a gapless semiconductor, with a completely filled valence band and an empty conduction band. It is really counterintuitive that in such a situation, at zero temperature, it has a finite conductivity, of the order of the conductance quantum e^2/h. This is one of the most striking consequences of its peculiar "ultrarelativistic" energy spectrum. Formally, the electron transport in undoped graphene is determined by zero modes of the Dirac operator, which are described by analytic functions with proper boundary conditions. Therefore, the whole power of complex calculus can be used here, just as in classical old-fashioned branches of mathematical physics such as two-dimensional hydrodynamics and electrostatics. These states cannot correspond to the waves propagating through the sample but, rather, are represented by evanescent waves. The transport via evanescent waves in undoped graphene is a completely new variety of electron transport in solids, being drastically different from all types known before (ballistic transport in nanowires and constrictions, diffusive transport in dirty metals, variable-range-hopping transport in Anderson insulators, etc.). Gaining a deeper understanding of these new quantum phenomena would seem to be a very important task.

4

The Klein paradox and chiral tunneling

4.1 The Klein paradox

Soon after the discovery of the Dirac equation, Oskar Klein (1929) noticed one of its strange properties, which was called afterwards the "Klein paradox." Klein considered the 4×4 matrix Dirac equation for a relativistic spin-$\frac{1}{2}$ particle propagating in three-dimensional space. To be closer to our main subject, we will discuss the 2×2 matrix equation for a particle propagating in two-dimensional space; the essence of the paradox remains the same. Thus, we will consider the stationary Schrödinger equation

$$\hat{H}\Psi = E\Psi \tag{4.1}$$

with the two-component spinor wave function

$$\Psi = \begin{pmatrix} \psi_1 \\ \psi_2 \end{pmatrix}$$

and the Hamiltonian

$$\hat{H} = -i\hbar c \vec{\sigma} \nabla + V(x, y)\hat{1} + mc^2 \hat{\sigma}_z. \tag{4.2}$$

Here c is the velocity of light, m is the mass of the particle, and $V(x, y)$ is a potential energy; we will explicitly write the identity matrix $\hat{1}$ to show the spinor structure of the Hamiltonian. Let us consider the one-dimensional case $V = V(x)$ and $\psi_i = \psi_i(x)$ (the latter means normal incidence). Eq. (4.1) now takes the form

$$-i\hbar c \frac{d\psi_2}{dx} = \left[E - mc^2 - V(x) \right]\psi_1,$$
$$\tag{4.3}$$
$$-i\hbar c \frac{d\psi_1}{dx} = \left[E + mc^2 - V(x) \right]\psi_2.$$

First consider just a jump of the potential:

$$V(x) = \begin{cases} 0, & x < 0, \\ V_0, & x > 0, \end{cases} \qquad (4.4)$$

with a positive V_0.

At the left side of the barrier, the solutions Ψ_1 and Ψ_2 have x-dependence as $\exp(\pm ikx)$, where the wave vector k satisfies the relativistic dispersion relation $E^2 = (\hbar c k)^2 + m^2 c^4$, or

$$k = \frac{\sqrt{E^2 - m^2 c^4}}{\hbar c}. \qquad (4.5)$$

The allowed energy values are $E > mc^2$ (electron states) or $E < -mc^2$ (hole, or positron, states). To be specific, we will consider the first case. Thus, using Eq. (4.3) with $V = 0$ one finds for the incident wave

$$\Psi_{in}(x) = \begin{pmatrix} 1 \\ \alpha \end{pmatrix} e^{ikx} \qquad (4.6)$$

and for the reflected wave

$$\Psi_r(x) = \begin{pmatrix} 1 \\ -\alpha \end{pmatrix} e^{-ikx}, \qquad (4.7)$$

where

$$\alpha = \sqrt{\frac{E - mc^2}{E + mc^2}}. \qquad (4.8)$$

We will assume a solution of the general form

$$\Psi(x) = \Psi_{in}(x) + r\Psi_r(x), \qquad (4.9)$$

where r is the reflection coefficient.

At the right side of the barrier, we have the dispersion relation $(E - V_0)^2 = \hbar^2 c^2 q^2 + m^2 c^4$ for the new wave vector q. We will consider the case of a potential jump that is strong enough:

$$V_0 > E + mc^2. \qquad (4.10)$$

In this case the solution

$$q = \frac{\sqrt{(V_0 - E)^2 - m^2 c^4}}{\hbar c} \qquad (4.11)$$

is real and the particle can also propagate on the right side of the barrier. However, this particle belongs to the lower (positron, or hole) continuum (see Fig. 4.1). It is in this situation that the paradox arises, so we will consider only this case.

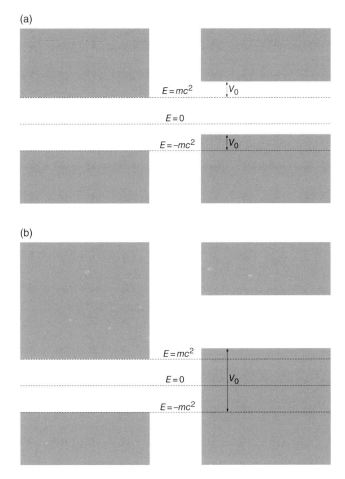

Fig. 4.1 Electron and positron states on the left and right sides of the barrier for the cases $V_0 < 2mc^2$ (a) and $V_0 > 2mc^2$ (b).

For smaller values of V_0, one has either the situation of propagating electrons on both sides of the barrier, if $V_0 < E - mc^2$, or evanescent waves at $x > 0$ if $E - mc^2 < V_0 < E + mc^2$ (Fig. 4.1(a)).

On solving the Schrödinger equation (4.3) for $x > 0$ one finds for the transmitted wave

$$\Psi_t(x) = \begin{pmatrix} 1 \\ -\dfrac{1}{\beta} \end{pmatrix} e^{iqx}, \tag{4.12}$$

where

$$\beta = \sqrt{\frac{V_0 - E - mc^2}{V_0 - E + mc^2}}. \tag{4.13}$$

One can find the reflection coefficient r and the transmission coefficient t, assuming that the wave function is continuous at $x = 0$, that is,

$$\Psi_{in} + r\Psi_r|_{x=-0} = t\Psi_t|_{x=+0} \tag{4.14}$$

or

$$1 + r = t,$$
$$\alpha(1 - r) = -\frac{1}{\beta}t. \tag{4.15}$$

We find straightforwardly

$$r = \frac{1 + \alpha\beta}{\alpha\beta - 1}. \tag{4.16}$$

Since for the case under consideration α and β are real, $0 < \alpha, \beta < 1$, one can see immediately that $r < 0$ and

$$R = |r|^2 = \left(\frac{1 + \alpha\beta}{1 - \alpha\beta}\right)^2 > 1. \tag{4.17}$$

However, R is nothing other than the reflection probability! Indeed, the current density

$$j_x = c\Psi^+\sigma_x\Psi = c(\psi_1{}^*\psi_2 + \psi_2{}^*\psi_1) \tag{4.18}$$

has the values $2\alpha c$ and $-2\alpha cR$ for the incident and reflected parts of the wave function (4.9), respectively. Thus, we have the very strange conclusion that, under the condition (4.10), the reflected current is larger than the incident one and the reflection probability is larger than unity. This was initially called the Klein paradox.

Our further discussion will follow Calogeracos and Dombey (1999) and Dombey and Calogeracos (1999). (A rather complete list of references can be found in Greiner and Schramm [2008].)

First, as was noticed by Pauli, there is a problem with the definition of the transmitted wave. For the case (4.10), the group velocity of the particle on the right side of the barrier

$$v_g = \frac{1}{\hbar}\frac{dE}{dq} = \frac{1}{\hbar}\left(\frac{dq}{dE}\right)^{-1} = \frac{\hbar qc^2}{E - V_0}, \tag{4.19}$$

is opposite to the direction of the wave vector q. This means that, formally speaking, the transmitted wave (4.12) describes the particle propagating to the left (for positive q), since the direction of propagation is determined by the direction of the group velocity, not by the momentum. So, at first sight, the formal paradox disappears (see also Vonsovsky & Svirsky, 1993).

However, it reappears in a more detailed view of the problem. Instead of the infinitely broad barrier (4.4), let us consider the finite one:

$$V(x) = \begin{cases} V_0, & |x| < a, \\ 0, & |x| > a. \end{cases} \qquad (4.20)$$

In this situation, there is no problem with the choice of the transmitted wave at the right side, it is just $t\Psi_{in}$; within the barrier region one has to consider the most general solution, with both parts, proportional to $\exp(\pm iqx)$. The calculations are simple and straightforward (see, e.g., Su, Siu, & Chou, 1993; Calogeracos & Dombey, 1999) and the results for the reflection and transmission probabilities R and T are

$$R = \frac{\left(1 - \alpha^2\beta^2\right)^2 \sin^2(2qa)}{4\alpha^2\beta^2 + \left(1 - \alpha^2\beta^2\right)^2 \sin^2(2qa)}, \qquad (4.21)$$

$$T = \frac{4\alpha^2\beta^2}{4\alpha^2\beta^2 + \left(1 - \alpha^2\beta^2\right)^2 \sin^2(2qa)}. \qquad (4.22)$$

There is no formal problem in the sense that $0 < R < 1$, $0 < T < 1$ and $R + T = 1$, as should be the case.

Now, the case of an infinitely broad barrier can be considered from Eq. (4.21) and (4.22) in the limit $a \to \infty$. We should be careful here, because of fast oscillations. If

$$qa = \frac{N\pi}{2} \qquad (4.23)$$

(N is an integer), then $\sin(2qa) = 0$, and we have complete transmission ($R = 0$, $T = 1$). If we just average over the fast oscillations in the limit $a \to \infty$, replacing $\sin^2(2qa)$ by its average value $\frac{1}{2}$, we will find the expressions

$$R_\infty = \frac{\left(1 - \alpha^2\beta^2\right)^2}{8\alpha^2\beta^2 + \left(1 - \alpha^2\beta^2\right)^2},$$

$$T_\infty = \frac{8\alpha^2\beta^2}{8\alpha^2\beta^2 + \left(1 - \alpha^2\beta^2\right)^2}. \qquad (4.24)$$

Thus, the paradox reappears in a different form. It is no longer a paradox in a logical or mathematical sense, it is just a physically counterintuitive behavior.

The well-known tunneling effect in quantum mechanics assumes that the particle can penetrate through a classically forbidden region with $E < V(x)$ but the probability of the penetration is exponentially small if the barrier is high and broad. In the semiclassical approximation, the transmission of the barrier between classical turning points $x_{1,2}$ satisfying the equation $E = V(x_{1,2})$ can be estimated as (Landau & Lifshitz, 1977)

$$T \approx \exp\left\{ -\frac{2}{\hbar} \int_{x_1}^{x_2} dx \sqrt{2m[V(x) - E]} \right\}, \qquad (4.25)$$

where m is the mass of the particle; the motion is supposed to be nonrelativistic. For the relativistic particle under the condition (4.10) the situation is dramatically different: In the limit $a \to \infty$ the penetration probability (4.24) remains finite and, in general, is not small at all. Even for an infinitely high barrier ($V_0 \to \infty$) one has $\beta = 1$ and

$$T_\infty = \frac{E^2 - m^2 c^4}{E^2 - \frac{1}{2} m^2 c^4}. \qquad (4.26)$$

This quantity is of the order of unity if $E - mc^2$ is of the order of mc^2. In the ultrarelativistic limit

$$E \gg mc^2, \qquad (4.27)$$

one has $T_\infty \approx 1$. The ability of quantum relativistic particles to penetrate with large enough probabilities through barriers with arbitrarily large height and width is the contemporary formulation of the Klein paradox (Calogeracos & Dombey, 1999).

A hand-waving explanation of the tunnel effect is based on the Heisenberg principle: Since one cannot know with arbitrary accuracy both the momentum and the position of a particle at a given instant one cannot accurately separate the total energy into a potential part and a kinetic part. Thus, the kinetic energy can be "a bit" negative.

In the relativistic regime, there is a much stronger restriction (Landau & Peierls, 1931). One cannot know even the position alone with accuracy better than $\hbar c/E$. This means that relativistic quantum mechanics cannot be *mechanics*, it can only be field theory (Berestetskii, Lifshitz, & Pitaevskii, 1971). It always contains particles and antiparticles, and to measure the position with an accuracy better than $\hbar c/E$ one needs to apply an energy so high that it will create particle-antiparticle pairs. The original particle whose position is supposed to be measured will be lost among the newly born particles since all electrons are identical.

This consideration is relevant for the Klein paradox since under the condition (4.10) both electron and positron states are explicitly involved.

The standard interpretation of the states with negative energy is based on the Dirac theory of holes (Bjorken & Drell, 1964; Berestetskii, Lifshitz, & Pitaevskii, 1971; Davydov, 1976). It is supposed that in the vacuum all the states with negative energy are occupied; antiparticles are the holes in this energy continuum. In the case (4.10), the tunneling of a relativistic particle happens from a state from the upper energy continuum ($x < 0$) to a state in the lower one ($x > 0$). In this situation the definition of the vacuum should be reconsidered. This reconstruction takes place necessarily when we switch on the potential and pass from the "normal" situation of small V to the "paradoxical" case (4.10).

Let us consider the case of a rectangular barrier (4.20) but for arbitrary V. If V is small enough, the bound states are formed in the gap, that is, with energies $|E| < mc^2$. A straightforward solution of this problem gives the following equation for the energy of the bound states (Calogeracos & Dombey, 1999):

$$\tan(qa) = \sqrt{\frac{(mc^2 - E)(mc^2 + E + V_0)}{(mc^2 + E)(E + V_0 - mc^2)}},$$

$$\tan(qa) = -\sqrt{\frac{(mc^2 + E)(mc^2 + E + V_0)}{(mc^2 - E)(E + V_0 - mc^2)}},$$

(4.28)

where

$$q = \frac{\sqrt{(E + V_0)^2 - m^2 c^4}}{\hbar c}$$

and we have made the replacement $V_0 \rightarrow -V_0$. When $qa = \pi/2$ and, thus,

$$V_0 = mc^2 + \sqrt{(mc)^2 + \frac{\pi^2 \hbar^2 c^2}{4a^2}},$$

(4.29)

the energy of one of the bound states reaches the boundary of the positron continuum, $E = -mc^2$ (Fig. 4.2). It is now energetically favorable to occupy this state, creating a hole in the negative energy continuum (positron emission). At $qa = \pi$ the next state reaches the continuum, and the vacuum state is reconstructed. This allows us to better understand the nature of the original Klein paradox. Despite the problem that a large enough barrier looks static, actually it is not. One needs to carefully study how this state is reached, and this process involves positron emission by the growing barrier. For a more detailed discussion of the role of the electron–positron pairs in the Klein paradox, see Krekora, Su, and Grobe

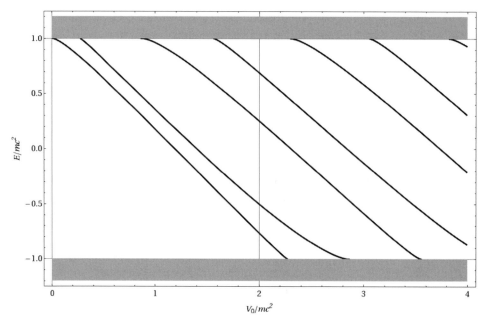

Fig. 4.2 Energies of the bound state found from Eq. (4.28) as functions of the height of the barrier; $a = 2\hbar/(mc)$.

(2005). We will come back to this issue later, when discussing supercritical charges in graphene (Chapter 8).

4.2 The massless case: the role of chirality

We are going to discuss the Klein paradox and related issues for the massless Dirac fermions in graphene (Katsnelson, Novoselov, & Geim, 2006). The case $m = 0$ is very special. If we put $m = 0$ in the results (4.21) and (4.22) we will have $T = 1$ and $R = 0$ for any parameters of the potential (one can see from Eq. (4.8) and (4.13) that $\alpha = \beta = 1$ for $m = 0$). This result is not related to a specific choice of the potential barriers (4.20).

For $m = 0$, the equations (4.3) can be very easily solved for arbitrary $V(x)$. Let us introduce a variable

$$w = \frac{1}{\hbar c} \int^{x} dx' [E - V(x')]. \tag{4.30}$$

Of course, we have to be careful: This change of variables is possible only for the intervals within which $E > V(x)$ or $E < V(x)$, so dw/dx never vanishes. Therefore, we will use (4.30) separately for each interval between two turning points (and for the intervals between $-\infty$ and the first turning point and between the last turning point and $+\infty$). There are two basic solutions for each such interval:

$$\Psi_> = \begin{pmatrix} 1 \\ 1 \end{pmatrix} \exp{(i|w|)} \tag{4.31}$$

and

$$\Psi_< = \begin{pmatrix} 1 \\ -1 \end{pmatrix} \exp{(-i|w|)}. \tag{4.32}$$

Both components of the spinor should be continuous at the turning points, so one can see immediately that the only way to match the solutions is to choose either $\Psi_>$ or $\Psi_<$ to be zero everywhere. One can never have a combination of incident and reflected waves, since propagation is only allowed in one direction (here one has to recall that we consider only the case of normal incidence; for two-dimensional problems with $\Psi(x, y)$ this is not the case, see the next section).

The point is that a massless Dirac particle can only propagate either along its (pseudo)spin direction or in the opposite direction. The scalar potential proportional to the identity matrix in the Hamiltonian (4.2) does not act on the pseudospin and therefore cannot change the direction of propagation of a massless particle with spin $\frac{1}{2}$ to the opposite.

This property has an analogue in more general two-dimensional and three-dimensional situations with $V = V(x, y)$ or $V = V(x, y, z)$: Backscattering is forbidden. This was found long ago for the scattering of ultrarelativistic particles in three dimensions (Yennie, Ravenhall, & Wilson, 1954; Berestetskii, Lifshitz, & Pitaevskii, 1971). Ando, Nakanishi, and Saito (1998) noticed an importance of this property for carbon materials. In particular, the absence of backscattering explains the existence of conducting channels in metallic carbon nanotubes; in a nonrelativistic one-dimensional system an arbitrarily small disorder leads to localization (Lifshitz, Gredeskul, & Pastur, 1988), so the conductive state of the nanotubes is not trivial.

The consideration of Ando, Nakanishi, and Saito (1998) is very instructive, since it shows explicitly the role of the Berry phase and time-reversal symmetry, but it is quite cumbersome. Here we present a somewhat simplified version of this proof. To this end, we consider the equation (Newton, 1966) for the scattering T-matrix

$$\hat{T} = \hat{V} + \hat{V}\hat{G}_0\hat{T}, \tag{4.33}$$

where $\hat{V}$ is the scattering potential operator,

$$\hat{G}_0 = \lim_{\delta \to +0} \frac{1}{E - \hat{H}_0 + i\delta} \tag{4.34}$$

is the Green function of the unperturbed Hamiltonian $\hat{H}_0$, and E is the electron energy (we will assume $E > 0$). For more details of this formalism, see Chapter 6. If $\hat{H}_0$ is the Dirac Hamiltonian for massless fermions (1.22), we have

$$\hat{G}_0(\vec{r}, \vec{r}') = \int \frac{d\vec{q}}{(2\pi)^2} \hat{G}_0(\vec{q}) \exp\left[i\vec{q}(\vec{r} - \vec{r}')\right], \tag{4.35}$$

where

$$\hat{G}_0(\vec{q}) = \frac{1}{E - \hbar v \vec{q}\vec{\sigma} + i\delta} = \frac{1}{\hbar v} \frac{k + \vec{q}\vec{\sigma}}{(k + i\delta)^2 - q^2} \tag{4.36}$$

with $k = E/(\hbar v)$. The probability amplitude of the backscattering can be found by iterations of Eq. (4.33) and is proportional to

$$T\left(-\vec{k}, \vec{k}\right) = \left\langle -\vec{k} \middle| V + V\hat{G}V + V\hat{G}V\hat{G}V + \cdots \middle| \vec{k} \right\rangle \equiv T^{(1)} + T^{(2)} + \cdots, \tag{4.37}$$

where $T^{(n)}$ is the contribution proportional to V^n.

Let us assume that $\vec{k} \| Ox$ (we can always choose the axes in such a way), then $\left|\vec{k}\right\rangle$ and $\left|-\vec{k}\right\rangle$ have spinor structures

$$\begin{pmatrix} 1 \\ 1 \end{pmatrix} \text{ and } \begin{pmatrix} 1 \\ -1 \end{pmatrix},$$

respectively (see Eq. (1.30)). Thus, if $\hat{T}$ is a 2×2 matrix

$$\hat{T} = T_0 + \vec{T}\hat{\vec{\sigma}}, \tag{4.38}$$

one has

$$T\left(-\vec{k}, \vec{k}\right) = \left\langle -\vec{k} \middle| T_z + iT_y \middle| \vec{k} \right\rangle. \tag{4.39}$$

Then, keeping in mind that V is proportional to the identity matrix, one can prove, term by term, that all contributions to $\left\langle -\vec{k} \middle| T_z \middle| \vec{k} \right\rangle$ and $\left\langle -\vec{k} \middle| T_y \middle| \vec{k} \right\rangle$ vanish by symmetry. Actually, this is just because $\vec{T}\left(\vec{k}\right) \propto \vec{k} \| Ox$. One cannot construct from the vectors $\vec{k}$ and $-\vec{k}$ anything with nonzero y- or z-components: For two nonparallel vectors $\vec{k}_1$ and $\vec{k}_2$, one of them has a nonzero y-component and $\vec{k}_1 \times \vec{k}_2 \| Oz$.

4.3 Klein tunneling in single-layer graphene

Keeping in mind electrons in quantum electrodynamics, it is not easy to create potential jumps larger than $2mc^2 \approx 1$ MeV. Similar phenomena take place in very high electric or gravitational fields (Greiner, Mueller, & Rafelski, 1985; Grib, Mamaev, & Mostepanenko, 1994; for a detailed list of references, see Greiner & Schramm, 2008), but the context is always quite exotic, such as collisions of ultraheavy ions or even black-hole evaporation. There were no experimental data that would require the Klein paradox for their explanation.

Soon after the discovery of graphene, it was realized that Klein tunneling (tunneling of Dirac fermions under the conditions of the Klein paradox) is one of the crucial phenomena for graphene physics and electronics (Katsnelson, Novoselov, & Geim, 2006). Soon after the theoretical prediction of Klein tunneling in graphene, it was confirmed experimentally (Stander, Huard, & Goldhaber-Gordon, 2009; Young & Kim, 2009).

In conventional terms of solid-state physics, Klein tunneling is nothing other than tunneling through a p-n-p (or n-p-n) junction when electrons are transformed into holes and then back to electrons (or vice versa) (Fig. 4.3). As we saw in the previous section, for massless Dirac fermions, the transmission at normal incidence is always 100%, irrespective of the height and width of the potential barrier. From the point of view of applications, this is very bad news: If one just copies the construction of a silicon transistor it will not work, since it is impossible to lock it. The gap opening is necessary. The good news is that, due to the Klein paradox, the unavoidable inhomogeneities of the electron density in graphene (see Section 13.1) do not lead to localization and, moreover, their effect on the electron mobility is not very great. We will come back to this important issue many times in this book.

Now consider, following Katsnelson, Novoselov, and Geim (2006), electron propagation through the barrier (4.20) for an arbitrary angle of incidence φ. The

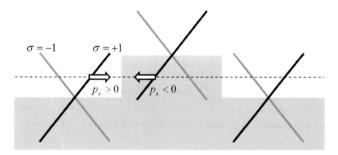

Fig. 4.3 Transformation of an electron to a hole under the potential barrier; the large arrows show directions of momenta, assuming that the group velocity is always parallel to the *Ox* axis. Black and gray lines show the dispersion of electronic states with opposite pseudospin projections.

energy $E = \hbar v k$ is supposed to be positive. There is a refraction of the electron wave at the potential jump, and the new angle θ is determined by the conservation of the y-component of the electron momentum (and, thus, of the wave vector):

$$k_y = k \sin \varphi = q_y = q \sin \theta, \tag{4.40}$$

where

$$q = \frac{|E - V_0|}{\hbar v} \tag{4.41}$$

is the length of the wave vector within the barrier. For massless Dirac fermions with energy E propagating at the angle φ to the x-axis, the components of the spinor wave functions are related by

$$\psi_2 = \psi_1 \exp(i\varphi) \operatorname{sgn} E \tag{4.42}$$

(see Eq. (1.30)). Thus, the wave function has the following form (cf. Eq. (3.13) for the case of zero energy):

$$\psi_1(x, y) = \begin{cases} [\exp(ik_x x) + r \exp(-ik_x x)] \exp(ik_y y), & x < -a, \\ [A \exp(iq_x x) + B \exp(-iq_x x)] \exp(ik_y y), & |x| < a, \\ t \exp(ik_x x + ik_y y), & x > a, \end{cases} \tag{4.43}$$

$$\psi_2(x, y) = \begin{cases} s[\exp(ik_x x + i\varphi) - r \exp(-ik_x x - i\varphi)] \exp(ik_y y), & x < -a, \\ s'[A \exp(iq_x x + i\theta) - B \exp(-iq_x x - i\theta)] \exp(ik_y y), & |x| < a, \\ st \exp(ik_x x + ik_y y + i\varphi), & x > a, \end{cases} \tag{4.44}$$

where

$$s = \operatorname{sgn} E, \quad s' = \operatorname{sgn}(E - V_0), \quad k_x = k \cos \varphi, \quad q_x = q \cos \theta \tag{4.45}$$

and we have taken into account that the reflected particle moves at the angle $\pi - \varphi$, $\exp[i(\pi - \varphi)] = -\exp(-i\varphi)$. The parameters r (the reflection coefficient), t (the transmission coefficient), A, and B should be found from the continuity of ψ_1 and ψ_2 at $x = \pm a$. Note that the Klein paradox situation is

$$ss' = -1 \tag{4.46}$$

(with opposite signs of the energy outside and inside the barrier). As a result, one finds

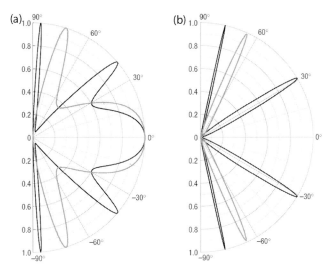

Fig. 4.4 Transmission probabilities through a 100-nm-wide barrier as a function of the angle of incidence for single-layer (a) and bilayer (b) graphene. The electron concentration n outside the barrier is chosen as 0.5×10^{12} cm^{-2} for all cases. Inside the barrier, hole concentrations p are 1×10^{12} and 3×10^{12} cm^{-2} for black and gray curves, respectively (such concentrations are most typical in experiments with graphene). This corresponds to Fermi energies E of incident electrons ≈ 80 and ≈ 17 meV for single-layer and bilayer graphene, respectively. The barrier heights V_0 are (a) 200 and (b) 50 meV (black curves) and (a) 285 and (b) 100 meV (gray curves).

(Reproduced with permission from Katsnelson, Novoselov, & Geim, 2006.)

$$r = 2 \exp{(i\varphi - 2ik_x a)} \sin{(2q_x a)}$$
$$\times \frac{\sin{\varphi} - ss' \sin{\theta}}{ss'[\exp{(-2iq_x a)}\cos{(\varphi + \theta)} + \exp{(2iq_x a)}\cos{(\varphi - \theta)}] - 2i\sin{(2q_x a)}}. \tag{4.47}$$

The transmission probability can be calculated as

$$T = |t|^2 = 1 - |r|^2. \tag{4.48}$$

The results are shown in Fig. 4.4. In agreement with the general consideration of the previous section, $r = 0$ at $\varphi = 0$ (this can be seen immediately from Eq. (4.47) and (4.40)).

There are also additional "magic angles" for which $r = 0$ and one has 100% transmission. They correspond to the condition $\sin{(2q_x a)} = 0$, or

$$q_x a = \frac{\pi}{2}N, \tag{4.49}$$

where $N = 0, \pm 1, \pm 2, \ldots$ Interestingly, this coincides with the condition (4.23) of complete transmission for the case of nonzero mass. These conditions correspond to the Fabry–Pérot resonances in optics (Born & Wolf, 1980). The same resonances can take place for a more general potential $V = V(x)$, as was shown in the semiclassical approximation by Shytov, Rudner, and Levitov (2008) (see also Shytov et al., 2009). At the same time, for some $V(x)$, these resonances *cannot* take place, and only full transmission for normally incident beam survives (Tudorovskiy, Reijnders, & Katsnelson, 2012; Reijnders, Tudorovskiy, & Katsnelson, 2013).

This issue will be considered in the next section.

4.4 Klein tunneling for a smooth potential barrier and the effect of magnetic fields

Strictly speaking, the Dirac-cone approximation itself does not work for the case of an atomically sharp potential since it will induce intervalley scattering, which can change the whole physical picture dramatically. The sharp potential jump considered in the previous sections means a sharpness in comparison with the electron wave length k^{-1} but not in comparison with the interatomic distance a. So, the typical spatial scale of the change of potential at the barrier d was assumed to satisfy the condition

$$a \ll d < \ll \frac{1}{k}. \tag{4.50}$$

The opposite limit case, that of a very smooth potential

$$kd \gg 1, \tag{4.51}$$

was first considered by Cheianov and Falko (2006). It turns out that in this case the region of high transmission near $\varphi \approx 0$ is pretty narrow:

$$T(\varphi) = \exp(-Ckd \sin^2 \varphi), \tag{4.52}$$

where C is a numerical factor depending on the specific shape of the potential, thus $T(\varphi) \approx 1$ if

$$|\varphi| \leq \frac{1}{\sqrt{kd}} \tag{4.53}$$

(the "Klein collimation"). The result (4.52) was obtained using both the exact solution of the Dirac equation in a constant electric field and the semiclassical approximation. Here we will present a simple derivation following Shytov, Gu, and Levitov (2007; see also Shytov et al., 2009).

Let us consider the Schrödinger equation (4.1) with the Hamiltonian (4.2) for the case when

$$V(x) = -eEx, \tag{4.54}$$

where E is the electric field. One can use the momentum representation for the coordinate x, $x \leftrightarrow k_x$. Then the coordinate $x \to i\partial/\partial k_x$ and the Schrödinger equation takes the form (with the replacement $c \to v$, keeping in mind the case of graphene)

$$-ieE\frac{\partial \Psi}{\partial k_x} = \hat{H}'\Psi, \tag{4.55}$$

where

$$\hat{H}' = \hbar v \vec{k}\vec{\sigma} - \varepsilon$$

(here we use the notation ε for the electron energy, in order not to confuse it with the electric field). The Eq. (4.55) is formally equivalent to the time-dependent Schrödinger equation with a time $t' = -\hbar k_x/(eE)$ and the Hamiltonian linearly dependent on the "time." This is nothing other than the problem of Landau–Zener breakdown, in which the term $\hbar v k_y \sigma_y$ plays the role of the gap in the Hamiltonian. Using the known solution of this problem (Vonsovsky & Katsnelson, 1989) one finds

$$T \approx \exp\left(-\frac{\pi \hbar v k_y^2}{|eE|}\right), \tag{4.56}$$

which coincides with Eq. (4.52), keeping in mind that $d \approx \hbar v k/|eE|$.

If we have crossed electric and magnetic fields E and B ($B \parallel Oz$), one can use the Lorentz transformation, similarly to what was done in Section 2.10 (see Eq. (2.196) and (2.197)). In the case

$$E > \frac{v}{c}B, \tag{4.57}$$

which is complementary to Eq. (2.198), one can exclude the *magnetic* field, and the electric field E is replaced in Eq. (4.56) by

$$E\sqrt{1 - \left(\frac{vB}{cE}\right)^2} = \sqrt{E^2 - \left(\frac{vB}{c}\right)^2}$$

(cf. Eq. (2.200)). The effects of disorder on the motion of an electron near a p-n junction were considered by Fogler et al. (2008).

Shytov, Rudner, and Levitov (2008) studied the case of a parabolic potential barrier

$$V(x) = ax^2 - \varepsilon \tag{4.58}$$

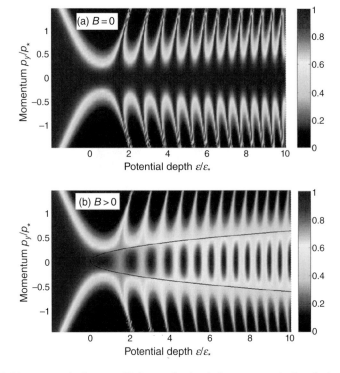

Fig. 4.5 The transmission coefficient, obtained from numerical solution of the Dirac equation with the potential (4.58), plotted as a function of the component of electron momentum p_y and potential depth. At zero magnetic field (a), transmission exhibits fringes with a phase that is nearly independent of p_y. At finite magnetic field (b), the fringe contrast reverses its sign on the parabola (black thin line). Here $\varepsilon^* = (a\hbar^2 v^2)^{1/3}$ and $p^* = \varepsilon^*/v$.
(Reproduced with permission from Shytov, Rudner, & Levitov, 2008.)

$(a, \varepsilon > 0)$, which creates p-n boundaries at

$$x = \pm x_\varepsilon = \pm\sqrt{\frac{\varepsilon}{a}}. \tag{4.59}$$

The magnetic field B is included in the Landau gauge, $A_x = 0$, $A_y = Bx$. Numerical solution of the Schrödinger equation gives the results shown in Fig. 4.5. One can see that a region of 100% transmission can exist not only for a rectangular barrier (see Eq. (4.49)) but also for a more general symmetric potential. At the same time, for *nonsymmetric* potentials $V(x) \neq V(-x)$, the side resonances with $\varphi \neq 0$ turn out to be suppressed (Tudorovskiy, Reijnders, & Katsnelson, 2012) as will be discussed later. The magnetic field modifies the picture of the transmission in a peculiar way. Oscillations of the conductance through the barrier as a function of the magnetic field were observed by Young and Kim (2009) (Fig. 4.6).

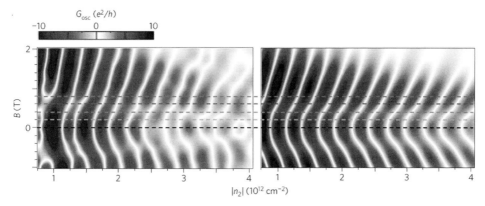

Fig. 4.6 The magnetic field and density dependences of the conductance of a p-n-p junction in graphene; left and right panels present experimental data and theoretical results, respectively.
(Reproduced with permission from Young & Kim, 2009.)

Now let us consider a general semiclassical theory for an arbitrary, smooth, one-dimensional potential $V(x)$; here we will follow the work by Reijnders, Tudorovskiy, and Katsnelson (2013).

First, let us introduce dimensionless units $x \to x/l, \hat{\vec{p}} \to \hat{\vec{p}}/p_0$, $E \to E/vp_0, V \to V/vp_0$, where l is a typical spatial scale of the change of potential and vp_0 is a typical energy scale of the difference $E - V$. Then, the Schrödinger equation for massless Dirac fermions in graphene takes the form

$$\left[-ih\hat{\sigma}_x \frac{d}{dx} + p_y\hat{\sigma}_y + U(x) \right] \Psi(x) = 0, \tag{4.60}$$

where $U(x) = V(x) - E$,

$$h = \hbar/(p_0 l), \tag{4.61}$$

and we try the solution in the form

$$\Psi(x, y) = \Psi(x) \exp\left(\frac{ip_y x}{h} \right) \tag{4.62}$$

(cf. Eq. (4.43), (4.44)). Note that semiclassical approximation is formally applicable if $h \ll 1$.

Similar to the transition from Eq. (2.39) to Eq. (2.42), we act by the operator $-ih\hat{\sigma}_x \frac{d}{dx} + p_y\hat{\sigma}_y - U(x)$ on Eq. (4.60). The result is

$$\left[-h^2 \frac{d^2}{dx^2} + p_y^2 - U^2(x) - ih\hat{\sigma}_x U'(x) \right] \Psi(x) = 0, \tag{4.63}$$

where $U'(x) = dU(x)/dx$. Since Eq. (4.63) contains only a single Pauli matrix, it can be diagonalized by the substitution

$$\Psi(x) = \begin{pmatrix} 1 \\ 1 \end{pmatrix} \eta_1(x) + \begin{pmatrix} 1 \\ -1 \end{pmatrix} \eta_2(x) \tag{4.64}$$

and we obtain (cf. Eq. (2.42), (2.43)):

$$\left[h^2 \frac{d^2}{dx^2} + U^2(x) \pm ihU'(x) - p_y^2 \right] \eta_{1,2}(x) = 0, \tag{4.65}$$

$$\eta_{2,1} = \frac{1}{p_y} \left(h \frac{d}{dx} \pm iU(x) \right) \eta_{1,2}. \tag{4.66}$$

Eq. (4.65) reminds the standard nonrelativistic Schrödinger equation with the effective potential $U^2(x) \pm ihU'(x)$ and the effective energy p_y^2. Just as in the conventional semiclassical approximation (Landau & Lifshitz, 1977) one can try the solution in the form

$$\eta_1(x) = A(x, h) \exp \left[\frac{iS(x, h)}{h} \right], \tag{4.67}$$

expanding the phase $S(x,h)$ and amplitude $A(x,h)$ functions in Taylor series in the parameter h (4.61). In the leading order approximation, we have

$$\eta_1(x) = A_+(x) \exp \left[\frac{iS_0(x)}{h} \right] + A_-(x) \exp \left[-\frac{iS_0(x)}{h} \right], \tag{4.68}$$

where

$$S_0(x) = \int_{x_0}^{x} dy \sqrt{p_y^2 - U^2(x)} \tag{4.69}$$

and x_0 is a constant. Eq. (4.68) describes incident and reflected waves in the classically allowed regions where $p_y^2 > U^2(x)$. In the classically forbidden regions $(p_y^2 < U^2(x))$ it describes evanescent waves, and near the turning points $(p_y^2 = U^2(x))$ the amplitude functions $A_\pm(x)$ are divergent, making the expression (4.68) inapplicable. The problem of how to match the semiclassical solutions in classically allowed and classically forbidden regions and how to build a "uniform asymptotics" valid in the vicinity of the turning points is discussed in detail by Tudorovskiy, Reijnders, and Katsnelson (2012) and Reijnders, Tudorovskiy, and Katsnelson (2013). Here we will show just some results.

　　Let us consider the case of n-p-n junction, with two smooth-enough junctions between electron and hole parts separated by a relatively long hole region (Fig. 4.7). In this situation, turning points $x_\pm$ and classically forbidden regions

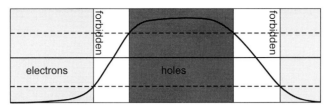

Fig. 4.7 A potential barrier for the case of n-p-n junction.

arise, and we have to solve the matching problem for each of them separately. The result for the transmission coefficient t_{npn} is (Shytov, Rudner, & Levitov, 2008; Tudorovskiy, Reijnders, & Katsnelson, 2012; Reijnders, Tudorovskiy, & Katsnelson, 2013):

$$t_{npn} = \frac{t_{np}t_{pn}e^{-iL/h}}{1 - r_{np}^* r_{pn}^* e^{-2iL/h}}, \tag{4.70}$$

where t_{np}, r_{np} are transmission and reflection coefficients for the left junction, t_{pn}, r_{pn} are the same for the right junctions, and

$$L = \int_{x_{1+}}^{x_{2-}} dy \sqrt{p_y^2 - U^2(x)}, \tag{4.71}$$

where the integral is taken over the classically allowed hole region. This is an analogue of the known expression describing Fabry–Pérot resonances in optics (Born & Wolf, 1980).

Keeping in mind that $|t|^2 + |r|^2 = 1$ for both n-p and p-n junctions, one can find that the maximum (resonant) value of the modulus of transmission coefficient (4.72) is equal to

$$\left| t_{npn} \right|_{res} = \frac{\left| t_{np} \right|\left| t_{pn} \right|}{1 - \sqrt{\left(1 - \left| t_{np} \right|^2\right)\left(1 - \left| t_{np} \right|^2\right)}}; \tag{4.72}$$

this value is equal to 1 only if $|t_{np}| = |t_{pn}|$ (symmetric barrier), otherwise we always have $|t_{npn}|_{res} < 1$.

In semiclassical approximation, one finds (Tudorovskiy, Reijnders, & Katsnelson, 2012; Reijnders, Tudorovskiy, & Katsnelson, 2013)

$$\left| t_{npn} \right|_{res} = \frac{1}{\cosh\left(\dfrac{K_{np} - K_{pn}}{h}\right)}, \tag{4.73}$$

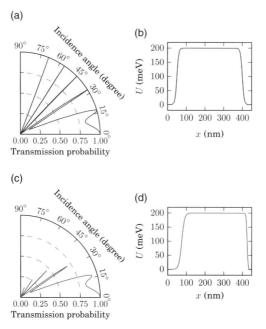

Fig. 4.8 Transmission probability for n-p-n junction for massless Dirac fermions; energy is 80 meV and the height of the potential is 200 meV. (a) The results for symmetric potential (shown in b). (c) The results for asymmetric potential (shown in d). (Reproduced with permission from Kleptsyn et al., 2015)

where

$$K = \int_{x_-}^{x_+} dy \sqrt{U^2(x) - p_y^2} \tag{4.74}$$

and the integral are taken over the corresponding (left or right) classically forbidden region. Therefore, one can see that for a generic, asymmetric one-dimensional barrier, the full transmission takes place only for $p_y = 0$, otherwise the suppression is exponentially strong in our formal small parameter (4.61).

Numerical results that illustrated suppression of the side resonances for the Dirac electrons are shown in Fig. 4.8 (Kleptsyn et al., 2015). This conclusion is also confirmed by numerical simulations on honeycomb lattice, that is, beyond Dirac approximation (Logemann et al., 2015).

4.5 Negative refraction coefficient and Veselago lenses for electrons in graphene

As was discussed in Section 4.1, the group velocity $\vec{v}_g$ is parallel to the wave vector $\vec{k}$ for particles (electrons) and antiparallel for antiparticles (holes). In the situation of the Klein paradox, the incident and transmitted waves propagate, by definition, in the same direction, and the propagation direction is determined by the group velocity. This means that the wave vectors for these waves are antiparallel. For massless particles with a linear dispersion, the group velocity is

$$\vec{v}_g = \pm v \frac{\vec{k}}{k}, \qquad (4.75)$$

where the signs $+$ and $-$ correspond to electrons and holes, respectively. The incident electron wave has the wave vector $\vec{k} = k(\cos\varphi, \sin\varphi)$ and the group velocity $\vec{v}_e = v(\cos\varphi, \sin\varphi)$. The reflected wave has the wave vector $\vec{k}' = k(-\cos\varphi, \sin\varphi)$ and the group velocity $\vec{v}'_e = v(-\cos\varphi, \sin\varphi)$. For the transmitted wave, in the situation of the Klein paradox (or for a p-n junction, using conventional semiconductor terminology) the group velocity $\vec{v}_h = v(\cos\theta', \sin\theta')$ and the wave vector $\vec{q} = -q(\cos\theta', \sin\theta')$, $\cos\theta' > 0$, q is determined by Eq. (4.41) and $\theta' = -\theta$. The refraction angle θ' is determined by the continuity of the y-component of the wave vector (see Eq. (4.40)), or

$$\frac{\sin\theta'}{\sin\varphi} = -\frac{k}{q} \equiv n \qquad (4.76)$$

with a negative refractive index n. This means that the p-n junction in graphene transforms a divergent electron beam into a collimated one, see Fig. 4.9 (Cheianov, Falko, & Altshuler, 2007).

In optics, such devices are known as Veselago lenses (Veselago, 1968), and materials with negative refractive indices are called left-handed materials, or metamaterials (Pendry, 2004). Creation of such a material for visual light is not an easy task. For electrons in graphene such a situation can be realized quite easily.

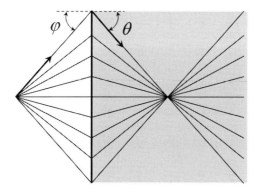

Fig. 4.9 A Veselago lens for the case of a negative refraction index.

For a detailed discussion of the relation between the negative refraction index and the Klein paradox, see Giiney and Meyer (2009).

Electron Veselago lensing in graphene was experimentally observed by Lee, Park, and Lee (2015) and by Chen et al. (2016). Bøggild et al. (2017) suggested a concept of "Dirac Fermion Microscope" where collimated electron beams in graphene in the ballistic regime are used to magnify atomic-scale inhomogeneities.

A detailed theory of Veselago lensing in graphene was developed by Reijnders and Katsnelson (2017a, 2017b). Here we will present only the main physical results of the theory.

First, for the massless Dirac fermions there is an intimate relation between propagation direction of the electron beam and the direction of the pseudospin if we use the beam with nonzero pseudospin polarization. The latter can be created, e.g., via electron injection from hexagonal boron nitride (Wallbank et al., 2016). In that case the sublattice symmetry is broken, as we will discuss in detail at the end of the book (Chapter 13). Numerical simulations as well as semiclassical theory (Reijnders & Katsnelson, 2017a) show that the pseudospin polarization can result into a splitting or asymmetric shift of the focus, see Fig. 4.10.

When we take into account the trigonal warping, we have different Hamiltonians for the different valleys, due to the τ_z term in Eq. (1.34). This leads to different trajectories for different valleys and to a valley splitting of the focus (Reijnders & Katsnelson, 2017b). Moreover, one can create a valley beam splitter based on n-p-n junction: the trigonal warping effects can essentially separate the K and K′ beam components (Garcia-Pomar, Cortijo, & Nieto-Vesperinas, 2008).

4.6 Klein tunneling and minimal conductivity

As was stressed in the previous chapter, the existence of a minimal conductivity of the order of e^2/h is one of the striking properties of graphene. We discussed this from the perspective of pure samples (the ballistic regime). It is instructive to consider the same problem from the opposite perspective of strong disorder (Katsnelson, Novoselov, & Geim, 2006).

First, it is worth recalling some basic ideas on the electronic structure of strongly disordered systems (Mott, 1974; Mott & Davis, 1979; Shklovskii & Efros, 1984; Lifshitz, Gredeskul, & Pastur, 1988). Let us start with the case in which typical fluctuations of the potential energy $V(x, y)$ are much stronger than the kinetic energy T. The electrons are locked into puddles restricted by the equipotential lines $E - V(x, y)$. There is a small probability of tunneling from one puddle to another, so some electrons are distributed among couples of puddles, fewer electrons among trios of puddles, etc. (Fig. 4.11). On increasing the ratio

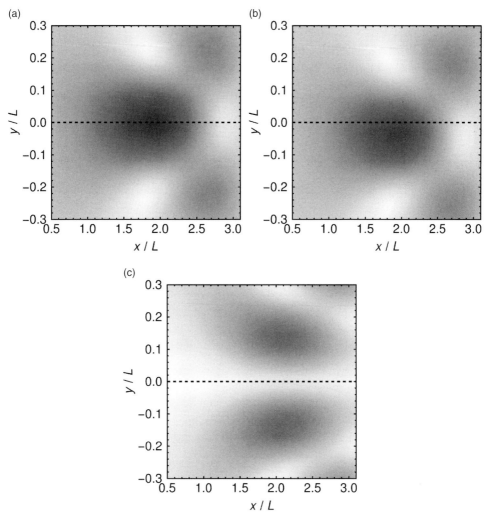

Fig. 4.10 The modulus of the electron wave function near the focus of the Veselago lens shown schematically in Fig. 4.9. Electron energy is 100 meV, the height of the potential barrier in 250 meV and the distance from the source to the lens is $L = 100$ nm. (a) The components of the spinor wave function are $(1, 1)/\sqrt{2}$, the electron density is symmetric about the x-axis. (b) The components of the spinor wave function are $(1, 0)$, the mirror symmetry is broken. (c) The components of the spinor wave function are $(1, -1)/\sqrt{2}$, the mirror symmetry is restored, but the central maximum has disappeared.
(Reproduced with permission from Reijnders & Katsnelson, 2017a.)

$|T/V|$ the tunneling probability increases, and at some point a percolation transition happens (Shklovskii & Efros, 1984), with the formation of an infinite cluster of regions connected by electron tunneling. This percolation is associated with the Mott–Anderson metal–insulator transition, although the latter involves more

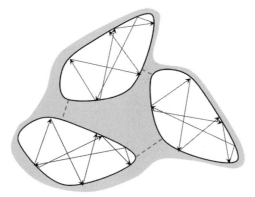

Fig. 4.11 A sketch of electronic states in conventional semiconductors with strong disorder; electrons tunnel, with a small probability, between classically allowed regions.

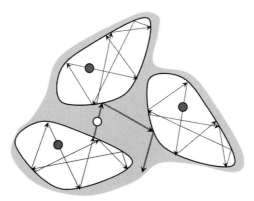

Fig. 4.12 A sketch of electronic states in graphene with strong disorder; due to Klein tunneling, electrons cannot be locked and penetrate through p-n boundaries, transforming into holes.

then just percolation, since phase relations between the electron wave functions are also important (Mott & Davis, 1979).

The Klein tunneling changes the situation dramatically. However small the kinetic energy is (or, equivalently, however high and broad the potential barriers are), the electrons cannot be locked into puddles (Fig. 4.12). Thus, their states cannot be localized.

In the absence of Anderson localization, the minimal conductivity can be estimated via Mott's considerations on the basis of the remark by Ioffe and Regel that for extended states the electron mean free path l cannot be smaller than the electron de Broglie wavelength (Mott, 1974; Mott & Davis, 1979). Here we apply this general consideration to graphene.

Let us start with Einstein's relation between the conductivity σ and the electron diffusion coefficient D (Zubarev, 1974).

$$\sigma = e^2 \frac{\partial n}{\partial \mu} D \qquad (4.77)$$

For a noninteracting degenerate (obeying Fermi statistics) electron gas

$$\frac{\partial n}{\partial \mu} = N(E_F) = \frac{2|E_F|}{\pi \hbar^2 v^2} = \frac{2k_F}{\pi \hbar v} \qquad (4.78)$$

(see Eq. (2.138) and Eq. (1.72)). For the two-dimensional case, the diffusion coefficient is

$$D = \frac{1}{2} v^2 \tau, \qquad (4.79)$$

where τ is the electron mean-free-path time. On substituting Eq. (4.78) and (4.79) into (4.77) one finds

$$\sigma = \frac{e^2}{\pi \hbar} k_F l = \frac{2e^2}{h} k_F l, \qquad (4.80)$$

where $l = v\tau$ is the mean free path. Assuming that the minimal possible value of $k_F l$ is of the order of unity, we have an estimation for the minimal conductivity of

$$\sigma_{\min} \sim \frac{e^2}{h} \qquad (4.81)$$

coinciding, in the order of magnitude, with the ballistic conductivity $e^2/(\pi h)$ per channel (see Eq. (3.16)).

This conclusion is very important, in the light of experimental observation of electron-hole puddles in graphene on a substrate in the vicinity of the neutrality point (Martin et al., 2008). Moreover, it was demonstrated theoretically that the puddles are unavoidable even for freely suspended graphene at room temperature since the inhomogeneities of electron density result from thermal bending fluctuations (Gibertini et al., 2010); this phenomenon will be considered in detail in Section 13.1. It is the Klein tunneling that protects electron states from localization and makes large-scale inhomogeneities rather irrelevant for electron transport.

The minimal conductivity was analyzed in terms of classical percolation by Cheianov et al. (2007). It follows from their analysis that the minimal conductivity is of the order of e^2/h if the number of electrons (holes) per puddle is of the order of 1.

4.7 Chiral tunneling in bilayer graphene

To elucidate which features of the anomalous tunneling in graphene are related to the linear dispersion and which features are related to the pseudospin and chirality of the Dirac spectrum, it is instructive to consider the same problem for bilayer graphene (Katsnelson, Novoselov, & Geim, 2006). We will restrict ourselves to the case of moderate electron energies, for which the parabolic approximation (1.46) works. This means that the energies are smaller than that of interlayer hopping, both outside and inside the barrier:

$$|E|, \ |E - V_0| \ll 2|\gamma_1| \tag{4.82}$$

and, at the same time, the trigonal warping effects are not important,

$$ka, qa > \left| \frac{\gamma_3 \gamma_1}{\gamma_0^2} \right| \tag{4.83}$$

(cf. Eq. (1.55)), where we assume that the potential barrier has the shape (4.20), and k and q are the wave vectors outside and inside the barrier, respectively:

$$k = \sqrt{\frac{2m^*|E|}{\hbar^2}},$$

$$q = \sqrt{\frac{2m^*|E - V_0|}{\hbar^2}}. \tag{4.84}$$

Assuming that the wave function propagates in the y-direction with the wave-vector component k_y, the two components of the spinor wave function are

$$\psi_1(x, y) = \psi_1(x) \exp\left(i k_y y\right),$$
$$\psi_2(x, y) = \psi_2(x) \exp\left(i k_y y\right), \tag{4.85}$$

where $\psi_i(x)$ satisfy the second-order equations

$$\left(\frac{d^2}{dx^2} - k_y^2 \right)^2 \psi_i = k^4 \psi_i \tag{4.86}$$

outside the barrier and

$$\left(\frac{d^2}{dx^2} - k_y^2 \right)^2 \psi_i = q^4 \psi_i \tag{4.87}$$

inside it. At the boundaries $x = \pm a$ one has to require that four conditions be fulfilled, namely continuity of ψ_1, ψ_2, $d\psi_1/dx$ and $d\psi_2/dx$. To satisfy them one has to include not only propagating but also evanescent solutions of Eq. (4.86) and (4.87) but, of course, without the terms growing exponentially at $x \to \pm\infty$.

Let us consider first the case $x < -a$. The two components of the wave function can be found from the equations

$$\left(\frac{d}{dx} + k_y\right)^2 \psi_2 = sk^2 \psi_1,$$
$$\left(\frac{d}{dx} + k_y\right)^2 \psi_1 = sk^2 \psi_2,$$

(4.88)

where $s = \text{sgn}\, E$ (cf. Eq. (4.45)). Thus, for this region one can try the solutions

$$\psi_1(x) = \alpha_1 \exp(ik_x x) + \beta_1 \exp(-ik_x x) + \gamma_1 \exp(\chi_x x),$$
$$\psi_2(x) = s[\alpha_1 \exp(ik_x x + 2i\varphi) + \beta_1 \exp(-ik_x x - 2i\varphi) - \gamma_1 h_1 \exp(\chi_x x)],$$

(4.89)

where φ is the angle of incidence,

$$k_y = k \sin\varphi,$$

$$k_x = k \cos\varphi,$$

(4.90)

$$\chi_x = \sqrt{k_x^2 + 2k_y^2} = k\sqrt{1 + \sin^2\varphi}$$

(4.91)

and

$$h_1 = \left(\sqrt{1 + \sin^2\varphi} - \sin\varphi\right)^2.$$

(4.92)

The coefficients α_1, β_1, and γ_1 are the amplitudes of the incident, reflected, and evanescent waves, respectively.

For the case $x > a$ there is no reflected wave:

$$\psi_1(x) = \alpha_3 \exp(ik_x x) + \delta_3 \exp(-\chi_x x),$$
$$\psi_2(x) = s\left[\alpha_3 \exp(ik_x x + 2i\varphi) - \frac{\delta_3}{h_1} \exp(-\chi_x x)\right];$$

(4.93)

the phase factor $\exp(2i\varphi)$ follows from Eq. (1.48). Finally, inside the barrier $|x| < a$ one has to use the most general solution with two propagating and two evanescent waves:

$$\psi_1(x) = \alpha_2 \exp(iq_x x) + \beta_2 \exp(-iq_x x) + \gamma_2 \exp(\chi_x' x) + \delta_2 \exp(-\chi_x' x),$$
$$\psi_2(x) = s'\left[\alpha_2 \exp(iq_x x + 2i\theta) + \beta_2 \exp(-iq_x x - 2i\theta) - \gamma_2 h_2 \exp(\chi_x' x) - \frac{\delta_2}{h_2} \exp(-\chi_x' x)\right],$$

(4.94)

where θ is the refraction angle,

$$q_y = q \sin \theta = k_y,$$

$$q_x = q \cos \theta, \tag{4.95}$$

$$\chi'_x = q\sqrt{1 + \sin^2 \theta}, \tag{4.96}$$

$$h_2 = \left(\sqrt{1 + \sin^2 \theta} - \sin \theta\right)^2 \tag{4.97}$$

and $s' = \mathrm{sgn}(E - V_0)$ (cf. Eq. (4.45)). The presence of the evanescent waves is a very interesting feature of bilayer graphene that is dramatically different both from the Dirac case and from the Schrödinger case.

Now we have to find the coefficients α_i, β_i, γ_i, and δ_i from eight conditions of continuity of $\psi_i(x)$ and $d\psi_i(x)/dx$ at $x = a$ and $x = -a$. In general, this can only be done numerically. Typical results for the "Klein" case $ss' = -1$ are shown in Fig. 4.4(b). Similarly to the case of single-layer graphene, there are "magic angles" with transmission probability equal to unity. A detailed mathematical analysis (Kleptsyn et al., 2015) shows, however, that contrary to the case of the single-layer graphene, where 100% transmission is protected by chirality, and for the case of symmetric potential, additional magic angles exist; for the case of bilayer, the magic angles are not necessary, and one can build a potential barrier with arbitrary, small transmission probability at *any* angle (for a given energy). It takes place for the potentials, which are oscillating rapidly enough (with a typical scale of the oscillations comparable with the de Broglie wavelength of the electrons), see Fig. 4.13.

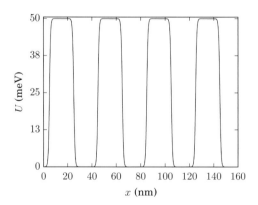

Fig. 4.13 An example of the fast-oscillating potential. Within the energy band from 20 to 30 meV the maximal transmission probability does not exceed 2×10^{-8}.

(Reproduced with permission from Kleptsyn et al., 2015.)

This provides a nice counterexample for a frequent statement that n-p-n (or p-n-p) junction cannot be locked in single-layer graphene due to the energy gap absence. In the case of bilayer, the gap is also absent but the junction *can* be locked! It is the chiral properties of electrons (conservation of pseudospin and, therefore, the propagation direction for the normally incident beam) rather than the gap absence. The difference can already be seen from our simple case of a rectangular barrier if we focus on the case of the normally incident beam.

For the case of normal incidence ($\varphi = 0$, $\theta = 0$) the problem can be solved analytically, and the result for the transmission coefficient is

$$t = \frac{\alpha_3}{\alpha_1} = \frac{4ikq \exp{(2ika)}}{(q+ik)^2 \exp{(-2qa)} - (q-ik)^2 \exp{(2qa)}}. \qquad (4.98)$$

In contrast with the case of single-layer graphene, $T = |t|^2$ decays exponentially with the height and the width of the barriers, as $\exp(-4qa)$ for $\varphi = 0$. This situation is sometimes called anti-Klein tunneling. This illustrates a drastic difference between the cases of chiral scattering with Berry phases π and 2π. For the latter case, the condition (1.49) does not fix the projection of the pseudospin to the direction of the motion (cf. Eq. (1.33)), so the conservation of the chirality does not forbid backscattering.

For the case $a \to \infty$ (which is just a potential step corresponding to a single p-n junction) $T = 0$ at $\varphi = 0$, which looks rather counterintuitive: There is a continuum of allowed states after the barrier but penetration there is forbidden. Furthermore, for a single p-n junction with $V_0 \gg E$, the following analytic solution for any φ has been found:

$$T = \frac{E}{V_0} \sin^2{(2\varphi)}, \qquad (4.99)$$

which, again, yields $T = 0$ for $\varphi = 0$. This behavior is in obvious contrast with that of single-layer graphene, where normally incident electrons are always perfectly transmitted.

The perfect reflection (instead of perfect transmission) can be viewed as another incarnation of the Klein paradox, because the effect is again due to the charge-conjugation symmetry. For single-layer graphene, an electron wave function at the barrier interface perfectly matches the corresponding wave function for a hole with the same direction of pseudospin, yielding $T = 1$. In contrast, for bilayer graphene, the charge conjugation requires a propagating electron with wave vector k to transform into a hole with wave vector ik (rather than $-k$), which is an evanescent wave inside a barrier.

For completeness, we compare the results obtained with those from the case of conventional nonrelativistic electrons. If a tunnel barrier contains no electronic

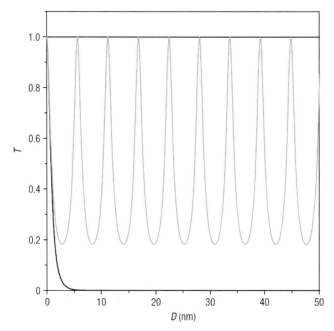

Fig. 4.14 The transmission probability T for normally incident electrons in single-layer and bilayer graphene and in a nonchiral, zero-gap semiconductor as a function of the width D of the tunnel barrier. The concentrations of charge carriers are chosen as $n = 0.5 \times 10^{12}$ cm^{-2} and $p = 1 \times 10^{13}$ cm^{-2} outside and inside the barrier, respectively, for all three cases. The transmission probability for bilayer graphene (the lowest line) decays exponentially with the barrier width, even though there are plenty of electronic states inside the barrier. For single-layer graphene it is always 1 (the upper line). For the nonchiral semiconductor it oscillates with the width of the barrier (the intermediate curve).
(Reproduced with permission from Katsnelson, Novoselov, & Geim, 2006.)

states, the difference is obvious: The transmission probability in this case is known to decay exponentially with increasing barrier width and height (Esaki, 1958), so that the tunnel barriers discussed previously would reflect electrons completely. However, both graphene systems are gapless, and it is more appropriate to compare them to gapless semiconductors with nonchiral charge carriers (such a situation can be realized in certain heterostructures (Meyer et al., 1995; Teissier et al., 1996)). In this case, we find

$$t = \frac{4k_x q_x \exp\left(2iq_x a\right)}{\left(q_x + k_x\right)^2 \exp\left(-2iq_x a\right) - \left(q_x - k_x\right)^2 \exp\left(2iq_x a\right)}, \tag{4.100}$$

where k_x and q_x are the x-components of the wave vector outside and inside the barrier, respectively. Again, similarly to the case of single-layer and bilayer graphene, there are cases of normal incidence ($\varphi = 0$), the resonance conditions

$2q_x a = \pi N$, $N = 0, \pm1$, at which the barrier is transparent. For the tunneling coefficient is then an oscillating function of the tunneling parameters and can exhibit any value from 0 to 1 (see Fig. 4.14). This is in contrast with graphene, for which T is always 1, and bilayer graphene, for which $T = 0$ for sufficiently wide barriers. This makes it clear that the drastic difference among the three cases is essentially due to the different chiralities or pseudospins of the quasiparticles involved rather than any other features of their energy spectra.

To summarize this chapter, the Klein paradox is a key phenomenon for electronic transport in graphene and for graphene-based electronics. On the one hand, it protects high electron mobility in inhomogeneous graphene and prevents Anderson localization. On the other hand, it is an essential obstacle to copying a "normal" transistor based on p-n-p (or n-p-n) junctions in conventional semiconductors. Usually, one can easily lock the transistor by applying a voltage to the potential barrier, which is impossible for the cases of both single-layer and bilayer graphene due to the Klein paradox. One needs to open a gap in the electron spectrum. One of the most natural ways to do this is the use of space quantization in graphene nanoribbons and nanoflakes, which will be one of the subjects of the next chapter.

5

Edges, nanoribbons, and quantum dots

5.1 The neutrino billiard model

Owing to the Klein paradox, the massless Dirac fermion cannot be confined in a restricted region by any configuration of a purely electrostatic (scalar) potential $V(x, y)$; one needs the gap opening. As discussed in Section 1.3, this requires a violation of the equivalence of the sublattices. Let us consider the Hamiltonian

$$\hat{H} = -i\hbar v \vec{\sigma} \nabla + \sigma_z \Delta(x, y),$$

where the last term represents a difference of potential energy between the A and B sites (or between (pseudo)spin up and (pseudo)spin down states). With $\Delta = $ constant the energy spectrum of the Hamiltonian (5.1) is

$$E(\vec{k}) = \pm\sqrt{\hbar^2 v^2 k^2 + \Delta^2},$$

where $\vec{k}$ is the wave vector and there is the energy gap $2|\Delta|$. For a given energy E, the regions where $|E| < |\Delta(x, y)|$ are classically forbidden; quantum mechanically, the probability of tunneling to these regions decays exponentially with the distance from the boundary. In particular, one can introduce the boundary condition

$$|\Delta(x, y)| = \pm\infty$$

at a line L; thus, only the region D restricted by the line L is allowed for the particle (Fig. 5.1). The line L is parameterized by the length s counted from some initial point:

$$x = x_L(s), \quad y = y_L(s)$$

We will assume

$$\Delta(x, y) = 0$$

within the region D.

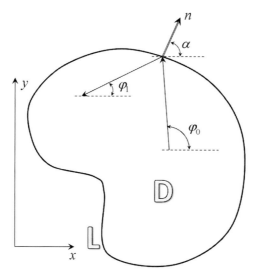

Fig. 5.1 The geometry of a "neutrino billiard." The particle moves within the region D restricted by the line L where the infinite energy gap opens.

This model was considered by Berry and Mondragon (1987) long before the discovery of graphene and was called the "neutrino billiard" (at that time it was assumed that the neutrino had zero mass). It is not sufficient to completely describe the edge effects and confinement in graphene nanoribbons and nanoflakes: As we will see further, the existence of two valleys is of crucial importance, thus, the single Dirac point approximation is not enough. However, it already contains some important physics, so it is convenient to start our consideration with this model.

An important property of the Hamiltonian (5.1) is that it is not invariant under the time-reversal symmetry operation $\hat{T}$. The latter can be represented (Landau & Lifshitz, 1977) as

$$\hat{T} = \hat{U}\hat{K}, \tag{5.6}$$

where

$$\hat{U} = i\hat{\sigma}_y = \begin{pmatrix} 0 & 1 \\ -1 & 0 \end{pmatrix} \tag{5.7}$$

and $\hat{K}$ is the complex conjugation. Under this operation the Hamiltonian $\hat{H}$ (5.1) is transformed into

$$\hat{H}' = \hat{U}\hat{H}^*\hat{U}^+ = -i\hbar v\vec{\sigma}\nabla - \sigma_z\Delta(x, y) \tag{5.8}$$

and differs from Eq. (5.1) by the sign of Δ. This means that there is no Kramers degeneracy (Landau & Lifshitz, 1977) of the energy levels of the

Hamiltonian (5.1). At the same time this means that the energy spectrum is insensitive to the sign of Δ: If

$$\Psi = \begin{pmatrix} \psi_1 \\ \psi_2 \end{pmatrix}$$

is an eigenstate of the Hamiltonian (5.1) with an energy E, the function

$$\Psi' = \hat{T}\Psi = \begin{pmatrix} \psi_2{}^* \\ -\psi_1{}^* \end{pmatrix} \tag{5.9}$$

corresponds to the same eigenvalue E for the Hamiltonian (5.8). Obviously, Ψ' is orthogonal to Ψ, since $(\Psi')^* \, \Psi = 0$.

The most general boundary condition for the Hamiltonian (5.1) and (5.5) follows from the requirement that it should be Hermitian (or, equivalently, its energy spectrum should be real). Using the Gauss theorem, one has

$$\iint_D dxdy\left(\psi^+\hat{H}\psi - \psi^+\hat{H}^+\psi\right) = -i\hbar v \iint_D dxdy[\psi^+\vec{\sigma}\nabla\psi + (\nabla\psi^+)\vec{\sigma}\psi]$$

$$= -i\hbar v \iint_D dxdy\nabla[\psi^+\vec{\sigma}\psi] = -i\hbar \oint_L d s\vec{n}(s)\vec{j}(s) = 0, \tag{5.10}$$

where $\vec{n}$ is the unit vector normal to the curve L and $\vec{j} = v\Psi^+\vec{\sigma}\Psi$ is the current density (cf. Eq. (3.2)).

The local boundary condition must ensure that there is no normal current to the boundary at any point. On introducing the angle α such that

$$\vec{n} = (\cos\alpha, \sin\alpha) \tag{5.11}$$

(see Fig. 5.1), one can write this condition as

$$\cos\alpha\,\mathrm{Re}\left(\psi_1^*\psi_2\right) + \sin\alpha\,\mathrm{Im}\left(\psi_1^*\psi_2\right) = 0 \tag{5.12}$$

or, equivalently,

$$\frac{\psi_2}{\psi_1} = iB\exp\left(i\alpha(s)\right), \tag{5.13}$$

where $B = B(s)$ is real.

To specify B, one can consider first the case of a flat boundary L $\parallel$ Oy. One can assume that $\Delta = 0$ at $x < 0$ and $\Delta = \Delta_0 = $ constant at $x > 0$, solve the Dirac equation explicitly as was done in the previous chapter, consider the reflection problem, and compare the result for $\psi_2(x = -0)/\psi_1(x = -0)$ with Eq. (5.13) at $\alpha = 0$. One can see that

$$B = \pm 1 \tag{5.14}$$

at $\Delta_0 \to \pm\infty$. We will call Eq. (5.13) with $B = \pm 1$ the *infinite-mass boundary condition* (Berry & Mondragon, 1987).

It is not surprising that this boundary condition is not invariant under the time-reversal operation. Indeed, it follows from Eq. (5.9) and (5.13) that

$$\frac{\psi'_2}{\psi'_1} = -\left(\frac{\psi_1}{\psi_2}\right)^* = -iB\exp\left(i\alpha(s)\right), \tag{5.15}$$

which differs from Eq. (5.13) by the sign (we have taken into account that $B^2 = 1$).

Confinement of electrons in a finite region leads to a discrete energy spectrum. Consider first the simplest case in which L is just a circle, $r = R$, where we pass to the polar coordinates

$$x = r\cos\varphi, \quad y = y\sin\varphi. \tag{5.16}$$

In these coordinates,

$$-i\vec{\sigma}\nabla = -i\left(\begin{array}{cc} 0 & e^{-i\varphi}\left(\dfrac{\partial}{\partial r} - \dfrac{i}{r}\dfrac{\partial}{\partial\varphi}\right) \\ e^{i\varphi}\left(\dfrac{\partial}{\partial r} + \dfrac{i}{r}\dfrac{\partial}{\partial\varphi}\right) & 0 \end{array}\right) \tag{5.17}$$

and the Schrödinger equation for the state with $E = \hbar v k$ takes the form

$$e^{-i\varphi}\left(\frac{\partial}{\partial r} - \frac{i}{r}\frac{\partial}{\partial\varphi}\right)\psi_2 = ik\psi_1,$$

$$e^{i\varphi}\left(\frac{\partial}{\partial r} + \frac{i}{r}\frac{\partial}{\partial\varphi}\right)\psi_1 = ik\psi_2. \tag{5.18}$$

One can try solutions of Eq. (5.18) of the form

$$\psi_1(r,\varphi) = \psi_1(r)\exp\left(il\varphi\right),$$

$$\psi_2(r,\varphi) = \psi_2(r)\exp\left[i(l+1)\varphi\right], \tag{5.19}$$

where l is integer. On substituting Eq. (5.19) into Eq. (5.18) one has

$$\begin{cases} \dfrac{d\psi_2}{dr} + \dfrac{l+1}{r}\psi_2 = ik\psi_1, \\ \dfrac{d\psi_1}{dr} - \dfrac{l}{r}\psi_1 = ik\psi_2. \end{cases} \tag{5.20}$$

By excluding ψ_1 (or ψ_2) from Eq. (5.20), one can find a second-order differential equation for the Bessel functions (Whittaker & Watson, 1927). The solutions regular at $r \to 0$ are

$$\psi_1(r) = J_l(kr),$$
$$\psi_2(r) = iJ_{l+1}(kr).$$
(5.21)

The energy spectrum $k = k_{nl}$ can be found from the boundary condition (5.13), keeping in mind that for the circle $\alpha = \varphi$. Thus, the quantization rule for the disc is

$$J_{l+1}(k_{nl}R) = BJ_l(k_{nl}R).$$
(5.22)

This leads to a discrete spectrum with a distance between neighboring energy levels with a given l of

$$\delta_l(E) \cong \frac{\pi\hbar v}{R}.$$
(5.23)

The density of states of the whole system is an extensive quantity proportional (in two dimensions) to the system area A. Therefore, the average energy distance (for an arbitrary shape of the billiard, not necessarily for the disc) can be estimated as

$$\delta(E) \approx \frac{1}{N(E)A},$$
(5.24)

where $N(E)$ is the density of states of the Dirac Hamiltonian per unit area:

$$N(E) = \frac{E}{2\pi\hbar^2 v^2} = \frac{k}{2\pi\hbar v}.$$
(5.25)

It differs from Eq. (1.72) by a factor of 4 (here we do not take into account the fourfold spin and valley degeneracy for graphene). The semiclassical estimation (5.24) (see Perenboom, Wyder, & Meier, 1981; Halperin, 1986; Stöckmann, 2000) is valid at

$$k\sqrt{A} \gg 1.$$
(5.26)

For the case of a circular disc that Eq. (5.20) gives, taking into account Eq. (5.23) through (5.25),

$$\delta(E) \approx \frac{\delta_l(E)}{kR} \propto \frac{1}{R^2}.$$
(5.27)

There is an important issue relating to the energy-level distribution in finite systems (Bohr & Mottelson, 1969; Perenboom, Wyder, & Meier, 1981; Stöckmann, 2000). In the case of integrable systems with regular classical motion of particles it is supposed that it follows the Poisson statistics. It was shown by Berry and Mondragon (1987) that this is indeed the case for the spectrum determined by Eq. (5.22). For a generic system with chaotic motion level repulsion takes place, and the probability of finding two very close energy levels is strongly suppressed. The main physical statement can be seen just from the two-level quantum-mechanical problem with a 2×2 Hamiltonian, for which the splitting of eigenvalues is

$$\Delta_{1,2} = \sqrt{(H_{11} - H_{22})^2 + 4|H_{12}|^2}. \tag{5.28}$$

If the Hamiltonian matrix is diagonal, the probability of degeneracy $\Delta_{1,2} = 0$ is equal to the probability that $H_{11} = H_{22}$; if the matrix is off-diagonal and real, it is the probability that $H_{11} = H_{22}$ and $H_{12} = 0$; if it is not real, it is the probability that $H_{11} = H_{22}$, and Re $H_{12} = 0$ and Im $H_{12} = 0$, which is obviously smaller.

For a generic chaotic system with time-reversal symmetry (this means that the basis exists in which the Hamiltonian is real) the distribution of the neighboring levels, $S = \Delta E / \delta(E)$, is given by the Gaussian orthogonal ensemble (GOE), with the probability function

$$P_{\text{GOE}}(S) = \frac{\pi S}{2} \exp\left(-\frac{\pi S^2}{4}\right), \tag{5.29}$$

whereas without time-reversal symmetry we have the Gaussian unitary ensemble (GUE), with

$$P_{\text{GUE}}(S) = \frac{32 S^2}{\pi^2} \exp\left(-\frac{4 S^2}{\pi}\right) \tag{5.30}$$

(Bohr & Mottelson, 1969; Perenboom, Wyder, & Meier, 1981; Stöckmann, 2000).

The numeral calculations of Berry and Mondragon (1987) demonstrate that the level distribution for neutrino billiards with chaotic classical motion obeys the GUE statistics (5.30). This is the consequence of violation of the time-reversal symmetry, which was discussed previously.

5.2 A generic boundary condition: valley mixing

As was discussed in Chapter 1, charge carriers in graphene can be described in the single Dirac-cone approximation only if all external inhomogeneities are smooth at the atomic scale. The edges of the terminated honeycomb lattice are sharp and can, in general, mix the electron states belonging to different valleys. So, one should use a more general, two-valley Hamiltonian (1.28) (we will use here the representation (1.27)). The current operator (cf. Eq. (3.3)) is

$$\hat{\vec{j}} = \frac{\delta \hat{H}}{\delta \vec{p}} = v \tau_0 \otimes \vec{\sigma}. \tag{5.31}$$

The most general restriction on the boundary condition generalizing Eq. (5.10) and (5.12) in the two-valley case is the absence of the normal component of the current through the boundary

$$\left\langle \Psi \left| \vec{n}(s)\hat{\vec{j}} \right| \Psi \right\rangle = 0, \tag{5.32}$$

at any s.

We will consider, following McCann and Fal'ko (2004) and Akhmerov and Beenakker (2008), the boundary conditions for the abruptly terminated honeycomb lattice, with zero probability of finding an electron outside the graphene flake. The simplest terminations, *zigzag* and *armchair* edges, are shown in Fig. 5.2.

Then the Schrödinger equation inside the flake reads

$$\left[-i\hbar v \tau_0 \otimes \vec{\sigma}\nabla + \hbar v \hat{M}' \delta(\vec{r} - \vec{r}_{\mathrm{B}}) \right] \Psi = E\Psi, \tag{5.33}$$

where $\vec{r} = \vec{r}_{\mathrm{B}}(s)$ in the equation of the boundary line L, and $\hat{M}'$ is an energy-independent Hermitian matrix. By integrating Eq. (5.33) along an infinitesimal line parallel to the normal $\vec{n}(s)$ to the boundary and taking into account that $\Psi = 0$ outside the flake, one finds the boundary condition

$$\hat{A}\Psi = i\hat{M}'\Psi \tag{5.34}$$

at $\vec{r} = \vec{r}_{\mathrm{B}}(s)$, where

$$\hat{A} = \vec{n}\hat{\tau}_0 \otimes \hat{\vec{\sigma}} = \frac{1}{v}\vec{n}.\hat{\vec{j}} \tag{5.35}$$

$\left(\hat{A}^2 = 1 \right)$. Equivalently, the condition (5.34) can be represented as

$$\Psi = \hat{M}\Psi \ (\vec{r} = \vec{r}_{\mathrm{B}}), \tag{5.36}$$

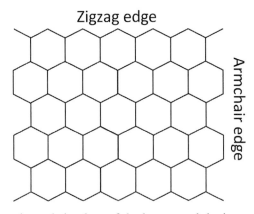

Fig. 5.2 Zigzag and armchair edges of the honeycomb lattice.

where

$$\hat{M} = i\hat{A}\hat{M}'. \tag{5.37}$$

On iterating Eq. (5.36) one can see that

$$\hat{M}^2 = 1. \tag{5.38}$$

If we require that the Hermitian matrices $\hat{A}$ and $\hat{M}'$ anticommute,

$$\{\hat{A}, \hat{M}'\} = 0, \tag{5.39}$$

the matrix (5.37) turns out to be Hermitian and, due to Eq. (5.38), also unitary:

$$\hat{M}^+ = \hat{M} = \hat{M}^{-1}. \tag{5.40}$$

It also anticommutes with the matrix $\hat{A}$:

$$\{\hat{A}, \hat{M}\} = i\hat{A}^2\hat{M}' + i(\hat{A}\hat{M}')\hat{A} = 0 \tag{5.41}$$

and the condition (5.32) is automatically satisfied in this case:

$$\Psi^+\hat{A}\Psi = \Psi^+\hat{M}^+\hat{A}\hat{M}\Psi = -\Psi^+\hat{A}\Psi = 0. \tag{5.42}$$

Thus, the boundary condition (5.36) with the most general matrix $\hat{M}$ satisfying the requirements (5.40) and (5.41) seems to be the most general form of the boundary conditions at the edges of terminated graphene flakes.

As was proven by Akhmerov and Beenakker (2008) the most general allowed matrix $\hat{M}$ can be represented as

$$\hat{M} = \sin\Lambda\hat{\tau}_0 \otimes \left(\vec{n}_1\hat{\vec{\sigma}}\right) + \cos\Lambda\left(\vec{v}\hat{\vec{\tau}}\right) \otimes \left(\vec{n}_2\hat{\vec{\sigma}}\right), \tag{5.43}$$

where Λ is an arbitrary real number and $\vec{v}, \vec{n}_1$ and $\vec{n}_2$ are three-dimensional unit vectors such that $\vec{n}_1$ and $\vec{n}_2$ are mutually orthogonal and also orthogonal to $\vec{n}$ ($\vec{v}$ is arbitrary).

One can assume that the boundary conditions for the graphene flake as a whole should be time-reversal symmetric. Formally, this follows from the fact that the tight-binding Hamiltonian for the honeycomb lattice in real space can be chosen as a *real* matrix. The time-reversal symmetry can be broken by spontaneous valley polarization at the edges or by spin polarization plus spin-orbit coupling. So far, there is no clear experimental evidence for such phenomena (as for the possible spin polarizaion at the edges, see Chapter 12).

On generalizing the definition of the time-reversal operation (5.6) to the case of two valleys one can write

$$\hat{T} = -\hat{\tau}_y \otimes \vec{\sigma}_y \cdot \hat{K}. \tag{5.44}$$

The matrix $\hat{M}$ (5.43) commutes with $\hat{T}$ only at $\Lambda = 0$; thus, for the time-reversal-invariant case

$$\hat{M} = \left(\vec{v}\hat{\vec{\tau}}\right) \otimes \left(\vec{m}\hat{\vec{\sigma}}\right), \vec{m} \perp \vec{n}. \qquad (5.45)$$

Further specification of the boundary conditions can be achieved by assuming the nearest-neighbor approximation (which is actually quite accurate for graphene, see Chapter 1). In this approximation there exist only hopping terms between sublattices, $\hat{H}_{AB}$, whereas intrasublattice terms vanish: $\hat{H}_{AA} = \hat{H}_{BB} = 0$ (see Eq. (1.14)). The Schrödinger equation for the two-component wave function (the components correspond to the sublattices)

$$\hat{H}_{AB}\psi_A = E\psi_B,$$

$$\hat{H}_{AB}^+\psi_B = E\psi_A \qquad (5.46)$$

has a rigorous electron–hole symmetry: $\psi_B \to -\psi_B$, $E \to -E$ transforms the equation to itself. In the limit of small energies $|E| << |t|$ this means that the operation $\hat{R} = \tau_z \otimes \sigma_z$ changes the sign of the Hamiltonian

$$\hat{R}\hat{H}\hat{R} = -\hat{H} \qquad (5.47)$$

or, equivalently (keeping in mind that $\hat{R}^2 = 1$),

$$\left\{\hat{H}, \hat{\tau}_z \otimes \hat{\sigma}_z\right\} = 0. \qquad (5.48)$$

This symmetry is an approximate one for real graphene, but this approximation is quite good due to the smallness of the second-neighbor hopping $|t'/t| \approx 0.1$ (see Section 1.2). If we require (5.48), there are only two classes of allowed boundary conditions: (1) $\vec{v}\|Oz, \vec{m}\|Oz$, for which

$$\hat{M} = \pm\hat{\tau}_z \otimes \hat{\sigma}_z; \qquad (5.49)$$

and (2) $v_z = m_z = 0$, for which

$$\hat{M} = \left(\cos\varphi\hat{\tau}_x + \sin\varphi\hat{\tau}_y\right) \otimes \sigma_x \qquad (5.50)$$

(we assume that the edge is along the x-axis $\vec{n}\|Oy$, and thus $\vec{m}\|Ox$).

Boundary conditions of the type (5.36) and (5.49) are called *zigzaglike*, whereas those of the type (5.36) and (5.50) are called *armchairlike*, for reasons that will be discussed in the next section. It is an important result (Akhmerov & Beenakker, 2008; Wimmer, Akhmerov, & Guinea, 2010) that zigzaglike boundary conditions are generic, whereas armchairlike boundary conditions occur only for some exceptional orientations of the edges.

5.3 Boundary conditions for a terminated honeycomb lattice

Here we present, following Akhmerov and Beenakker (2008), a microscopic derivation of the boundary conditions for a terminated honeycomb lattice in the nearest-neighbor approximation. The geometry of our problem is clear from Fig. 5.3. The translation vector along the boundary is

$$\hat{T} = n\vec{R}_1 + m\vec{R}_2, \tag{5.51}$$

where

$$\hat{R}_{1,2} = \frac{a}{2}\left(\sqrt{3}, \mp 1\right) \tag{5.52}$$

are elementary translation vectors and n and m are integers. The number N of missing sites and the number N' of dangling bonds per period are larger than or equal to $n + m$. Fig. 5.3(d) shows a *minimal* boundary where $N = N' = n + m$.

The Schrödinger equation for the tight-binding model in the nearest-neighbor approximation reads

$$\psi_B(\vec{r}) + \psi_B(\vec{r} - \vec{R}_1) + \psi_B(\vec{r} - \vec{R}_2) = \varepsilon\psi_A(\vec{r}),$$
$$\psi_A(\vec{r}) + \psi_A(\vec{r} - \vec{R}_1) + \psi_A(\vec{r} - \vec{R}_2) = \varepsilon\psi_B(\vec{r}), \tag{5.53}$$

where $\varepsilon = E/t$ is the dimensionless energy and subscripts A and B label sublattices.

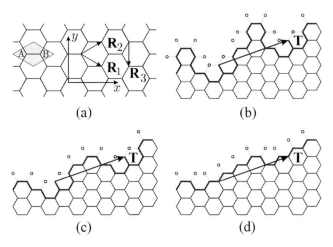

(a)   (b)   (c)   (d)

Fig. 5.3 (a) A honeycomb lattice constructed from a unit cell (gray rhombus) containing two atoms (labelled A and B), translated over lattice vectors $\mathbf{R}_1$ and $\mathbf{R}_2$. Panels (b)–(d) show three different periodic boundaries with the same period $\mathbf{T} = n\mathbf{R}_1 + m\mathbf{R}_2$. Atoms on the boundary (connected by thick solid lines) have dangling bonds (thin gray line segments) to empty neighboring sites (open circles). The number N of missing sites and the number N' of dangling bonds per period are $n + m$. Panel (d) shows a minimal boundary, for which $N = N' = n + m$. (Reproduced with permission from Akhmerov & Beenakker, 2008.)

The angle between the translation vector $\vec{T}$ and the armchair orientation (the direction Ox in Fig. 5.3(a)) is

$$\varphi = \arctan\left(\frac{1}{\sqrt{3}}\frac{n-m}{n+m}\right). \tag{5.54}$$

Owing to symmetry with respect to rotations at $\pm\pi/3$ we can restrict ourselves to the case $|\varphi| < \pi/6$ only.

The boundary condition is the requirement that the wave function vanishes at the empty sites. One can assume that it depends smoothly on the energy ε. We are interested in the case of small ε (the states close to the Dirac points) and, thus, can put $\varepsilon = 0$ in Eq. (5.53). So, as a first step one can find zero-energy modes for the terminated honeycomb lattice. Owing to the translational invariance along the boundary, one can use the Bloch theorem and require that

$$\psi_{A,B}(\vec{r}+\vec{T}) = e^{ik}\psi_{A,B}(\vec{r}) \tag{5.55}$$

with a real $0 \le k < 2\pi$.

For the behavior normal to the boundary, we assume that

$$\psi_{A,B}(\vec{r}+\vec{R}_3) = \lambda\psi_{A,B}(\vec{r}), \tag{5.56}$$

where $\vec{R}_3 = \vec{R}_1 - \vec{R}_2$ is antiparallel to the y-axis in Fig. 5.3(a). This lattice vector has a nonzero component $a\cos\varphi > a\sqrt{3}/2$ perpendicular to $\vec{T}$. For the states localized at the edge $|\lambda| < 1$ and for propagating states $|\lambda| = 1$; of course, the case $|\lambda| > 1$ is meaningless, since the corresponding wave function cannot be normalized. If $|\lambda| < 1$, the solution satisfying Eq. (5.56) has a decay length in the direction normal to $\vec{T}$ of

$$l = -\frac{a\cos\varphi}{\ln|\lambda|}. \tag{5.57}$$

Taking into account that $\vec{R}_1 = \vec{R}_2 + \vec{R}_3$, one can rewrite Eq. (5.53) at $\varepsilon = 0$ as

$$\psi_B(\vec{r}) + \psi_B(\vec{r} - \vec{R}_2 - \vec{R}_3) + \psi_B(\vec{r} - \vec{R}_2) = 0,$$
$$\psi_A(\vec{r}) + \psi_A(\vec{r} - \vec{R}_2 - \vec{R}_3) + \psi_A(\vec{r} - \vec{R}_2) = 0. \tag{5.58}$$

On substituting Eq. (5.56) into Eq. (5.58) one finds

$$\psi_B(\vec{r}+\vec{R}_2) = -\frac{1}{1+\lambda}\psi_B(\vec{r}),$$
$$\psi_A(\vec{r}+\vec{R}_2) = -(1+\lambda)\psi_A(\vec{r}). \tag{5.59}$$

Using Eq. (5.56) and (5.55) together, we have, for any integer p and q,

$$\psi_B\left(\vec{r} + p\vec{R}_2 + q\vec{R}_3\right) = \lambda^q(-1-\lambda)^{-p}\psi_B(\vec{r}),$$

$$\psi_A\left(\vec{r} + p\vec{R}_2 + q\vec{R}_3\right) = \lambda^q(-1-\lambda)^{-p}\psi_A(\vec{r})$$

(5.60)

Now we have to recall the Bloch theorem (5.55) for

$$\vec{T} = n\left(\vec{R}_2 + \vec{R}_3\right) + m\vec{R}_2 = (n+m)\vec{R}_2 + n\vec{R}_3.$$

(5.61)

Thus, we have two equations relating k and λ:

$$(-1-\lambda)^{m+n} = e^{ik}\lambda^n$$

(5.62)

for the sublattice A and

$$(-1-\lambda)^{m+n} = e^{ik}\lambda^m$$

(5.63)

for the sublattice B. One needs to find all solutions λ of Eq. (5.62) and (5.63) for a given k satisfying the conditions $|\lambda| \leq 1$.

A general zero-energy state can be represented as

$$\psi_A = \sum_{p=1}^{M_A} a_p \psi_p$$

$$\psi_B = \sum_{p=1}^{M_B} a_p' \psi_p',$$

(5.64)

where M_A and M_B are the numbers of solutions of Eq. (5.62) and (5.63) within the unit circle, respectively, and ψ_p and ψ_p' are the corresponding eigenstates. The coefficients a_p and a_p' should be chosen in such a way that ψ_A and ψ_B vanish at missing sites from the sublattices A and B.

The Dirac limit corresponds to the case of small k. Explicit calculations for the case $k = 0$ give the result (Akhmerov & Beenakker, 2008)

$$M_A = \frac{2n+m}{3}$$

$$M_B = \frac{2m+n}{3} + 1.$$

(5.65)

These solutions include also the values

$$\lambda_\pm = \exp\left(\pm\frac{2\pi i}{3}\right),$$

(5.66)

corresponding to the propagating modes; for all other modes $|\lambda| < 1$, so they are localized at the edge. The corresponding eigenstate is $\exp(\pm i\vec{K}\vec{r})$, with

$$\vec{K} = \frac{4\pi}{3a^2}\vec{R}_3. \tag{5.67}$$

Thus, the general zero-energy mode at $k = 0$ can be represented as

$$\psi_A = \psi_1 \exp(i\vec{K}\vec{r}) + \psi_4 \exp(-i\vec{K}\vec{r}) + \sum_{p=1}^{M_A-2} \alpha_p \psi_p,$$

$$\psi_B = \psi_2 \exp(i\vec{K}\vec{r}) + \psi_3 \exp(-i\vec{K}\vec{r}) + \sum_{p=1}^{M_B-2} \alpha'_p \psi'_p. \tag{5.68}$$

The four amplitudes $(\psi_1, -i\psi_2, i\psi_3, -\psi_4)$ correspond to the four components of the wave function (1.27) in the Dirac limit; ψ_1 and ψ_2 are associated with the valley K, ψ_3 and ψ_4 with the valley K'.

At the same time, there are N_A conditions $\psi_A = 0$ at the missing sites belonging to the sublattice A and N_B conditions $\psi_B = 0$ at the missing sites belonging to the sublattice B (N_A and N_B are the numbers of missing sites belonging to the corresponding sublattice).

For the minimal boundary, $N_A = n$ and $N_B = m$. At the same time, for $n > m$ one has $M_A < n$ conditions $\psi_A = 0$ at some sites. The only way to satisfy them is to require that $\psi_A = 0$ on the whole boundary, including $\psi_1 = \psi_4 = 0$. At the same time, $M_B > m + 2$, so ψ_2 and ψ_3 remain undetermined.

This corresponds to the zigzag boundary conditions Eq. (5.49), with the minus sign. Similarly, for $n < m$ one has the zigzag boundary conditions with the plus sign. Only at $n = m$ does one have $M_A = M_B = n + 1 > n$, such that one has the same condition for sublattices A and B. All ψ_i are nonzero in this case, with

$$|\psi_1| = |\psi_4|, \quad |\psi_2| = |\psi_3| \tag{5.69}$$

(armchair boundary conditions (5.50)).

So, at least for the case of minimal edges, one can prove that the armchair boundary conditions are exceptional whereas the zigzag ones are generic. This result also seems to be correct for nonminimal edges, as well as for the case of disorder at the edges (Martin & Blanter, 2009; Wimmer, Akhmerov, & Guinea, 2010).

For the case $n > m$, the number of independent zero-energy modes per unit length is (Akhmerov & Beenakker, 2008; Wimmer, Akhmerov, & Guinea, 2010)

$$\rho = \frac{M_A - n}{|\vec{T}|} = \frac{|m - n|}{3a\sqrt{n^2 + nm + m^2}} = \frac{2}{3a}|\sin\varphi|. \tag{5.70}$$

At $\varphi = 0$ (armchair boundaries) there are no such states. The existence of the zero-energy modes and the corresponding sharp peak in the density of states at zigzag edges was first found numerically by Nakada et al. (1996). It will be analyzed in more detail in the next sections.

Akhmerov and Beenakker (2008) have demonstrated that the infinite-mass boundary condition (5.13) with $B = \pm 1$ can be obtained in the limit of an infinite staggered field (difference of on-site energies between sublattices A and B at the edge). The sign of B is determined by the sign of this staggered field.

5.4 Electronic states of graphene nanoribbons

The previous consideration was a bit formal, but the result is quite simple. For the case of pure zigzag edges *all* missing atoms belong to sublattice A only (or sublattice B only), thus the corresponding components of the wave function for the two valleys, K and K′, should vanish at the boundary. If the numbers of missing atoms belonging to A and B are not equal, the boundary conditions remain the same, depending on the majority of the atoms: "The winner takes all." Only in the exceptional case, in which the numbers of missing atoms from A and B coincide exactly (armchair edges), are all four components of the Dirac spinors finite at the edge, satisfying the two relations (5.69).

If we have a nanoribbon of a constant width $L(|y| \leq L/2)$ with zigzag edges, one edge corresponds to the missing atoms A and the other to the missing atoms B. The boundary conditions are

$$u\left(y = -\frac{L}{2}\right) = 0,$$
$$v\left(y = \frac{L}{2}\right) = 0,$$

(5.71)

where u is ψ_1 or ψ_4 and v is ψ_2 or ψ_3. In this case the valleys are decoupled, so in the Dirac approximation we can consider them independently. For the valley K, the Schrödinger equation reads

$$\left(\frac{\partial}{\partial x} + i\frac{\partial}{\partial y}\right)u(x, y) = ikv(x, y),$$
$$\left(\frac{\partial}{\partial x} - i\frac{\partial}{\partial y}\right)v(x, y) = iku(x, y),$$

(5.72)

where $k = E/(\hbar v)$. For the valley K′, the signs before $\partial/\partial y$ are exchanged. The analytic solution of Eq. (5.72) with the boundary conditions (5.71) has been found by Brey and Fertig (2006). Let us try the solutions as

$$u(x, y) = \exp(ik_x x)u(y),$$

$$v(x, y) = \exp(ik_x y)v(y), \tag{5.73}$$

where u and v satisfy a system of two linear ordinary differential equations with constant coefficients:

$$\left(k_x + \frac{d}{dy}\right)u(y) = kv(y),$$

$$\left(k_x - \frac{d}{dy}\right)v(y) = ku(y). \tag{5.74}$$

The solution can be tried as

$$u(y) = A\exp(zy) + B\exp(-zy),$$

$$v(y) = C\exp(zy) + D\exp(-zy), \tag{5.75}$$

where

$$z = \sqrt{k_x^2 - k^2} \tag{5.76}$$

can be either real (for evanescent waves) or imaginary (for propagating waves). On substituting Eq. (5.75) into Eq. (5.74) and taking into account Eq. (5.71), one finds a dispersion relation for the waves in the nanoribbon:

$$\varphi(z) = \frac{k_x - z}{k_x + z} = \exp(-2Lz). \tag{5.77}$$

Graphical solution of Eq. (5.77) (Fig. 5.4) shows that a real solution (other than the trivial one, $z = 0$) exists if

$$k_x > \frac{1}{L}. \tag{5.78}$$

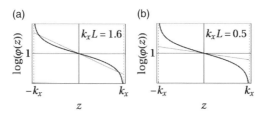

Fig. 5.4 Graphical solution of Eq. (5.77) (the logarithm of both sides is taken). If the condition (5.78) is satisfied, there is a nontrivial ($z \neq 0$) solution (a); otherwise, $z = 0$ is the only solution (b).

Indeed, at this condition $\varphi(z) \approx 1 - 2z/k_x$ is larger than $\exp(-2Lz) \approx 1 - 2Lz$ at small z. At the same time, $\varphi(k_x) = 0 < \exp(-2Lk_x)$, thus the curves should cross. Otherwise, there are no solutions.

Eq. (5.78) is the condition of existence of the edge state; for the semispace ($L \to \infty$) it always exists, with the decay decrement $z = k_x$, in agreement with the consideration of the previous section. For a finite width L, those states with energies

$$E_s = \pm \hbar v \sqrt{k_x^2 - z^2} \qquad (5.79)$$

are linear combinations of the states localized on the left and right edges of the ribbon. There are no solutions at $k_x < 0$, so, for a given valley, these edge states can propagate only in one direction. Conversely, for the valley K' the solutions exist only for $k_x < 0$. Numerical calculations for honeycomb-lattice nanoribbons (Brey & Fertig, 2006; Peres, Castro Neto, & Guinea, 2006) show that these edge states connect the valleys K and K' (Fig. 5.5).

For the case of purely imaginary $z = ik_y$ Eq. (5.77) can be rewritten as

$$k_x = k_y \cot (k_y L), \qquad (5.80)$$

which gives "bulk" standing waves with discrete values of k_y and energy

$$E_b = \pm \hbar v \sqrt{k_x^2 + k_y^2}. \qquad (5.81)$$

For the case of armchair nanoribbons, the amplitudes of the components of wave functions belonging to different valleys are the same but the phases can differ (see Eq. (5.69)). A detailed analysis (Brey & Fertig, 2006) results in the following boundary conditions:

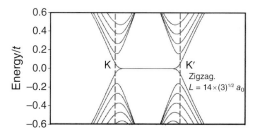

Fig. 5.5 The energy spectrum for zigzag-terminated graphene nanoribbon with 56 atoms per unit cell.
(Reproduced with permission from Brey & Fertig, 2006.)

$$u\left(-\frac{L}{2}\right) = u'\left(-\frac{L}{2}\right),$$

$$v\left(-\frac{L}{2}\right) = v\left(-\frac{L}{2}\right),$$

$$u\left(\frac{L}{2}\right) = \exp(2\pi i v)u'\left(\frac{L}{2}\right),$$ (5.82)

$$v\left(\frac{L}{2}\right) = \exp(2\pi i v)v'\left(\frac{L}{2}\right),$$

where the functions with (without) primes correspond to the states from valley K' (K) and $v = 0, \pm\frac{2}{3}$, depending on the number of rows in the nanoribbons; $v = 0$ if this number is $3p$ (p is an integer) and $v = \pm\frac{2}{3}$ if it is $3p \pm 1$. In this case there are no edge states with real z, the wave functions of the bulk states are very simple, namely

$$u_j(y) = -iv_j(y) = \frac{1}{\sqrt{2L}} \exp(ik_j y),$$
$$u'_j(y) = -iv'_j(y) = \frac{1}{\sqrt{2L}} \exp(-ik_j y),$$ (5.83)

and k_j is discrete:

$$k_j = \frac{(j + v)\pi}{L}, \quad j = 0, \pm 1, \dots.$$ (5.84)

5.5 Conductance quantization in graphene nanoribbons

For the case of zigzag edges, electron motion along the edges is coupled with that in the perpendicular direction; see Eq. (5.80). This coupling leads to interesting consequences for the electron transport in nanoribbons with varying width, such as those with nanoconstrictions (Fig. 5.6).

Let us consider a ribbon with a slowly varying width $L(x)$, assuming that

$$\left|\frac{dL}{dx}\right| \ll 1.$$ (5.85)

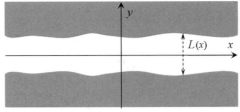

Fig. 5.6 A sketch of a graphene nanoribbon with a smoothly varying width.

For simplicity, we will assume mirror symmetry so that the edges correspond to $y = \pm L(x)/2$ (Fig. 5.6). For the case of the Schrödinger equation for conventional nonrelativistic electrons

$$-\frac{\hbar^2}{2m}\nabla^2\Psi(x, y) = E\Psi(x, y) \tag{5.86}$$

with boundary conditions

$$\Psi\left(y = \pm\frac{L(x)}{2}\right) = 0 \tag{5.87}$$

(impenetrable walls), the electron states can be considered in the adiabatic approximation (Glazman et al., 1988; Yacoby & Imry, 1990). Owing to the condition (5.87), one can try having the wave function as

$$\Psi(x, y) = \chi(x)\varphi_x(y), \tag{5.88}$$

where

$$\varphi_x(y) = \sqrt{\frac{2}{L(x)}}\sin\left(\frac{\pi n[2y + L(x)]}{L(x)}\right) \tag{5.89}$$

is the standing wave of transverse motion satisfying the boundary condition (5.87) and depending on x as a parameter via $L(x)$. It can be proven (Glazman et al., 1988; Yacoby & Imry, 1990) that the wave function of longitudinal motion satisfies the Schrödinger equation

$$\frac{d^2\chi_n(x)}{dx^2} + \left(k^2 - k_n^2(x)\right)\chi_n(x) = 0, \tag{5.90}$$

where $k^2 = 2mE/\hbar^2$ and

$$k_n(x) = \frac{\pi n}{L(x)}. \tag{5.91}$$

Owing to Eq. (5.85), one can use the semiclassical approximation (Landau & Lifshitz, 1977). At $k > k_n(x)$, the solutions of Eq. (5.90) are propagating waves with an exponentially small probability of reflection, whereas for the classically forbidden regions $k < k_n(x)$, the electron states decay quickly. This means that the electron transport in the adiabatic approximation is determined by the minimal width of the constriction $L_{\min}$: All states with

$$n < \frac{kL_{\min}}{\pi} \tag{5.92}$$

have transmission coefficients close to unity, and all states with larger n do not contribute to the electron transmission at all. According to the Landauer formula (see Section 3.2) the conductance in the adiabatic regime should be quantized, with an exponential accuracy of

$$G = \frac{2e^2}{h} n, \tag{5.93}$$

where n is an integer and the factor of 2 is due to spin degeneracy. Each transverse mode corresponds to an independent channel of transmission.

For the case of graphene nanoribbons the situation is more complicated (Katsnelson, 2007b). Here we will only consider the case of zigzag boundary conditions, since they are generic for inhomogeneous nanoribbons as discussed earlier.

Thus, one can solve the equations (5.72) with x-dependent boundary conditions (5.71):

$$u\left(x, y = -\frac{L(x)}{2}\right) = 0,$$

$$v\left(x, y = \frac{L(x)}{2}\right) = 0. \tag{5.94}$$

Following Katsnelson (2007b) we expand a general solution in the standing waves with $k_x = 0$. For this case,

$$k_y = k_j = \frac{\pi j}{L}, \; j = \pm\frac{1}{2}, \; \pm\frac{3}{2}, \; \ldots \tag{5.95}$$

(cf. Eq. (5.80)), and the eigenfunctions can be written explicitly:

$$u_j(y) = \frac{1}{\sqrt{L}} \cos\left[k_j\left(y - \frac{L}{2}\right)\right],$$

$$v_j(y) = -\frac{1}{\sqrt{L}} \sin\left[k_j\left(y - \frac{L}{2}\right)\right]. \tag{5.96}$$

Instead of Eq. (5.88), let us use the most general expansion

$$u(x, y) = \sum_j c_j(x) u_j^{(x)}(y),$$

$$v(x, y) = \sum_j c_j(x) v_j^{(x)}(y), \tag{5.97}$$

where $u^{(x)}$ and $v^{(x)}$ are the functions (5.96) with the replacement $L \to L(x)$:

$$u_j^{(x)}(y) = \frac{1}{\sqrt{L(x)}} \cos\left[\pi j\left(\frac{y}{L(x)} - \frac{1}{2}\right)\right],$$

$$v_j^{(x)}(y) = -\frac{1}{\sqrt{L(x)}} \sin\left[\pi j\left(\frac{y}{L(x)} - \frac{1}{2}\right)\right].$$

(5.98)

The functions (5.98) satisfy by construction the boundary conditions. On substituting the expansion (5.97) into Eq. (5.72) and multiplying the first equation by $\langle v_j|$ and the second one by $\langle u_j|$ one finds

$$\sum_{j'}\left[\frac{dc_{j'}}{dx}\langle v_j|v_{j'}\rangle + c_{j'}\left\langle v_j\left|\frac{dv_{j'}}{dx}\right.\right\rangle\right] = i\sum_{j'}(k - k_{j'})c_{j'}\langle v_j|u_{j'}\rangle,$$

$$\sum_{j'}\left[\frac{dc_{j'}}{dx}\langle u_j|u_{j'}\rangle + c_{j'}\left\langle u_j\left|\frac{du_{j'}}{dx}\right.\right\rangle\right] = i\sum_{j'}(k - k_{j'})c_{j'}\langle u_j|v_{j'}\rangle. \qquad (5.99)$$

These equations are formally exact. As a first step to the adiabatic approximation, one should neglect the terms with

$$\left\langle v_j\left|\frac{dv_{j'}}{dx}\right.\right\rangle \text{and} \left\langle u_j\left|\frac{du_{j'}}{dx}\right.\right\rangle,$$

which is justified by the smallness of dL/dx, as in the case of nonrelativistic electrons (Yacoby & Imry, 1990).

To proceed further, we need to calculate the overlap integrals

$$\langle \phi_1|\phi_2\rangle = \int_{-L/2}^{L/2} dy \phi_1^* \phi_2$$

for different basis functions:

$$\langle u_j|u_{j'}\rangle = \frac{1}{2}\left(\delta_{jj'} + \delta_{j,-j'}\right),$$

$$\langle v_j|v_{j'}\rangle = \frac{1}{2}\left(\delta_{jj'} + \delta_{j,-j'}\right),$$

$$\langle u_j|v_{j'}\rangle = \langle v_{j'}|u_j\rangle = \begin{cases} -\dfrac{1}{\pi(j'-j)}, & j'-j = 2n+1, \\[2mm] -\dfrac{1}{\pi(j'+j)}, & j'-j = 2n, \end{cases}$$

(5.100)

where n is an integer. On substituting Eq. (5.100) into Eq. (5.99) and neglecting the nonadiabatic terms within the matrix elements of the operator d/d_x, we obtain after simple transformations

$$\frac{dc_j}{dx} = -\frac{2i}{\pi} \sum_{j'} \frac{k - k_{j'}(x)}{j + j'} c_{j'}(x),$$ (5.101)

where the sum is over all j' such that $j' - j$ is even.

Until now we have employed transformations and approximations that are identical to those used in the case of nonrelativistic electrons. However, we still have a coupling between different standing waves, so we cannot prove that the electron transmission through the constriction is adiabatic. To prove this we need one more step, namely a transition from the discrete variable j to a continuous one and a replacement of the sums on the right-hand side of Eq. (5.101) by integrals: $\sum_{j'} \ldots \rightarrow \frac{1}{2} P \int dy \ldots$, where P is the symbol of principal value. This step is justified by assuming that $kL \gg 1$, i.e., it is valid only for highly excited states. For low-lying electron standing waves it is difficult to see any way to appreciably simplify the set of equations (5.101) for the coupled states.

For any function $f(z)$ that is analytic in the upper (lower) complex half-plane one has

$$\int_{-\infty}^{\infty} dx f(x) \frac{1}{x - x_1 \pm i0} = 0$$ (5.102)

or, equivalently,

$$\int_{-\infty}^{\infty} dx f(x) \frac{P}{x - x_1} = \pm i\pi f(x_1).$$ (5.103)

Assuming that $c_j(x)$ is analytic in the lower half-plane as a function of the complex variable j one obtains, instead of Eq. (5.101),

$$\frac{dc_j(x)}{dx} = [k + k_j(x)] c_{-j}(x).$$ (5.104)

Similarly, taking into account that $c_{-j}(x)$ is analytic in the upper half-plane as a function of the complex variable j we have

$$\frac{dc_{-j}(x)}{dx} = [k_j(x) - k] c_j(x).$$ (5.105)

Finally, on differentiating Eq. (5.104) with respect to x, neglecting the derivatives of $k_j(x)$ due to the smallness of dL/dx and taking into account Eq. (5.105) we find

$$\frac{d^2 c_j(x)}{dx^2} + [k^2 - k_j^2(x)] c_j(x) = 0.$$ (5.106)

Further analysis completely follows that for the nonrelativistic case. The potential is semiclassical for the case of smoothly varying $L(x)$. Therefore, the transmission coefficient is very close to unity if the electron energy exceeds the energy of the j'th level in the narrowest place of the constriction and is exponentially small otherwise. Standard arguments based on the Landauer formula prove the conductance quantization in this situation.

At the same time, for the lowest energy levels the replacement of sums by integrals in Eq. (5.101) cannot be justified, and thus the states with different js are in general coupled even for a smooth constriction (5.85). Therefore, electron motion along the strip is strongly coupled with that in the perpendicular direction and different electron standing waves are essentially entangled. In this situation there is no general reason to expect sharp jumps and well-defined plateaux in the energy dependence of the conductance. This means that the criterion of the adiabatic approximation is more restrictive for the case of Dirac electrons than it is for nonrelativistic ones. The formal reason is an overlap between components of the wave functions with different pseudospins or, equivalently, between the hole component of the state j and the electron component of the state $j' \neq j$. This conclusion (Katsnelson, 2007b) seems to be confirmed by the numerical simulations of Muños-Rojas et al. (2008).

5.6 The band gap in graphene nanoribbons with generic boundary conditions

One has to keep in mind that the terminated honeycomb lattice is a special case of graphene edges. Density-functional calculations show that the reconstructed "5–7" edge (Fig. 5.7) has an energy lower than those of both armchair and zigzag edges (Koskinen, Malola, & Häkkinen, 2008). The reconstruction to this low-energy state requires the overcoming of energy barriers, so the zigzag edges are metastable (Kroes et al., 2011), but under some circumstances it will definitely happen. Zigzag edges are very chemically active, so they will bind hydrogen, oxygen, or hydroxyl groups (see, e.g., Boukhvalov & Katsnelson, 2008; Bhandary et al., 2010). Lastly, that the density of states peaks due to zero-energy modes means ferromagnetic instability (Fujita et al., 1996; Son, Cohen, & Louie, 2006a; see also Section 12.3).

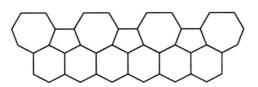

Fig. 5.7 A sketch of a reconstructed 5–7 zigzag edge.

All of this will substantially modify the boundary conditions. The most general form is given by Eq. (5.36) and (5.45). It assumes only time-reversal symmetry. Time-reversal symmetry can be broken by ferromagnetic ordering; however, the latter can exist in one-dimensional systems at zero temperature only. At finite temperatures one has, instead, a superparamagnetic state with a finite correlation length ξ which is just several interatomic distances at room temperature (Yazyev & Katsnelson, 2008). If all essential sizes of the problem (e.g., the width of nanoribbons L) are larger than ξ then the system should be considered time-reversal invariant.

The most general boundary conditions for the nanoribbons are therefore

$$
\Psi\left(x, y = -\frac{L}{2}\right) = \left(\vec{v}_1 \cdot \vec{\tau}\right) \otimes \left(\vec{n}_1 \cdot \vec{\sigma}\right) \Psi\left(x, y = -\frac{L}{2}\right),
$$

$$
\Psi\left(x, y = \frac{L}{2}\right) = \left(\vec{v}_2 \cdot \vec{\tau}\right) \otimes \left(\vec{n}_2 \cdot \vec{\sigma}\right) \Psi\left(x, y = \frac{L}{2}\right),
$$

(5.107)

where $\vec{v}_i$ are three-dimensional unit vectors (no restrictions) and $\vec{n}_i$ are three-dimensional unit vectors perpendicular to the y-axis:

$$
\vec{n}_1 = (\cos\theta_1, 0, \sin\theta),
$$

$$
\vec{n}_2 = (\cos\theta_2, 0, \sin\theta_2).
$$

(5.108)

Valley symmetry implies that only the relative directions of the vectors $\vec{v}_1$ and $\vec{v}_2$ are essential. Thus, the problem is characterized by three angles: θ_1, θ_2, and the angle γ between $\vec{v}_1$ and $\vec{v}_2$.

The most general dispersion relation $E = E(k)$ for the propagating waves

$$
\Psi(x, y) \propto \exp(ikx + iqy),
$$

(5.109)

satisfying the boundary conditions (5.107) has been obtained by Akhmerov and Beenakker (2008). It reads

$$
\cos\theta_1 \cos\theta_2 \left(\cos\omega - \cos^2\Omega\right) + \cos\omega \sin\theta_1 \sin\theta_2 \sin^2\Omega
$$
$$
- \sin\Omega[\sin\Omega\cos\gamma + \sin\omega\sin(\theta_1 - \theta_2)] = 0,
$$

(5.110)

where

$$
\omega^2 = 4L^2\left[\frac{E^2}{\hbar^2 v^2} - k^2\right]
$$

and

$$
\cos\Omega = \frac{\hbar v k}{E}.
$$

(5.111)

Different solutions of Eq. (5.110) correspond to different standing waves with discrete q_n. Analysis of this equation shows that there is a gap in the energy spectrum if $\gamma = 0, \pi$ (which means that valleys are coupled at the boundaries) or at $\gamma = \pi$, $\sin \theta_1 \sin \theta_2 > 0$, or at $\gamma = 0$, $\sin \theta_1 \sin \theta_2 < 0$ (Akhmerov & Beenakker, 2008). One can see that the case of zigzag-terminated edges when states with arbitrarily small energy, up to $E = 0$, exist is very exceptional. For generic boundary conditions, the gap is of the order of

$$\Delta \cong \frac{\hbar v}{L}.$$

(5.112)

A detailed analysis of the gap, both in a tight-binding model and in realistic density-functional calculations, was carried out by Son, Cohen, and Louie (2006b) (see also, e.g., Wassmann et al., 2008).

The gap opening in nanoribbons is very important for applications. It allows one to overcome restrictions due to Klein tunneling and build a transistor that can really be locked by a gate voltage (Han et al., 2007; Wang et al., 2008; Han, Brant, & Kim, 2010).

5.7 Energy levels in graphene quantum dots

Nanoribbons are restricted in one dimension, therefore their electron spectra consist of bands $E_n(k)$. It is possible to make graphene devices in which electrons are confined in two dimensions – graphene quantum dots (Ponomarenko et al., 2008; Stampfer et al., 2008; Güttinger et al., 2009; Molitor et al., 2010; Zhang et al., 2010). Fig. 5.8 (Ponomarenko et al., 2008) shows an example of such a device, together with the voltage dependence of the differential conductance G through the device. Oscillations of G are due to the discreteness of the electron energy spectrum in the dot. First of all, there is a classical electrostatic effect, namely the dependence of the energy on the total charge Q,

$$E_C(Q) = \frac{Q^2}{2C},$$

(5.113)

where C is the capacitance of the dot. When the electron tunnels to the dot or from the dot, the charge Q, is changed by $\pm e$. This effect is known as Coulomb blockade; see Kouwenhoven, Marcus, and McEuen (1997). Apart from this, there is a discreteness of the single-electron energy spectrum superimposed on the Coulomb-blockade peaks. The sharp dependence of G on the gate voltage allows one to use the device as a single-electron transistor (Ponomarenko et al., 2008; Stampfer et al., 2008). The data extracted from the measurements clearly show the effect of level repulsion, which was discussed in Section 5.1; this means that the

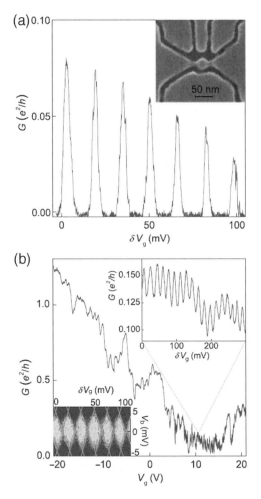

Fig. 5.8 (a) A graphene-based, single-electron transistor. The conductance G of a device shown in the insert in the upper right corner is given as a function of the gate voltage, at temperature $T = 0.3$ K. Two panels in (b) show the picture with different resolutions.
(Reproduced with permission from Ponomarenko et al., 2008.)

single-electron spectrum of real graphene quantum dots is certainly chaotic (De Raedt & Katsnelson, 2008; Ponomarenko et al., 2008). The function $P(S)$ (cf. Eq. (5.29) and (5.30)) extracted from the experimental data by Ponomarenko et al. (2008) for a 40-nm graphene quantum dot is shown in Fig. 5.9. Its decrease at small S is a manifestation of the level repulsion. At the same time, it is difficult to distinguish between the cases of orthogonal and unitary ensembles. Theoretically, the distinction depends on the probability of intervalley scattering. If it is large enough, then, due to atomic-scale inhomogeneity at the edges, the system is time-

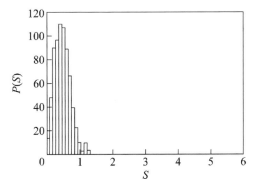

Fig. 5.9 The level-spacing distribution extracted from experimental data on a graphene 40-nm quantum dot.
(Reproduced with permission from De Raedt & Katsnelson, 2008.)

reversal invariant, and one should expect the behavior typical for the Gaussian orthogonal ensemble, Eq. (5.29). This is obvious already from the fact that, in the absence of a magnetic field, the tight-binding Hamiltonian can be chosen to be real. At the same time, if the inhomogeneities at the edges are smooth enough and intervalley scattering is therefore weak, the situation should be close to the case of a neutrino billiard (Section 5.1), and a unitary ensemble is to be expected. This can indeed be the case, since for chemical passivation of the edges the electronic structure changes smoothly within a rather broad strip near the edges (Boukhvalov & Katsnelson, 2008). Theoretical discussions of the energy-level statistics in graphene quantum dots can be found in Wurm et al. (2009), Libisch, Stampfer, and Burgdörfer (2009), Wimmer, Akhmerov, and Guinea (2010), and Huang, Lai, and Grebogi (2010).

5.8 Edge states in magnetic fields and the anomalous quantum Hall effect

Now we can come back to the physics of the half-integer quantum Hall effect discussed in Chapter 2. Our analysis in Section 2.9 was based on the solution of the quantum-mechanical problem for bulk graphene. There is an alternative approach to the quantum Hall effect that is based on the analysis of the edge states of electrons in a magnetic field (Halperin, 1982; MacDonald & Středa, 1984).

Let us start with the classical picture of electron motion in a magnetic field. In two dimensions, the electron orbits are closed circles (Larmor rotation). Depending on the direction of the magnetic field, all electrons in the bulk rotate either clockwise or counterclockwise. However, for the electrons with the centers of their orbits close enough to the boundary, reflections form a completely different kind of trajectory, skipping orbits (Fig. 5.10). They possess a magnetic moment

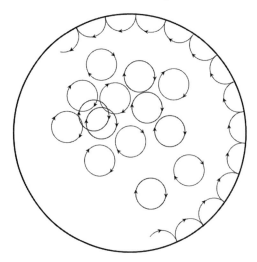

Fig. 5.10 Skipping orbits of electrons due to the combination of Larmor rotation in a magnetic field and reflection from the edges.

opposite to that of the "bulk" orbits and, actually, exactly compensate for the latter, so that, in agreement with a general theorem, the *classical* system of electrons can be neither paramagnetic nor diamagnetic (Vonsovsky & Katsnelson, 1989). In quantum theory, the skipping orbits are associated with the edge states localized near the boundary and carrying the current. These states are chiral, since only one direction of propagation is allowed. Therefore, they are protected against localization by disorder; the situation is similar to the Klein tunneling and forbidden backscattering for massless Dirac fermions (Chapter 4). Simply speaking, there are no other states with the same energy for electrons to be scattered to. Thus, if one assumes that all bulk states are localized there is still a current being carried by the skipping electrons, with a contribution to the conductance of e^2/h per spin (complete transmission). This gives an alternative explanation of the quantum Hall effect (Halperin, 1982; MacDonald & Středa, 1984).

A topological analysis shows that the number of edge states at the border between a quantum Hall insulator and vacuum is equal to the integer in (2.181) and, thus, "bulk" and "edge" approaches to the quantum Hall effect give the same results for σ_{xy} (Hatsugai, 1993; Kellendonk & Schulz-Baldes, 2004; Prodan, 2009).

The counting of the edge states is therefore an alternative way to explain the anomalous ("half-integer") quantum Hall effect in graphene (Abanin, Lee, & Levitov, 2006; Hatsugai, Fukui, & Aoki, 2006). Here we will use the approach of the first of these works, which is based on a solution of the Dirac equation in a magnetic field (the second one uses an analysis of the geometry of the honeycomb lattice).

Let us assume that graphene fills the semispace $x < 0$. The solutions of the Dirac equation for the valley K satisfying the conditions $\psi_i(x) \to 0$ at $x \to -\infty$ are given by Eq. (2.45) and (2.46),

$$\psi_1(X) = D_n(-X),$$
$$\psi_2(X) = i\varepsilon D_{n-1}(-X), \tag{5.114}$$

where $n = \varepsilon^2$ and X is given by Eq. (2.26) and (2.41). For the valley K' the results are the same but with the replacement $\psi_1 \to \psi_2', \psi_2 \to \psi_1'$ (see Eq. (1.27) and (1.28)), thus,

$$\psi_1'(X) = i\varepsilon D_{n-1}(-X),$$
$$\psi_2'(X) = D_n(-X). \tag{5.115}$$

The eigenenergy ε can be found from the boundary conditions. For example, for the armchair-terminated edge, one needs to put

$$\psi_1(x = 0) = \psi_1'(x = 0),$$
$$\psi_2(x = 0) = \psi_2'(x = 0). \tag{5.116}$$

For the case of zigzag-terminated edges, the valleys are decoupled, and the conditions are

$$\psi_1(x = 0) = 0,$$
$$\psi_1'(x = 0) = 0 \tag{5.117}$$

(for the zigzag edge with missing A atoms). Then, Eq. (5.114) and (5.115) give the energy (2.26) depending on the coordinate of the center of the orbit x_0 (2.40) or, equivalently, on the wave vector k_y along the edge.

It is easier to analyze these solutions after transformation of the original problem to the Schrödinger equation for a double-well potential (Abanin, Lee, & Levitov, 2006; Delplace & Montambaux, 2010). The Hamiltonian $\hat{H}^2$ (2.33) can be represented as

$$\hat{H}^2 = \frac{2\hbar|e|Bv^2}{c}\hat{Q}, \tag{5.118}$$

where

$$\hat{Q} = -\frac{1}{2}\frac{d^2}{dx^2} + \frac{1}{2}(x - x_0)^2 - \frac{1}{2}\sigma_z\tau_z,$$

where x and x_0 are in units of the magnetic length l_B and $\sigma_z = +1$ for components corresponding to the sublattice A and $\sigma_z = -1$ for components corresponding to the sublattice B, with $\tau_z = \pm 1$ for the valley K and K', respectively.

For the case of zigzag edges, the valleys and sublattices are decoupled. The eigenvalues of the operators $\hat{Q}$ for the valleys K and K' differ by 1. The sublattices are also decoupled, but the edge states for the B sublattice are associated with another edge.

The eigenstates of the problem

$$\hat{Q}\psi(x) = \varepsilon^2\psi(x) \tag{5.119}$$

with the boundary condition (5.117) are the same as the antisymmetric eigenstates for the symmetric potential

$$\hat{Q} = -\frac{1}{2}\frac{d^2}{dx^2} + V(x), \tag{5.120}$$

$$V(x) = \frac{1}{2}(|x| - x_0)^2 \mp \frac{1}{2} \tag{5.121}$$

with $\mp$ signs for the valleys K and K', respectively (see Fig. 5.11).

If $|x_0| \gg 1$, the potential wells are well separated and the probability of tunneling between the wells is exponentially small, for

$$\varepsilon^2 \leq \frac{1}{2}x_0^2. \tag{5.122}$$

Then, in zeroth-order approximation, the eigenvalues are the same as for independent walls

$$\varepsilon_n^2 = n + \frac{1}{2} \mp \frac{1}{2}, \tag{5.123}$$

where ($n = 0, 1, 2, \ldots$). Tunneling leads to the splitting of each eigenvalue for symmetric and antisymmetric states

$$\delta\varepsilon_n^2 = \pm\Delta_n \tag{5.124}$$

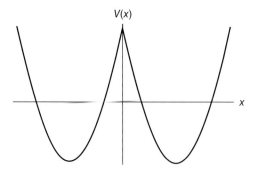

Fig. 5.11 The effective potential (5.121) (for the case of the minus sign).

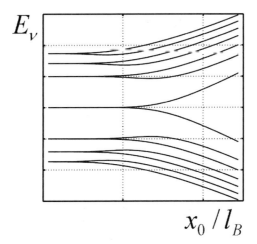

Fig. 5.12 A sketch of the energy spectrum for magnetic edge states.

with

$$\Delta_n \propto \exp\left[-\int_{-x_1}^{x_2} dx \sqrt{V(x) - \varepsilon_n^2}\right], \tag{5.125}$$

where $x_{1,2}$ are the classical turning points: $V(x_{1,2}) = \varepsilon_n^2$. One needs to choose the plus sign in Eq. (5.124) corresponding to the antisymmetric eigenfunctions.

For the minus sign in Eq. (5.121) (valley K) one has some growing dependence of E_n on the function $|x_0|/l_B$ (the larger $|x_0|$ the smaller the shift) starting from $E = 0$. Starting from the first Landau level, the second valley K' also contributes, but Δ_n for the same energy corresponds to another value of $n(n \to n - 1)$ and, thus, will be different. As a result, we have the picture of the energy levels shown schematically in Fig. 5.12. An almost zero-energy Landau *band* (originating from the zero-energy Landau level for an infinite system) corresponds, for a given edge, to the states from a single valley; the states from the second valley are associated with another edge.

For the case of armchair edges, the boundary conditions (5.116) lead to the Schrödinger equations (5.119) and (5.120), but with the potential

$$V(x) = \frac{1}{2}(|x| - x_0)^2 - \frac{1}{2}\mathrm{sgn}x \tag{5.126}$$

(see Fig. 5.13). Indeed, one can define formally

$$\psi_1'(x) \equiv \psi_1(-x),$$

$$\psi_2'(x) \equiv \psi_2(-x), \tag{5.127}$$

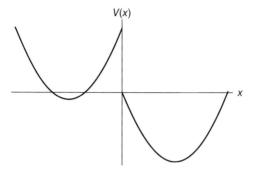

Fig. 5.13 The effective potential (5.126).

so that Eq. (5.116) is nothing other than the condition of continuity of the wave function ψ_1 and its derivative $d\psi_1/dx$ (which is related to ψ_2 by the Dirac equation) at $x = 0$. The qualitative dependence $E_n(x_0/l)$ remains the same as that shown in Fig. 5.12. A more detailed analysis of the problem in the semiclassical approximation was performed by Delplace and Montambaux (2010).

To calculate the Hall conductivity one just needs to count the occupied edge states for a given Fermi energy, with each state contributing e^2/h per spin. One can immediately see from Fig. 5.12 that the lowest-energy Landau band always produces *one* edge electron (for $E > 0$) or hole ($E < 0$) state and all other bands produce two such states. This immediately gives Eq. (2.167) for σ_{xy}, with $g_v = 2$ and $g_s = 1$ (Abanin, Lee, & Levitov, 2006).

5.9 Spectral flow for massless Dirac fermions

In Chapter 2, we discussed nontrivial topological properties of massless Dirac fermions in graphene; the existence of topologically protected zero-energy Landau levels (Section 2.3) and related to them half-integer quantization of Hall conductivity (Section 2.9) is probably the most important one. Here we consider the other nontrivial topological effect, namely, nonvanishing spectral flow of the Dirac Hamiltonian (5.1), (5.5) in quantum dots which are not simply connected (that is, with holes; Fig. 5.14). The effect was considered by Prokhorova (2013) and Katsnelson and Nazaikinskii (2012); our presentation will follow the latter paper.

Let us consider the situation with magnetic fluxes Φ_i entering ith inner hole; for the external boundary we will assume, by definition,

$$\Phi_1 = -\sum_{i=2}^{m} \Phi_i \tag{5.128}$$

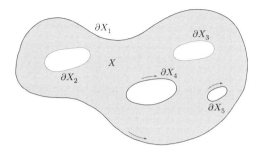

Fig. 5.14 Example of a quantum dot with $m = 5$ boundaries. When calculating the flux associated to each boundary the sign is positive for the inner boundaries (∂X_2, ∂X_3, ∂X_4, ∂X_5) when going around the boundary clockwise and for the external boundary (∂X_1) when going counterclockwise, as shown by arrows. If B is positive for X_1, X_4, and X_5 and negative for X_2 and X_3, $\partial^+ X = \partial X_1 \cup \partial X_4 \cup \partial X_5$.

where $m > 1$ is the number of boundaries. Due to Aharonov–Bohm effect (Aharonov & Bohm, 1959; Olariu & Popescu, 1985; see Section 3.4) the phase of the wave function when going around ith hole is changed by $2\pi\mu_i$, $\mu_i = \Phi_i/\Phi_0$ where Φ_0 is the flux quantum (2.52) (cf. Eq. (3.43)). If all μ_i are integer, the system is equivalent to the system without magnetic field and, in particular, should have the same eigenenergies E_α. For the conventional nonrelativistic electron gas it would mean that each eigenenergy separately is a periodic function of the magnetic fluxes:

$$E_\alpha(\mu_i) = E_\alpha(\mu_i + n_i) \tag{5.129}$$

when all n_i are integer.

However, the Dirac Hamiltonian is not semibounded, it can have arbitrarily large negative and positive eigenvalues; this is an unavoidable consequence of its electron–hole symmetry. In this situation, periodicity of the *spectrum* $\{E_\alpha\}$ as a whole does not mean periodicity of each eigenvalue separately. Indeed, if we consider a transformation $n \to n + 1$, then the set of all integers Z transforms to itself, despite that each number is shifted up; if we consider only the set of *positive* integers, nothing similar is possible. For unbound operators, one can introduce a concept of *spectral flow* (Atiyah, Patodi, & Singer, 1976). For our problem, the latter can be defined as follows. Fix any real value of energy E. Let us consider the transformation of the Hamiltonian $\hat{H}$ to itself (such as $\mu_i \to \mu_i + n_i$ in Eq. (5.129)) due to a continuous change of parameters (such as increase or decrease of magnetic field through the holes). Then some eigenvalues ($N_<$ of them) will cross the value E from up to down and some eigenvalues ($N_>$ of them) will cross the value E from down to up. The spectral flow of the operator $\hat{H}$ is

$$sf\{\hat{H}\} = N_> - N_< \tag{5.130}$$

It is easy to see that it cannot be dependent on the value of E. In particular, it means that if the spectral flow of the Dirac operator is not zero, then, under a smooth increase of magnetic fluxes, some energy levels will cross zero, which means a creation of electron–hole pairs from vacuum.

Let us assume Berry–Mondragon boundary condition (5.13) for each boundary. Then, as was proven by Prokhorova (2013) and Katsnelson and Nazaikinskii (2012) the spectral flow is dependent on the signs of the constants B (that is, on the signs of gap functions Δ at the boundaries). The result is:

$$sf\{\hat{H}\} = \sum_{\partial^+ X} \mu_i = \sum_{\partial^+ X} \frac{\Phi_i}{\Phi_0}. \tag{5.131}$$

Here $\partial^+ X$ is the sum of all boundaries for which B is positive, see Fig. 5.14. In particular, if the sign of B is the same for all boundaries, $sf\{\hat{H}\} = 0$ due to Eq. (5.128).

There are two Dirac cones in graphene, and for a given configuration of magnetic fluxes their spectral flows should be opposite, to have the total spectral flow of the lattice Hamiltonian equal to zero (similar to the index; see Section 2.3). It means that if we would create a configuration with different signs of the gap functions at different boundaries, then, at a smooth increase of the magnetic field, some energy levels will move up for one valley and down for the other one.

Until now, there is still no experimental confirmation of this interesting prediction. The concept of spectral flow of the Dirac operator turns out to be useful in the physics of vortices in superfluid helium-3, where it leads to the appearance of additional hydrodynamic forces acting on moving vortices (Kopnin forces; Kopnin, 2002; Volovik, 2003).

6

Point defects

6.1 Scattering theory for Dirac electrons

Here we discuss quantum relativistic effects in the electron scattering by a radially symmetric potential $V(r)$. This will give us a feeling for the peculiar properties of charge carriers in *imperfect* graphene, in comparison with the conventional two-dimensional electron gas with impurities (Ando, Fowler, & Stern, 1982). Further, we will consider a more realistic model of defects in a honeycomb lattice, beyond the Dirac approximation. In this section we follow the papers by Katsnelson and Novoselov (2007), Hentschel and Guinea (2007), Guinea (2008), and Novikov (2007). It is instructive to compare the scattering theory developed in those works with the two-dimensional scattering theory for the Schrödinger equation (Adhikari, 1986).

Let us start with the equation

$$\left(-i\hbar v \hat{\vec{\sigma}} \nabla + V(r) \right) \begin{pmatrix} \psi_1 \\ \psi_2 \end{pmatrix} = E \begin{pmatrix} \psi_1 \\ \psi_2 \end{pmatrix}, \tag{6.1}$$

where the potential $V(r)$ is supposed to be isotropic, that is, dependent only on the modulus $r = \sqrt{x^2 + y^2}$. We have to pass to the radial coordinates (see Eq. (5.16) through Eq. (5.19)). Then Eq. (6.1) is transformed to the couple of ordinary differential equations

$$\frac{dg_l(r)}{dr} - \frac{l}{r} g_l(r) - \frac{i}{\hbar v} [E - V(r)] f_1(r) = 0,$$

$$\frac{df_l(r)}{dr} + \frac{l+1}{r} f_l(r) - \frac{i}{\hbar v} [E - V(r)] g_1(r) = 0, \tag{6.2}$$

where $l = 0, \pm 1, \pm 2, \ldots$ is the angular-momentum quantum number and we try the solution in the following form (cf. Eq. (5.19):

$$\psi_1(\vec{r}) = g_l(r)\exp{(il\varphi)},$$
$$\psi_2(\vec{r}) = f_l(r)\exp{(i(l+1)\varphi)}.$$
(6.3)

To be specific, we will further consider the case of electrons with $E = \hbar v k > 0$. In two dimensions, the incident electron plane wave has the expansion

$$\exp{(i\vec{k}\vec{r})} = \exp{(ikr\cos\varphi)} = \sum_{l=-\infty}^{\infty} i^l J_l(kr)\exp{(il\varphi)},$$
(6.4)

where $J_l(z)$ are the Bessel functions (Whittaker & Watson, 1927). At large values of the argument $(kr \gg 1)$, they have asymptotics

$$J_l(kr) \cong \sqrt{\frac{2}{\pi kr}}\cos\left(kr - \frac{l\pi}{2} - \frac{\pi}{4}\right).$$
(6.5)

The radial Dirac equation (6.2) for free space $(V(r) = 0)$ has, for a given l, two independent solutions, which are proportional to the Bessel and Neumann functions, $J_l(kr)$ and $Y_l(kr)$, the latter having the asymptotics $(kr \gg 1)$

$$Y_l(kr) \cong \sqrt{\frac{2}{\pi kr}}\sin\left(kr - \frac{l\pi}{2} - \frac{\pi}{4}\right),$$
(6.6)

but the functions $Y_l(kr)$ are divergent at $r \to 0$. Instead, one can use Hankel functions

$$H_l^{(1,2)}(kr) = J_l(kr) \pm iY_l(kr)$$
(6.7)

with the asymptotics, at $kr \gg 1$,

$$H_l^{(1,2)}(kr) \cong \sqrt{\frac{2}{\pi kr}}\exp\left[\pm i\left(kr - \frac{l\pi}{2} - \frac{\pi}{4}\right)\right].$$
(6.8)

Thus, the function $H_l^{(1)}$ describes the scattering wave $H_l^{(2)}$ and describes the wave falling at the center.

If we have a potential of finite radius R $(V(r > R) = 0)$, the solution of Eq. (6.2) at $r > R$ can be represented in the form

$$g_l(r) = A\left[J_l(kr) + t_l H_l^{(1)}(kr)\right],$$
$$f_l(r) = iA\left[J_{l+1}(kr) + t_l H_{l+1}^{(1)}(kr)\right],$$
(6.9)

where the terms proportional to Bessel (Hankel) functions describe incident (scattering) waves. The complex factors t_l in Eq. (6.9) are scattering amplitudes.

One can represent them in a more conventional way, via scattering phases δ_l (Newton, 1966; Adhikari, 1986). The latter are determined via the asymptotics of radial solutions at $kr \gg 1$,

$$g_l(kr) \propto \frac{1}{\sqrt{kr}} \cos\left[kr - \frac{l\pi}{2} - \frac{\pi}{4} + \delta_l\right]. \tag{6.10}$$

Taking into account Eq. (6.5) through Eq. (6.8), Eq. (6.10) can be represented as

$$g_l(r) \propto \cos\delta_l J_l(kr) - \sin\delta_l Y_l(kr)$$
$$= \exp(-i\delta_l)\left[J_l(kr) + i\sin\delta_l \exp(i\delta_l)H_l^{(1)}(kr)\right]. \tag{6.11}$$

On comparing Eq. (6.9) and (6.11) one finds

$$t_l(k) = i\,\sin\,\delta_l(k)\exp[i\delta_l(k)] = \frac{\exp[2i\delta_l(k)] - 1}{2}. \tag{6.12}$$

It follows from Eq. (6.12) that

$$|t_l(k)| \leq 1, \tag{6.13}$$

which means, as we will see later, that the scattering current cannot be larger than the incident one.

Let us now calculate the scattering cross-section. For the incident wave propagating along the x-axis we have

$$\Psi^{(0)} = \frac{1}{\sqrt{2}}\begin{pmatrix} 1 \\ 1 \end{pmatrix}\exp(ikx), \tag{6.14}$$

where the numerical factor provides normalization of the incident current:

$$j_x^{(0)} = [\Psi^{(0)}]^{\dagger}\sigma_x\Psi^{(0)} = 1. \tag{6.15}$$

Thus, one can choose $A = 1/\sqrt{2}$ in Eq. (6.9). Taking into account Eq. (6.9) and (6.8), one finds for the asymptotics of the scattering waves at large distances

$$\Psi_{sc} \approx \frac{1}{\sqrt{\pi kr}}\exp\left(ikr - \frac{i\pi}{4}\right)\sum_{l=-\infty}^{\infty} t_l\begin{pmatrix} \exp[i(l+1)\varphi] \\ \exp(il\varphi) \end{pmatrix}. \tag{6.16}$$

The current operator in the direction $\vec{n} = \vec{r}/r$ is

$$\hat{j}_n = \vec{n}\hat{\vec{\sigma}} = \begin{pmatrix} 0 & e^{-i\varphi} \\ e^{i\varphi} & 0 \end{pmatrix}, \tag{6.17}$$

which gives us for the scattering current

$$j^{(sc)} = \Psi^+_{sc} \hat{j}_n \Psi_{sc} = \frac{2}{\pi k r} |F(\varphi)|^2, \tag{6.18}$$

where

$$F(\varphi) = \sum_{l=-\infty}^{\infty} t_l \exp(il\varphi). \tag{6.19}$$

Eq. (6.18) gives for the differential cross-section

$$\frac{d\sigma}{d\varphi} = \frac{2}{\pi k} |F(\varphi)|^2. \tag{6.20}$$

The Dirac equation (6.2) for the massless case has an important symmetry with respect to the replacement $f \leftrightarrow g$, $l \leftrightarrow -l-1$, which leads to the result

$$t_l(k) = t_{-l-1}(k). \tag{6.21}$$

Taking into account Eq. (6.21), the equation (6.20) can be rewritten in the final form (Katsnelson & Novoselov, 2007)

$$\frac{d\sigma}{d\varphi} = \frac{8}{\pi k} \left| \sum_{l=0}^{\infty} t_l \cos\left[\left(l + \frac{1}{2}\right)\varphi \right] \right|^2. \tag{6.22}$$

It follows immediately from Eq. (6.22) that $d\sigma/d\varphi = 0$ at $\varphi = \pi$, that is, backscattering is absent. This is in agreement with the general considerations of Section 4.2.

If we have a small concentration of point defects n_{imp}, then, according to the standard semiclassical Boltzmann theory (Shon & Ando, 1998; Ziman, 2001; see also later, Chapter 11), their contribution to the resistivity is

$$\rho = \frac{2}{e^2 v^2 N(E_F)} \frac{1}{\tau(k_F)}, \tag{6.23}$$

where $\tau(k_F)$ is the mean-free-path time and

$$\frac{1}{\tau(k_F)} = n_{imp} v \sigma_{tr}, \tag{6.24}$$

where

$$\sigma_{tr} = \int_0^{2\pi} d\varphi \frac{d\sigma}{d\varphi} (1 - \cos \varphi) \tag{6.25}$$

is the *transport cross-section*. The applicability of the semiclassical Boltzmann theory to quantum relativistic particles in graphene is not clear, a priori. This issue will be considered in detail in Chapter 11, and the answer will be that, yes, we can use this theory, except in the very close vicinity of the neutrality point, where the minimal conductivity is a purely quantum phenomenon (see Chapter 3). On substituting Eq. (6.20) into Eq. (6.25) one finds

$$\sigma_{tr} = \frac{4}{k} \sum_{l=0}^{\infty} \sin^2(\delta_l - \delta_{l+1}).$$

(6.26)

Note that Eq. (6.23), for the case of graphene, coincides with Eq. (4.80), where $l = v\tau(k_F)$ is the mean free path.

6.2 Scattering by a region of constant potential

Let us apply a general theory from the previous section to the simplest case of a rectangular potential well (or hump)

$$V(r) = \begin{cases} V_0, & r < R, \\ 0, & r > R. \end{cases}$$

(6.27)

Then, the asymptotic expression (6.9) gives us an exact solution for $r > R$. At $r < R$, k should be replaced by

$$q = \frac{E - V_0}{\hbar v},$$

(6.28)

and only Bessel functions $J_l(qr)$ are allowed (otherwise, the solution will not be normalizable, due to divergence $Y_l(z) \sim z^{-l}$ at $z \to 0$):

$$g_l(r) = BJ_l(qr)$$
$$f_l(r) = iBJ_{l+1}(qr)$$

(6.29)

at $r < R$. One needs to add the conditions of continuity of the functions $g_l(r)$ and $f_l(r)$ at $r = R$. The result is (Hentschel & Guinea, 2007; Katsnelson & Novoselov, 2007)

$$t_l(k) = \frac{J_l(qR)J_{l+1}(kR) - J_l(kR)J_{l+1}(qR)}{H_l^{(1)}(kR)J_{l+1}(qR) - J_l(qR)H_{l+1}^{(1)}(kR)}.$$

(6.30)

Let us consider first the case of a short-range potential

$$kR \ll 1;$$

(6.31)

then $q = -V_0/(\hbar v)$ can be considered an energy-independent quantity. At $z \to 0$,

$$J_l(z) \approx \frac{1}{l!} \left(\frac{z}{2}\right)^l,$$

$$H_l^{(1)}(z) \approx -\frac{i}{\pi} \left(\frac{2}{z}\right)^l (l-1)! \quad (l \neq 0), \tag{6.32}$$

$$H_0^{(1)}(z) \approx \frac{2i}{\pi} \ln z.$$

On substituting Eq. (6.32) into Eq. (6.30), one finds

$$t_l(k) \approx \frac{\pi i}{(l!)^2} \frac{J_{l+1}(qR)}{J_l(qR)} \left(\frac{kR}{2}\right)^{2l+1} \tag{6.33}$$

and, thus, the s-scattering ($l = 0$) dominates

$$t_0(k) \propto \delta_0(k) \propto kR. \tag{6.34}$$

Substituting Eq. (6.33) and (6.34) into Eq. (6.26),

$$\sigma_{tr} \propto k \tag{6.35}$$

and the contribution to the resistivity (6.23), (4.80) for the short-range scatterers, can be estimated as

$$\rho \cong \frac{h}{e^2} n_{\mathrm{imp}} R^2. \tag{6.36}$$

We will see later (see the detailed analysis in Chapter 11) that this contribution is negligible.

The results (6.34) and (6.35) are quite clear, keeping in mind an analogy with optics (Born & Wolf, 1980). The dispersion relation for massless Dirac fermions is the same as for photons, but for the latter case we know that obstacles with geometrical sizes much smaller than the wavelength are very inefficient scatterers.

There is a special case, however, if

$$J_0(qR) = 0. \tag{6.37}$$

Then, the expression (6.33) does not work at $l = 0$, and higher-order terms should be taken into account. The result is

$$t_0(k) \cong \frac{\pi i}{2} \frac{1}{\ln (kR)} \tag{6.38}$$

and

$$\sigma_{\text{tr}} = \frac{\pi^2}{k \ln^2(kR)}. \tag{6.39}$$

Therefore, instead of (6.36) we have a much larger contribution to the resistivity (Ostrovsky, Gornyi, & Mirlin, 2006; Katsnelson & Novoselov, 2007):

$$\rho \cong \frac{h}{e^2} \frac{n_{\text{imp}}}{n} \frac{1}{\ln^2(k_F R)}, \tag{6.40}$$

where $n = k_F^2/\pi$ is the charge-carrier concentration.

The condition (6.37) corresponds to the case of resonance, for which a virtual bound state in the well lies close to the neutrality point. Later in this chapter we will consider more realistic models of such resonant scatterers, namely vacancies and adatoms. It is interesting to see, however, that the effect already exists in the Dirac approximation.

If we were to repeat the same calculations for a nonrelativistic electron gas (Adhikari, 1986), then, instead of continuity of two components of the spinor wave function at $r = R$, we would have conditions of continuity of the single-component wave function and its derivative. The result is

$$t_l(k) = \frac{\left(\dfrac{k}{q}\right) J_l(qR) J_{l+1}(kR) - J_l(kR) J_{l+1}(qR)}{H_l^{(1)}(kR) J_{l+1}(qR) - \left(\dfrac{k}{q}\right) J_l(qR) H_{l+1}^{(1)}(kR)}, \tag{6.41}$$

where k and q are, again, wave vectors outside and inside the potential region. In this case $t_0(k) \sim 1/\ln(kR)$ (cf. Eq. (6.38)) for general values of the parameters, and the contribution to the resistivity takes the form (6.40). One can say that for the two-dimensional nonrelativistic electron gas *any* potential scattering should be considered resonant. This agrees with the fact that the perturbation theory does not work in such a situation and an arbitrarily weak potential leads to the formation of a bound state (Landau & Lifshitz, 1977).

The opposite limit

$$kR \gg 1 \tag{6.42}$$

is relevant for the problem of electron scattering by clusters of charge impurities (Katsnelson, Guinea, & Geim, 2009; see also Chapter 11). On substituting the asymptotics (6.5) and (6.8) into Eq. (6.30) one finds

$$t_l(k) \approx \frac{1}{2} \left[\exp\left(\frac{2iV_0 R}{\hbar v}\right) - 1 \right]. \tag{6.43}$$

The summation in Eq. (6.19) should be taken up to $|l| \le l_{\max} \approx kR$, thus

$$\frac{d\sigma}{d\varphi} = \frac{2}{\pi k} \sin^2\left(\frac{V_0 R}{\hbar v}\right) \left| \sum_{l=-l_{max}}^{l_{max}} e^{il\varphi} \right|^2 = \frac{2}{\pi k} \sin^2\left(\frac{V_0 R}{\hbar v}\right) \frac{\sin^2\left(\frac{(2l_{max}+1)\varphi}{2}\right)}{\sin^2\left(\frac{\varphi}{2}\right)}.$$

(6.44)

The expression (6.44) has sharp maxima at the angles

$$\varphi = \pi \frac{2n+1}{2l_{max}+1}, \quad n = 0, \pm 1, \cdots,$$

which can be related to periodic classical trajectories of electrons within the potential well (for more details, see Katsnelson, Guinea, & Geim, 2009). On substituting Eq. (6.44) into Eq. (6.25) one finds

$$\sigma_{tr} \cong \frac{4}{k} \sin^2\left(\frac{V_0 R}{\hbar v}\right).$$

(6.45)

Interestingly, the cross-section (6.45) is small in comparison with the geometrical size of the potential region R. Indeed, the region is transparent, due to Klein tunneling. The corresponding contribution to the resistivity is

$$\rho \cong \frac{h}{e^2} \frac{n_{imp}}{n} \sin^2\left(\frac{V_0 R}{\hbar v}\right).$$

(6.46)

Thus, long-range potential scattering leads to a contribution to the resistivity proportional to $1/n$.

6.3 Scattering theory for bilayer graphene in the parabolic-band approximation

We saw in the previous section that the scattering of massless Dirac fermions in graphene (chiral states, a linear dispersion relation) is essentially different from that of nonrelativistic electrons (nonchiral states, a parabolic dispersion relation) in a two-dimensional electron gas. To better understand the role of chirality and of dispersion relations, it is instructive to consider the case of chiral states with a parabolic dispersion relation, that is, the case of bilayer graphene in the parabolic-band approximation (1.46). The corresponding scattering theory was developed by Katsnelson (2007c).

To solve the Schrödinger equation for the Hamiltonian (1.46) with the addition of a radially symmetric potential $V(r)$, one has to use, instead of Eq. (6.3), the angular dependences of the two components of the spinor wave function

$$\psi_1(\vec{r}) = g_l(r) \exp(il\varphi),$$
$$\psi_2(\vec{r}) = f_l(r) \exp(i(l+2)\varphi), \tag{6.47}$$

where $l = 0, \pm 1, \ldots$ The radial components satisfy the equations

$$\left(\frac{d}{dr} - \frac{l+1}{r}\right)\left(\frac{d}{dr} - \frac{l}{r}\right)g_l = \left(k^2 - \frac{2m^*V}{\hbar^2}\right)f_l,$$
$$\left(\frac{d}{dr} + \frac{l+1}{r}\right)\left(\frac{d}{dr} + \frac{l+2}{r}\right)f_l = \left(k^2 - \frac{2m^*V}{\hbar^2}\right)g_l, \tag{6.48}$$

where, to be specific, we consider the case of electrons with $E = \hbar^2 k^2/(2m^*) > 0$.

The problem of scattering for this case is essentially different from both the Dirac theory and the Schrödinger theory, since evanescent waves are unavoidably involved (cf. the discussion of Klein tunneling for the case of bilayer graphene, Section 4.7). This means that, beyond the radius of action of the potential, Bessel functions of imaginary arguments have to be added to Eq. (6.9). More specifically, we mean the Macdonald function $K_l(kr)$ (Whittaker & Watson, 1927) with the asymptotics

$$K_l(kr) \approx \sqrt{\frac{\pi}{2kr}} \exp(-kr) \tag{6.49}$$

at $kr \gg 1$; the Bessel functions $I_l(kr)$ grow exponentially at large r and cannot be used, due to the normalization condition for the wave function. Thus, one should try for the solution at large distances

$$g_l(r) = A\left[J_l(kr) + t_l H_l^{(1)}(kr) + c_l K_l(kr)\right],$$
$$f_l(r) = A\left[J_{l+2}(kr) + t_l H_{l+2}^{(1)}(kr) + c_l K_{l+2}(kr)\right]. \tag{6.50}$$

One can check straightforwardly that the functions (6.50) satisfy the equations (6.48) at $V(r) = 0$ for any A, t_l and c_l.

The terms proportional to $J_l(kr)$ are related to the incident wave (see Eq. (6.4)), with those proportional to $H_l^{(1)}(kr)$ to the scattering waves and those proportional to $K_l(kr)$ to the evanescent waves. The coexistence of scattering and evanescent waves at the same energy makes the case of bilayer graphene really peculiar.

The normal component of the current operator

$$\hat{j}_n = \vec{n}\frac{\delta\hat{H}}{\delta\vec{k}}, \tag{6.51}$$

where $\vec{n} = \vec{r}/r$ and $\hat{H}$ is the Hamiltonian (1.46), has the form (cf. Eq. (6.17))

$$\hat{j}_n = \frac{\hbar k}{m^*} \begin{pmatrix} 0 & \exp(-2i\varphi) \\ \exp(2i\varphi) & 0 \end{pmatrix}. \tag{6.52}$$

By further calculating the scattering cross-section, as in the previous section, we find the same expression (6.19) and (6.20) formally, as for the case of single-layer graphene. However, the symmetry properties of Eq. (6.48) are different. Namely, they are invariant under the replacement $f \leftrightarrow g$, $l \leftrightarrow -l - 2$. As a result, instead of Eq. (6.21) we have

$$t_l(k) = t_{-l-2}(k). \tag{6.53}$$

Substituting Eq. (6.53) into Eq. (6.19), we rewrite Eq. (6.20) as

$$\frac{d\sigma}{d\varphi} = \frac{2}{\pi k} \left| t_{-1} + 2 \sum_{l=0}^{\infty} t_l \cos\left[(l + 1)\varphi\right] \right|^2, \tag{6.54}$$

which gives us a general solution of the scattering problem.

To find the scattering amplitudes t_l one needs to specify $V(r)$. For simplicity, we will use the expression (6.27) (a region of constant potential). Then, for the solution of Eq. (6.48) at $r < R$ that is regular as $r \to 0$ one can try

$$g_l(r) = \alpha_l J_l(qr) + \beta_l I_l(qr),$$
$$f_l(r) = \sigma[\alpha_l J_{l+2}(qr) + \beta_l I_{l+2}(qr)], \tag{6.55}$$

where

$$\sigma = \mathrm{sgn}(E - V_0),$$
$$q = \sqrt{\frac{2m^* |E - V_0|}{\hbar^2}}. \tag{6.56}$$

Eq. (6.48) are now satisfied identically, and the coefficients α_l, β_l, t_l, and c_l should be found from continuity of $g_l(r)$, $f_l(r)$, $dg_l(r)/dr$, and $df_l(r)/dr$ at $r = R$.

Further, we will consider only the case of a short-range potential, $kR \ll 1$. For the case $l = -1$, taking into account the identities $K_1(z) = K_{-1}(z)$, $I_1(z) = I_1(-z)$, $J_1(z) = -J_{-1}(z)$, and $H_1^{(1)}(z) = -H_{-1}^{(1)}(z)$, one can prove immediately that $c_{-1} = 0$ and

$$t_{-1} \propto (kR)^2. \tag{6.57}$$

Also, taking into account the asymptotics of the Macdonald and Hankel functions for $l > 2$, $z \to 0$ (we need here next-order terms, in comparison with Eq. (6.32)),

$$K_l(z) \approx \frac{1}{2} \left(\frac{2}{z}\right)^l (l-1)! \frac{1}{2} \left(\frac{2}{z}\right)^{l-2} (l-2)!,$$

$$H_l^{(1)}(z) \approx -\frac{i}{\pi} \left(\frac{2}{z}\right)^l (l-1)! - \frac{i}{\pi} \left(\frac{2}{z}\right)^{l-2} (l-2)!,$$

(6.58)

one can prove that for $l \geq 1$ and $kR \to 0$ both t_l and c_l are of the order of $(ka)^{2l}$ or smaller and thus only the s-channel ($l = 0$) contributes to the scattering cross-section, so that Eq. (6.54) can be rewritten as

$$\frac{d\sigma}{d\varphi} = \frac{8}{\pi k} |t_0(k)|^2 \cos^2 \varphi.$$

(6.59)

For single-layer graphene, $d\sigma/d\varphi \sim \cos^2(\varphi/2)$ (see Eq. (6.22)) and backscattering is forbidden. For the case of bilayer graphene, there is a strong suppression of the scattering at $\varphi \approx \pi/2$. This reflects a difference of the chiral properties of electron states in these two situations.

For the case $l = 0$, the wave functions at $r > R$ (but for $kR << 1$), Eq. (6.50), have the forms

$$g_0(r) = A \left[1 + t_0 + \tau_0 \ln \left(\frac{kr}{2}\right) + \gamma \right] + O\left[(kr)^2 \ln(kr) \right],$$

$$f_0(r) = A \left[-\frac{2i}{\pi} t_0 - \tau_0 \left(\frac{2}{(kr)^2} - \frac{1}{2} \right) \right] + O\left[(kr)^2 \ln(kr) \right],$$

(6.60)

where $\gamma \approx 0.577 \ldots$ is the Euler constant,

$$\tau_0 = \frac{2it_0}{\pi} - c_0.$$

(6.61)

It follows from the continuity of $df_0(r)/dr$ at $r = R$ that

$$\tau_0 = \frac{k^2 R^3}{4A} \left. \frac{df_0(r)}{dr} \right|_{r=R}.$$

(6.62)

and, thus,

$$\left. \frac{dg_0}{dr} \right|_{r=R} \propto k^2.$$

In the limit $k \to 0$ one has the condition

$$\left. \frac{dg_0}{dr} \right|_{r=R} = 0,$$

(6.63)

which gives us a ratio of β_0/α_0. As a result, for $r < R$

$$g_0(r) = \alpha_0 \left[J_0(qr) - I_0(qr) \frac{J_0'(qR)}{I_0'(qR)} \right],$$

$$f_0(r) = \sigma\alpha_0 \left[J_2(qr) - I_2(qr) \frac{J_0'(qR)}{I_0'(qR)} \right], \tag{6.64}$$

where prime means d/dR. Thus, we have two equations for the constant α_0 and A,

$$g_0(R) = A(1 + t_0),$$

$$f_0(R) + \frac{R}{2} \frac{df_0(R)}{dR} = -\frac{2iA}{\pi} t_0, \tag{6.65}$$

which gives us the final expression for t_0.

It is clear that t_0 does not depend on k in the limit $kR \to 0$. It takes the value with the maximum possible modulus, $t_0 = -1$ (the *unitary limit*), when

$$\frac{d}{dR} \frac{J_0(qR)}{I_0(qR)} = 0. \tag{6.66}$$

This behavior is dramatically different from both that of massless Dirac fermions and that of conventional nonrelativistic electrons, for which $t_0(k) \to 0$ at $k \to 0$ (either linearly or $\sim 1/|\ln k|$).

As a result, for the case of short-range scattering in bilayer graphene (in the parabolic-band approximation)

$$\sigma_{tr} \propto \frac{1}{k}, \tag{6.67}$$

and the corresponding contribution to the resistivity is

$$\rho \approx \frac{h}{e^2} \frac{n_{\text{imp}}}{n}. \tag{6.68}$$

Within the perturbation theory, this concentration dependence was obtained by Koshino and Ando (2006).

We will postpone further discussion of these results until Chapter 11, where we will discuss electronic transport in graphene; here we restrict ourselves to the quantum-mechanical problem.

6.4 General theory of defects in a honeycomb lattice

In general, the continuum medium approximation used earlier is not sufficient for discussing short-range scattering centers in graphene, since they induce intervalley transitions (Shon & Ando, 1998). To study these effects, we pass here to consideration of defects in a honeycomb lattice (Peres, Guinea, & Castro Neto, 2006;

Wehling et al., 2007; Basko, 2008; Wehling, Katsnelson, & Lichtenstein, 2009a). We will use the *T*-matrix formalism, which has already been mentioned in Section 4.2 (see Eq. (4.33) and (4.34)), but here we will present it in a more systematic way (see Lifshitz, Gredeskul, & Pastur, 1988; Vonsovsky & Katsnelson, 1989).

Let us consider a general, single-particle Hamiltonian

$$\hat{H} = \hat{H}_0 + \hat{V} \tag{6.69}$$

defined on a crystal lattice, $\hat{H}_0$ being the Hamiltonian of the ideal lattice and $\hat{V}$ the perturbation created by defects. The local density of states at site i is determined by the expression

$$N_i(E) = \langle i | \delta(E - \hat{H}) | i \rangle, \tag{6.70}$$

which can also be represented as

$$N_i(E) = -\frac{1}{\pi} \operatorname{Im} \hat{G}_{ii}(E), \tag{6.71}$$

where

$$\hat{G}(E) = \lim_{\delta \to +0} \frac{1}{E - \hat{H} + i\delta} \tag{6.72}$$

is the Green function (resolvent) of the operator $\hat{H}$. It follows immediately from Eq. (6.69) that

$$\hat{G}^{-1} = \hat{G}_0^{-1} - \hat{V}, \tag{6.73}$$

where $\hat{G}_0$ is the Green function of the unperturbed problem Eq. (4.34). By multiplying Eq. (6.73) by operators $\hat{G}$ from the right side and $\hat{G}_0$ from the left side we derive the *Dyson equation*

$$\hat{G}(E) = \hat{G}_0(E) + \hat{G}_0(E)\hat{V}\hat{G}(E). \tag{6.74}$$

Its formal solution can be written as

$$\hat{G}(E) = \hat{G}_0(E) \left[1 - \hat{V}\hat{G}_0(E) \right]^{-1}, \tag{6.75}$$

which is a compact notation for the infinite series

$$\hat{G}(E) = \hat{G}_0(E) + \hat{G}_0(E)\hat{V}\hat{G}_0(E) + \hat{G}_0(E)\hat{V}\hat{G}_0(E)\hat{V}\hat{G}_0(E) + \cdots \tag{6.76}$$

Alternatively, the series (6.76) can be written as

$$\hat{G}(E) = \hat{G}_0(E) + \hat{G}_0(E)\hat{T}(E)\hat{G}_0(E), \tag{6.77}$$

where $\hat{T}$ is the *T-matrix* satisfying Eq. (4.33). Its formal solution can be represented as

$$\hat{T}(E) = \left[1 - \hat{V}\hat{G}_0(E)\right]^{-1}\hat{V}. \tag{6.78}$$

The change of the spectral density can be expressed in terms of the *T*-matrix. The total density of states

$$N(E) = \text{Tr } \delta\left(E - \hat{H}\right) = -\frac{1}{\pi}\text{Tr Im } \hat{G}(E) \tag{6.79}$$

can be written, due to Eq. (6.72) and (6.75), as

$$N(E) = \frac{1}{\pi}\frac{\partial}{\partial E}\text{Tr Im ln } \hat{G}(E) = \frac{1}{\pi}\frac{\partial}{\partial E}\text{Tr Im}\left[\ln \hat{G}_0(E) - \ln\left(1 - \hat{V}\hat{G}_0(E)\right)\right] \tag{6.80}$$

since

$$\hat{G}(E) = -\frac{\partial}{\partial E}\ln \hat{G}(E). \tag{6.81}$$

At the same time, due to Eq. (6.78),

$$\ln \hat{T}(E) = -\ln\left[1 - \hat{V}\hat{G}_0(E)\right] + \ln \hat{V}, \tag{6.82}$$

the last term being energy-independent. As a result, the change of the density of states due to the perturbation $\hat{V}$ can be presented as

$$\Delta N(E) = N(E) - N_0(E) = \frac{1}{\pi}\frac{\partial}{\partial E}\text{Im Tr ln } \hat{T}(E). \tag{6.83}$$

Finally, using the operator identity

$$\text{Tr ln } \hat{A} = \ln \text{ det } \hat{A}, \tag{6.84}$$

one can represent Eq. (6.83) in the form

$$\Delta N(E) = -\frac{1}{\pi}\text{Im}\frac{\partial}{\partial E}\ln \text{ det}\left[1 - \hat{G}_0(E)\hat{V}\right], \tag{6.85}$$

which is more convenient for real calculations.

The contribution of point defects to the resistivity can be also expressed in terms of the *T*-matrix, see Chapter 11.

If the perturbation $\hat{V}$ is localized on one site $i = 0$ only

$$V_{ij} = V\delta_{i0}\delta_{j0}, \tag{6.86}$$

then one can see from Eq. (6.78) that the $\hat{T}$-matrix is also localized on the same site:

$$T_{ij}(E) = T_{00}(E)\delta_{i0}\delta_{j0}, \tag{6.87}$$

where

$$T_{00}(E) = \frac{V}{1 - VG_{00}^{(0)}(E)} \tag{6.88}$$

and $G_{00}^{(0)}(E)$ is the matrix element of the Green function for the ideal crystal lattice at site 0. For the lattice without basis,

$$G_{00}^{(0)}(E) = \lim_{\delta \to +0} \sum_{\vec{k}} \frac{1}{E - t(\vec{k}) + i\delta}. \tag{6.89}$$

However, for the case of a honeycomb lattice the Hamiltonian $\hat{H}_0$ is a 2×2 matrix, which has, in the nearest-neighbor approximation, the form (1.14). By inverting the matrix $E - \hat{H}_0$ one finds the Green function $\hat{G}_0$ in the k representation:

$$\hat{G}_0\left(E, \vec{k}\right) = \lim_{\delta \to +0} \frac{1}{(E + i\delta)^2 - \left|t\left(\vec{k}\right)\right|^2} \begin{pmatrix} E & t\left(\vec{k}\right) \\ t^*\left(\vec{k}\right) & E \end{pmatrix}, \tag{6.90}$$

where $t\left(\vec{k}\right) = tS\left(\vec{k}\right)$. Thus, instead of Eq. (6.89) we have, for the on-site Green function

$$\begin{aligned}
G_{00}^{(0)}(E) &= \lim_{\delta \to +0} \sum_{\vec{k}} \frac{E}{(E + i\delta)^2 - \left|t\left(\vec{k}\right)\right|^2} \\[2mm]
&= \frac{1}{2} \lim_{\delta \to +0} \sum_{\vec{k}} \left(\frac{1}{E + i\delta - \left|t\left(\vec{k}\right)\right|} + \frac{1}{E + i\delta + \left|t\left(\vec{k}\right)\right|} \right)
\end{aligned} \tag{6.91}$$

for which it does not matter whether the site 0 belongs to sublattice A or sublattice B. At $|E| \ll |t|$

$$N_0(E) = -\frac{1}{\pi} \operatorname{Im} G_{00}^{(0)}(E) = \frac{1}{\pi} \frac{|E|}{\hbar^2 v^2} \tag{6.92}$$

(cf. Eq. (1.72); our quantity is smaller by a factor of 2, since here we do not take into account the spin degeneracy). To find the real part of $G_{00}^{(0)}$ one can use Kramers–Kronig relations:

$$\operatorname{Re} G_{00}^{(0)}(E) = P \int_{-\infty}^{\infty} dE' \frac{N_0(E')}{E - E'}, \tag{6.93}$$

where P is the symbol for the principal value. We can also just guess the answer, keeping in mind that $G_{00}^{(0)}(E)$ is a regular function of energy in the upper complex half-plane.

Notice that $|E| = E\operatorname{sgn} E = E[1 - 2\theta(-E)]$, where $\theta(x > 0) = 1$, $\theta(x < 0) = 0$ and

$$\theta(-E) = \frac{1}{\pi} \operatorname{Im} \ln(E + i\delta).$$

This means that

$$|E| = E - \frac{2}{\pi} E \operatorname{Im} \ln(E + i\delta) \tag{6.94}$$

and, thus, the term $|E|$ in $-(1/\pi)\operatorname{Im} G_{00}^{(0)}(E)$ corresponds to $2E \operatorname{Re}\ln(E + i\delta) = 2E \ln|E|$ in $\operatorname{Re} G_{00}^{(0)}(E)$. Taking into account also that, by symmetry,

$$G_{00}^{(0)}(E = 0) = 0, \tag{6.95}$$

one finds

$$\operatorname{Re}G_{00}^{(0)}(E) \cong \frac{2}{\pi} \frac{E \ln\left(\dfrac{|E|}{D}\right)}{\hbar^2 v^2}, \tag{6.96}$$

where we introduce within the logarithm a factor D of the order of the bandwidth. For the accurate calculation of this factor, see Basko (2008). A general theory of scattering by short-range defects in graphene, including group-theory analysis, can also be found in that paper.

The contributions of various types of defects to the transport properties will be considered in detail in Chapter 11. Here we will just give some simple estimations, in order to establish relations between this section and the previous ones.

For the case of a weak enough potential V, the scattering rate (6.24) can be estimated, according to the Fermi golden rule, as

$$\frac{1}{\tau(k_F)} = \frac{2\pi}{\hbar} n_{imp} |V|^2 N_0(E_F). \tag{6.97}$$

For the case of a small concentration of defects but strong scattering, one can prove rigorously (Luttinger & Kohn, 1958) that the potential V should be replaced by the T-matrix:

$$\frac{1}{\tau(k_F)} = \frac{2\pi}{\hbar} n_{imp} |T_{00}(E_F)|^2 N_0(E_F) \tag{6.98}$$

(for the case of graphene, see Robinson et al., 2008; Wehling et al., 2010a).

6.5 The case of vacancies

As a specific application of the general theory described previously, consider first the case of vacancies (Peres, Guinea, & Castro Neto, 2006). Vacancies are not naturally present in graphene, due to their very high formation energy of about 7.5 eV; see Kotakoski, Krasheninnikov, and Nordlund (2006). However, they can be created by ion bombardment (Chen et al., 2009).

The simplest way to simulate the vacancy is just to put $V = \infty$ in the expression (6.88), thus making the site $i = 0$ unavailable for electrons. In this case,

$$T_{00}(E) = -\frac{1}{G_{00}^{(0)}(E)}. \tag{6.99}$$

On substituting Eq. (6.99) into Eq. (6.77) one finds that $G_{00}(E) = 0$, as it should be.

For small energies $|E| << D$ one finds from Eq. (6.96) and (6.99)

$$T_{00}(E) = -\frac{\pi \hbar^2 v^2}{E} \frac{1}{2 \ln\left(\frac{|E|}{D}\right) - i\pi \, \text{sgn} \, E}. \tag{6.100}$$

The change of the density of states, according to Eq. (6.83), is

$$\begin{aligned} \Delta N(E) &= \frac{1}{\pi} \frac{\partial}{\partial E} \, \text{Im} \ln T_{00}(E) \\ &= -\frac{1}{\pi} \frac{\partial}{\partial E} \, \text{Im} \ln \left[2 \ln \frac{|E|}{\Delta} - i\pi \text{sgn} E \right] \approx -\frac{2}{|E| \ln^2\left(\frac{D}{|E|}\right)} \end{aligned} \tag{6.101}$$

This contribution is negative since the vacancy changes the total number of sites in the system by one, thus

$$\int_{-\infty}^{\infty} dE \, \Delta N(E) = -1. \tag{6.102}$$

It is singular at $E \to 0$.

By substituting Eq. (6.100) into Eq. (6.98) and (6.23), one can estimate the vacancy contribution to the momentum relaxation rate and, thus, to the resistivity:

$$\rho \approx \frac{h}{e^2} \frac{n_{\text{imp}}}{n} \frac{1}{\ln^2(k_F a)}, \tag{6.103}$$

coinciding with Eq. (6.40). Thus, the vacancy is a resonant scatterer, contributing essentially to the resistivity (Hentschel & Guinea, 2007; Chen et al., 2009).

Qualitatively, this result can also be obtained within the continuum model. Let us consider the Dirac equation for the empty space with the radial wave functions (6.9). Let us assume that the disc $r < R$ is just cut from the sample. To be specific, let us assume boundary conditions of zigzag type $\psi_A = 0$, that is

$$g_l(R) = 0 \tag{6.104}$$

(the case $\psi_B = 0$ can be derived just by the replacement $l \rightarrow -l - 1$, as was explained in Section 6.1). Taking into account the behavior of Bessel and Hankel functions at $kr << 1$ (Eq. 6.32), one finds immediately that

$$t_0(k) = -\frac{J_0(kR)}{H_0^{(1)}(kR)} \approx \frac{\pi i}{2 \ln (kR)}, \tag{6.105}$$

coinciding with Eq. (6.38). As we have seen in Section 6.2, this gives the estimation (6.103) for the resistivity (Hentschel & Guinea, 2007).

Consider now the asymptotics of the perturbed density of states

$$\Delta N_i(E) = -\frac{1}{\pi} \mathrm{Im} \left[G_{i0}^{(0)}(E) T_{00}(E) G_{0i}^{(0)}(E) \right] \tag{6.106}$$

(see Eq. (6.77)) at $R_i \rightarrow \infty$. The asymptotics of the Green function

$$G_{i0}^{(0)}(E) = \sum_{\vec{k}} \exp \left(i\vec{k}\vec{R}_i \right) G_0 \left(E, \vec{k} \right), \tag{6.107}$$

where $G_0 \left(E, \vec{k} \right)$ is defined by Eq. (6.90), is determined by the region of $\vec{k}$ close to one of the conical points, K or K'. For a generic perturbation V the result is (Bena & Kivelson, 2005; Lin, 2005, 2006; Wehling et al., 2007)

$$\Delta N_i(E) \sim \frac{1}{R_i} \tag{6.108}$$

at $\frac{|E|R_i}{\hbar v} >> 1$. For the case of a vacancy ($V = \infty$) we have, instead of Eq. (6.108),

$$\Delta N_i(E) \sim \frac{1}{R_i^2} \tag{6.109}$$

(Pereira et al., 2006).

Finally, consider the case of a finite concentration of vacancies. The singularity of the scattering amplitude, Eq. (6.100) and (6.105), results in the formation of *mid-gap states*, or *vacancy bands* (Pereira et al., 2006; Yuan, De Raedt & Katsnelson, 2010a, 2010b). Fig. 6.1 shows the total density of states (in the small-energy region) obtained numerically for a large (about 10^7 nodes) piece of honeycomb lattice with periodic boundary conditions, with different concentrations of

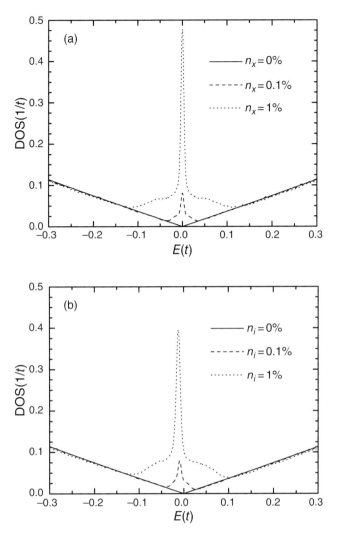

Fig. 6.1 The density of states of graphene with a small concentration of vacancies (a) or hydrogen atoms (that is, adatoms with the parameters (6.114)) (b). Solid lines, pure graphene; dashed lines, 0.1% of defects; dotted lines, 1% of defects. (Reproduced with permission from Yuan, De Raedt, & Katsnelson, 2010a.)

randomly distributed vacancies (Yuan, De Raedt, & Katsnelson, 2010a). The vacancy-induced states form a peak at $E = 0$ which was observed experimentally by Ugeda et al. (2010). In the continuum-medium model (see Eq. (6.104)) these states are associated with the edge states at the boundary of the void (Pereira et al., 2006). Note, however, that the latter model is valid only qualitatively, since the atomically sharp disorder induces intervalley processes, which should be taken into account (Basko, 2008).

6.6 Adsorbates on graphene

Adsorbed atoms and molecules are probably the most important examples of point defects in the physics of graphene. Owing to the outstanding strength of the carbon honeycomb lattice it is very difficult to introduce any defects into the lattice itself. At the same time, some contamination of graphene is unavoidable. A systematic study of adsorbates on graphene was started by Schedin et al. (2007), who discovered an extreme sensitivity of the electric properties of graphene to gaseous impurities; even the adsorption of a single molecule can be detected. The case of NO_2 was studied in detail, both theoretically and experimentally, by Wehling et al. (2008b). Optimized structures and electron densities of states for the NO_2 monomer and dimer are shown in Fig. 6.2. One can see that for the latter case (N_2O_4) there is a peak in the density of states that is reminiscent of the vacancy-induced mid-gap states. Chemical functionalization of graphene, leading, in particular, to the derivation of new two-dimensional crystals, such as graphane, CH (Elias et al., 2009), and fluorographene, CF (Nair et al., 2010), starts with chemisorption of the corresponding adatoms or admolecules (for a review, see Boukhvalov & Katsnelson, 2009a). Last but not least, scattering by adatoms and admolecules seems to be one of the most important factors limiting electron mobility in graphene (Wehling et al., 2010a; Ni et al., 2010); for more details, see Chapter 11.

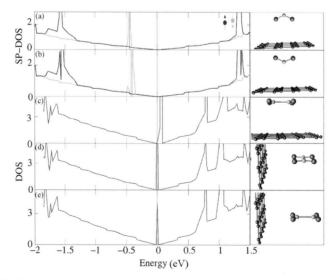

Fig. 6.2 Left: the spin-polarized density of states of graphene with adsorbed NO_2 (the black line is for spin up and the gray line is for spin down), (a) and (b), and the density of states for N_2O_4, (c)–(e), in various adsorption geometries. Right: adsorption geometries obtained from the calculations.
(Reproduced with permission from Wehling et al., 2008b.)

The simplest single-electron model describing adsorbates is the hybridization model with the Hamiltonian (Robinson et al., 2008; Wehling et al., 2010a)

$$\hat{H} = \sum_{ij} t_{ij} \hat{c}_i^+ \hat{c}_j + \sum_{ij} \gamma_{ij} \left(\hat{c}_i^+ \hat{d}_j + \hat{d}_j^+ \hat{c}_i \right) + E_d \sum_i \hat{d}_i^+ \hat{d}_i, \qquad (6.110)$$

where the operators $\hat{c}_i$ and $\hat{d}_i$ annihilate electrons on the ith carbon atom and ith atom of adsorbate, respectively, t_{ij} are the hopping parameters for the carbon honeycomb lattice, E_d is the electron energy for the adsorbate atoms (which are assumed to be identical), and γ_{ij} are hybridization parameters between the ith carbon atom and jth adsorbed atom. The d electron subsystem can be rigorously excluded by projection to c subspace only; the effective Hamiltonian for c electrons has the form (6.69), where $\hat{H}_0$ is the first term on the right-hand side of Eq. (6.110) (the band Hamiltonian for graphene), and $\hat{V}$ is the *energy-dependent* perturbation

$$V_{ij} = \frac{\sum_l \gamma_{il} \gamma_{lj}}{E - E_d}. \qquad (6.111)$$

If we consider the case of a single adatom ($i = 0$) and assume, for simplicity, that $\gamma_{ij} = \gamma \delta_{ij}$, we pass to the problem (6.86) with

$$V(E) = \frac{\gamma^2}{E - E_d}. \qquad (6.112)$$

Further, we can simply use the theory developed in the previous section.

If the condition

$$\gamma^2 >> |E_d||t| \qquad (6.113)$$

is satisfied, then, at energies close enough to the Dirac point ($|E| << |t|$), the potential (6.112) is very strong, and an adatom is effectively equivalent to a vacancy.

To understand this very important point, let us consider the hydrogen atom as an example. It is attached to one of the carbon atoms, transforming locally its state from sp^2 bonded to sp^3 bonded; distortions make the angles between the bonds and bond lengths locally similar to those in diamond (Boukhvalov, Katsnelson, & Lichtenstein, 2008), see Fig. 6.3. This means that the carbon atom bonded with hydrogen is almost unavailable for p_z electrons, since their energies are locally shifted too strongly. This makes it similar to a vacancy. *Ab initio* calculations (Wehling et al., 2010a) show that the local electronic structure for the case of a hydrogen adatom can be quite accurately fitted by the hybridization model with the parameters

$$\gamma \approx 2|t|, \quad E_d \approx -\frac{|t|}{16}, \qquad (6.114)$$

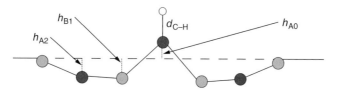

Fig. 6.3 Atomic displacements around a hydrogen atom attached to one of the carbon atoms in graphene. Carbon atoms belonging to sublattices A and B are shown in dark gray and light gray, respectively; $h_{A0} = 0.257$ Å, $h_{B1} = -0.047$ Å, $h_{A2} = -0.036$ Å, and $d_{C-H} = 1.22$ Å. (Reproduced with permission from Boukhvalov, Katsnelson, & Lichtenstein, 2008.)

so the inequality (6.113) is satisfied with high precision. This means that hydrogen atoms form mid-gap states, which are, however, slightly shifted with respect to the Dirac point, because $E_d < 0$ (Wehling et al., 2010a; Yuan, De Raedt, & Katsnelson, 2010a), see Fig. 6.1.

Interestingly, approximately the same parameters (6.114) describe the case of various organic groups, such as CH_3, C_2H_5, and CH_2OH, attached to carbon atoms via the carbon–carbon chemical bond (Wehling et al., 2010a). One can assume that such bonds can be formed in real graphene with organic contaminants, which, therefore, can be responsible for the appearance of strongly "resonant" scatterers (Ni et al., 2010; Wehling et al., 2010a).

The position of the impurity peak corresponds to the pole of the T-matrix

$$1 = V\left(E_{\text{imp}}\right) G_{00}^{(0)}\left(E_{\text{imp}}\right) \tag{6.115}$$

(see Eq. (6.88)). With the parameters (6.114) we find $E_{\text{imp}} \approx -0.03$ eV, in agreement with the results of straightforward *ab initio* calculations (Wehling, Katsnelson, & Lichtenstein, 2009b). For the case of fluorine, F, and the hydroxyl group, OH, the latter parameters give, respectively, $E_{\text{imp}} \approx -0.67$ eV and $E_{\text{imp}} \approx -0.70$ eV, so these impurities are weaker scatterers than hydrogen or a vacancy (Wehling, Katsnelson, & Lichtenstein, 2009b).

Further discussion will be presented in Chapters 11 and 12, in relation to the effects of adatoms on electronic transport in graphene and their magnetic properties, respectively.

6.7 Scanning tunneling microscopy of point defects on graphene

Scanning tunneling microscopy (STM) allows us to probe the electronic properties of conducting materials with atomic-scale spatial resolution (Binnig & Rohrer, 1987). Being a local probe, it is especially suitable for studying the electronic structures of various types of defects and defect-induced features, including many-body effects (Li et al., 1998; Madhavan et al., 1998; Balatsky, Vekhter, & Zhu,

2006). In particular, it was used locally to probe vacancies in the top (graphene) layer of graphite (Ugeda et al., 2010) and a magnetic adatom (Co) on graphene (Brar et al., 2011). Here we will discuss some general peculiarities of the STM spectra of graphene (Uchoa et al., 2009; Saha, Paul, & Sengupta, 2010; Wehling et al., 2010b).

Assuming that the tunneling between the sample and the STM tip is weak enough, one can derive, to lowest order in the tunneling amplitude M, the following expression for the current–voltage (I–V) characteristic (Tersoff & Hamann, 1985; Mahan, 1990):

$$I(V) = \frac{\pi e}{\hbar} \sum_{nv\sigma} |M_{nv}^{\sigma}|^2 \int dE \, N_n^{\sigma}(E) N_n^{\sigma}(E - eV)[f(E - eV) - f(E)], \quad (6.116)$$

where $f(E)$ is the Fermi distribution function, σ is the spin projection, Greek (Latin) indices label electron eigenstates for the sample (tip) $\psi_{v\sigma}$ and $\psi_{n\sigma}$,

$$M_{nv}^{\sigma} = \frac{\hbar^2}{2m} \int d\vec{S} \left(\psi_{n\sigma}^* \nabla \psi_{v\sigma} - \psi_{v\sigma} \nabla \psi_{n\sigma}^* \right) \quad (6.117)$$

is the current-matrix element, m is the free-electron mass, and the surface integral in Eq. (6.117) is taken over arbitrary area between the tip and sample. The spectral densities

$$N_v^{\sigma}(E) = -\frac{1}{\pi} \text{Im} \, G_v^{\sigma}(E) \quad (6.118)$$

for the sample and a similar quantity $N_v^{\sigma}(E)$ for the tip determine the intensity of tunneling. If one neglects the spin polarization, assumes that the spectral density of the tip is a smooth function, and uses a semiclassical approximation (Ukraintsev, 1996), one can demonstrate that, at low enough temperatures ($T \ll |eV|$),

$$\frac{dI}{dV} \propto -\frac{1}{\pi} \text{Im} \, G_{ii}(E = eV), \quad (6.119)$$

where i is the site index for the atom of the sample nearest to the tip. This means that, using STM, one can probe the spatial distribution of the electron density around the defect (see Wehling et al., 2007).

Let us assume that the adatom situated at the site $i = 0$ has a resonant state which can be of single-electron or many-body origin (e.g., the Kondo effect). The expression (6.116) and, thus, (6.119) are correct, anyway, assuming that the tunneling amplitude M is small enough and the lowest-order perturbation theory in M works (Mahan, 1990).

The resonance at $E = E_d$ is manifested in this situation via two contributions, namely, the direct contribution of d electrons to tunneling and the contribution of c electrons to the tunneling, via c-d hybridization. This leads to the *Fano (anti) resonance effect* (Madhavan et al., 2001). For simplicity, we can assume that d states are more localized than c states and, thus, only the second effect is important. In this situation, we can use Eq. (6.119), assuming that G is the Green function of c electrons. Its change due to the presence of an impurity is determined by Eq. (6.77). On putting $i = 0$ one finds

$$\text{Im}\left[G_{00}(E) - G_{00}^{(0)}(E)\right] = \text{Im}\left\{\left[G_{00}^{(0)}(E)\right]^2 T_{00}(E)\right\}$$

$$= \left\{\text{Re}\left[G_{00}^{(0)}(E)\right]^2 - \text{Im}\left[G_{00}^{(0)}(E)\right]^2\right\}\text{Im } T_{00}(E) \quad (6.120)$$

$$+ 2\left\{\text{Im}\left[G_{00}^{(0)}(E)\right]\text{Re}\left[G_{00}^{(0)}(E)\right]\right\}\text{Re } T_{00}(E).$$

In the case of resonance,

$$T_{00}(E) \sim \frac{1}{E - E_d + i\Delta}, \quad (6.121)$$

where Δ is the halfwidth of the resonance, thus

$$-\text{Im } T_{00}(E) \sim \frac{\Delta}{(E - E_d)^2 + \Delta^2} \quad (6.122)$$

has a maximum at $E = E_d$ and

$$\text{Re } T_{00}(E) \sim \frac{E - E_d}{(E - E_d)^2 + \Delta^2} \quad (6.123)$$

changes sign. Assuming that $G_{00}^{(0)}(E)$ is smoothly dependent on the energy at the energy scale $|E - E_d| \approx \Delta$ and substituting Eq. (6.120) through (Eq. (6.123)) into Eq. (6.119), one finds

$$\frac{dI}{dV} \propto \frac{q^2 - 1 + 2q\varepsilon'}{1 + \varepsilon'^2}, \quad (6.124)$$

where

$$\varepsilon' = \frac{eV - E_d}{\Delta} \quad (6.125)$$

and the quantity

$$q = -\frac{\text{Re } G_{00}^{(0)}(E_d)}{\text{Im } G_{00}^{(0)}(E_d)} \quad (6.126)$$

is called the *Fano asymmetry factor* (which should not be confused with the Fano factor (3.17) – the usual problem when a particular scientist made essential contributions to various fields!). If q is large then the *resonance* should be observed, whereas for small q one will observe rather the *antiresonance* (a dip in dI/dV instead of a peak).

For graphene, due to Eq. (6.92) and (6.96), the Fano factor at $|E| \ll \Delta$,

$$q = \frac{2}{\pi} \ln \left| \frac{\Delta}{E_d} \right|, \tag{6.127}$$

is very large (Wehling et al., 2010b).

For a more detailed analysis, see Uchoa et al. (2009), Wehling et al. (2010b) and Saha, Paul, and Sengupta (2010).

6.8 Long-range interaction between adatoms on graphene

Consider now the *energetics* of point defects and their clusters. On substituting Eq. (6.83) for the change of the total density of states into the expression for the thermodynamic potential of noninteracting fermions Eq. (2.134), one finds

$$\Delta\Omega = -\frac{T}{\pi} \operatorname{Im} \operatorname{Tr} \int_{-\infty}^{\infty} dE \, \ln \left[1 + \exp \left(\frac{\mu - E}{T} \right) \right] \frac{\partial}{\partial E} \ln \hat{T}(E)$$

$$= \frac{1}{\pi} \operatorname{Im} \operatorname{Tr} \int_{-\infty}^{\infty} dE \, f(E) \ln \hat{T}(E) \tag{6.128}$$

$$= \frac{1}{\pi} \operatorname{Im} \int_{-\infty}^{\infty} dE \, f(E) \, \ln \, \det \left[1 - \hat{G}_0(E) \hat{V} \right]$$

(see Eq. (6.85)).

This expression can be used, for example, to study the effects of interactions between impurities. Let us assume that

$$V_{ij} = V_1 \delta_{i1} \delta_{j1} + V_2 \delta_{i2} \delta_{j2}, \tag{6.129}$$

which means two defects with local potential at sites $i = 1$ and $i = 2$ (cf. Eq. (6.86)). Then,

$$\det \left[1 - \hat{G}_0 V \right] = \left[1 - \hat{G}_{11}^{(0)} V_1 \right] \left[1 - \hat{G}_{12}^{(0)} V_2 \right] - V_1 \hat{G}_{12}^{(0)} V_2 \hat{G}_{21}^{(0)}. \tag{6.130}$$

To find the interaction energy one needs to substitute Eq. (6.130) into Eq. (6.128) and subtract the same expression with $\hat{G}_{12}^{(0)} = 0$, which corresponds to the case of noninteracting defects. As a result, we obtain

$$\Omega_{\text{int}} = \frac{1}{\pi} \text{Im} \int\limits_{-\infty}^{\infty} dE\, f(E) \ln \left[1 - T_{11}^{(0)}(E) G_{12}^{(0)}(E) T_{22}^{(0)}(E) G_{21}^{(0)}(E)\right], \qquad (6.131)$$

where $T_{ii}^{(0)}(E)$ are the single-site T-matrices (6.88). Keeping in mind that the functions $G^{(0)}(E)$ and $T(E)$ are analytic at Im $E > 0$, that the Fermi function has poles at

$$E = \mu + i\varepsilon_n, \qquad (6.132)$$

where

$$\varepsilon_n = \pi T(2n + 1),$$

with the residues $-T$, and recalling that

$$\text{Im}\, A(E + i0) = \frac{1}{2i}[A(E + i0) - A(E - i0)], \qquad (6.133)$$

one can rewrite the expression (6.131) as

$$\Omega_{\text{int}} = -T \sum_{\varepsilon_n} \ln \left[1 - T_{11}^{(0)}(i\varepsilon_n + \mu) G_{12}^{(0)}(i\varepsilon_n + \mu) T_{22}^{(0)}(i\varepsilon_n + \mu) G_{21}^{(0)}(i\varepsilon_n + \mu)\right]$$

$$(6.134)$$

(Shytov, Abanin, & Levitov, 2009). One can use this expression to calculate the interaction energy for two resonant impurities, such as vacancies or hydrogen adatoms, when Eq. (6.99) can be used for the T-matrix.

To calculate the asymptotics of the interaction energy at large distances, one can assume that G_{12} is small and only take into account the first term in the Taylor expansion of Eq. (6.134):

$$\Omega_{\text{int}} \approx T \sum_{\varepsilon_n} T_{11}^{(0)}(i\varepsilon_n + \mu) G_{12}^{(0)}(i\varepsilon_n + \mu) T_{22}^{(0)}(i\varepsilon_n + \mu) G_{21}^{(0)}(i\varepsilon_n + \mu). \qquad (6.135)$$

Later we will consider the case of undoped graphene ($\mu = 0$).

Using this expression, one can prove that the sign of the interaction is different for impurities belonging to the same sublattice and to a different sublattice. In the latter case, there is *attraction* between the impurities, decaying as

$$U_{AB}(r) \propto -\frac{1}{r \ln (r/a)} \qquad (6.136)$$

($r \gg a$), whereas for the former case there is *repulsion*

$$U_{AA}(r) \propto \frac{1}{r^2 \ln (r/a)}.$$ (6.137)

This means that the resonant impurities would prefer to sit in different sublattices (Shytov, Abanin, & Levitov, 2009). This consideration is valid only at large distances. Interestingly, first-principles electronic-structure calculations (Boukhvalov, Katsnelson, & Lichtenstein, 2008; Boukhvalov & Katsnelson, 2009a) show that the same happens for the nearest-neighbor, next-nearest-neighbor, etc. distances: The resonant impurities *always* prefer to sit in different sublattices.

7

Optics and response functions

7.1 Light absorption by Dirac fermions: visualization of the fine-structure constant

In this chapter we will discuss electromagnetic properties of graphene related to electron–photon interaction. The discussion of optical properties related to phonons (infrared adsorption, the Raman effect) will be postponed until Section 9.8.

Massless Dirac fermions in two dimensions have an amazing property: their optical response is universal and expressed only in terms of the fine-structure constant (Ando, Zheng, & Suzuura, 2002; Gusynin, Sharapov, & Carbotte, 2006; Nair et al., 2008)

$$\alpha = \frac{e^2}{\hbar c} \approx \frac{1}{137.036}. \tag{7.1}$$

Experiments on light absorption of graphene can, literally, visualize this fundamental constant (Nair et al., 2008). To see this, let us determine the electric field of the light via the vector potential $\vec{A}(t) = \vec{A} \exp(-i\omega t)$,

$$\vec{E}(t) = -\frac{1}{c}\frac{\partial \vec{A}}{\partial t} = \frac{i\omega}{c}\vec{A}. \tag{7.2}$$

This is more convenient for optics than the representation via the scalar potential $\vec{E} = -\vec{\nabla}\varphi$, but is, of course, equivalent to it due to gauge invariance. Thus, the Hamiltonian of Dirac electrons in the presence of an electric field is (cf. Eq. (2.20) and (2.24))

$$\hat{H} = v\vec{\sigma}\left(\hat{\vec{p}} - \frac{e}{c}\vec{A}\right) = \hat{H}_0 + \hat{H}_{\text{int}}, \tag{7.3}$$

where

$$\hat{H}_{\text{int}} = -\frac{ve}{2c}\vec{\sigma}\vec{A} = \frac{iev}{2\omega}\vec{\sigma}\vec{E} \tag{7.4}$$

168

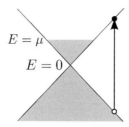

Fig. 7.1 A schematic representation of direct optical transitions in graphene.

is the Hamiltonian of the electron–photon interaction. The factor $\frac{1}{2}$ in Eq. (7.4) is necessary, since the standard expression for the complex field is

$$\vec{E}(t) = \mathrm{Re}\left[\vec{E}\,\exp\left(-i\omega t\right)\right] = \frac{1}{2}\left[\vec{E}\,\exp\left(-i\omega t\right) + \vec{E}^*\exp\left(-i\omega t\right)\right] \qquad (7.5)$$

and we take into account only the first term. This interaction induces transitions from the occupied hole states $\psi_h\left(\vec{k}\right)$ to the empty electron states $\psi_e\left(\vec{k}\right)$ with the same wave vector $\vec{k}$ (see Eq. (1.30)), the intraband transitions being forbidden by the momentum conservation (Fig. 7.1). The matrix element of the Hamiltonian (7.4) is

$$\left\langle\psi_h|\hat{H}_{\mathrm{int}}|\psi_e\right\rangle = \frac{ev}{2\omega}\left(E_y\cos\,\varphi\mp E_x\sin\,\varphi\right), \qquad (7.6)$$

where the $-$ and $+$ signs correspond to K and K′ valleys. It only depends on the polar angle φ of the $\vec{k}$ vector, not on its length. On averaging the square matrix element over φ one finds

$$|M|^2 = \overline{\left|\left\langle\psi_h|\hat{H}_{\mathrm{int}}|\psi_e\right\rangle\right|^2} = \frac{e^2v^2}{8\omega^2}\left|\vec{E}\right|^2, \qquad (7.7)$$

where we assume that the photon propagates perpendicular to the graphene plane and, thus, the vector $\vec{E} = \left(E_x, E_y, 0\right)$ lies within the plane. The absorption probability per unit time, to the lowest order of perturbation theory, is (Landau & Lifshitz, 1977)

$$P = \frac{2\pi}{\hbar}|M|^2N\left(\varepsilon = \frac{\hbar\omega}{2}\right), \qquad (7.8)$$

where $N(\varepsilon)$ is the density of states (1.72) (we take into account the spin and valley degeneracy) and the energy of the final states is $\hbar\omega/2$ as is obvious from Fig. 7.1. On substituting Eq. (1.72) and (7.7) into (7.8) we find

$$P = \frac{e^2}{4\hbar^2\omega}\left|\vec{E}\right|^2. \qquad (7.9)$$

Thus, the absorption energy per unit time is

$$W_a = P\hbar\omega = \frac{e^2}{4\hbar}|\vec{E}|^2. \tag{7.10}$$

At the same time, the incident energy flux is (Jackson, 1962)

$$W_i = \frac{c}{4\pi}|\vec{E}|^2. \tag{7.11}$$

The absorption coefficient is, therefore,

$$\eta = \frac{W_a}{W_i} = \frac{\pi e^2}{\hbar c} \approx 2.3\% \tag{7.12}$$

and is universal, assuming that $\hbar\omega > 2|\mu|$. Otherwise, the transitions are forbidden by the Pauli principle (see Fig. 7.1) and $\eta = 0$. For visible light, $\hbar\omega \approx 1 - 2$ eV is much higher than the Fermi energy in graphene. Moreover, it is much higher than the energy of electron hopping between layers in multilayer graphene or graphite. Therefore, the absorption for N-layer graphene is just $N\eta$. This behavior was observed experimentally for single-layer and bilayer graphene (Nair et al., 2008) and for graphite (Kuzmenko et al., 2008). According to Eq. (7.12), graphene is quite transparent. At the same time, one should keep in mind that this is an absorption coefficient of more than 2% *per single atomic layer*, which is a huge value. Thus, the interaction of Dirac electrons with photons is actually very strong.

In the first work (Novoselov et al., 2004) single-layer graphene on SiO_2 was first detected just by the human eye, via a conventional (optical) microscope. It was a lucky coincidence that the contrast due to light absorption in graphene was strongly enhanced by interference phenomena in the SiO_2 layer with appropriate thickness. The optics of the visibility of graphene on a substrate was considered by Blake et al. (2007) and Abergel, Russel, and Fal'ko (2007).

7.2 The optics of Dirac fermions: the pseudospin precession formalism

The optical properties of Dirac fermions can be studied in a physically transparent way using the equations of motion for the density matrix (Katsnelson, 2008). It has the form (2.173). For the Hamiltonian one can use Eq. (7.3); however, it is more instructive to change the gauge and write

$$\hat{H}_{int} = -e\vec{E}(t)\hat{\vec{r}} = -ie\vec{E}(t)\vec{\nabla}_{\vec{k}} \tag{7.13}$$

(see Eq. (2.178)). We will show explicitly that the result (7.12) can be derived within this representation as well. Thus, the equation (2.173) reads

$$i\hbar\frac{\partial\hat{\rho}_{\vec{k}}}{\partial t} = \hbar v\vec{k}\left[\hat{\vec{\sigma}},\hat{\rho}_{\vec{k}}\right] - ie\left(\vec{E}(t)\cdot\vec{\nabla}_{\vec{k}}\right)\hat{\rho}_{\vec{k}},\tag{7.14}$$

where $\hat{\rho}_{\vec{k}}$ is the 2×2 pseudospin matrix

$$\left(\hat{\rho}_{\vec{k}}\right)_{\alpha\beta} = \left\langle\psi^+_{\vec{k}\beta}\psi_{\vec{k}\alpha}\right\rangle\tag{7.15}$$

(cf. Eq. (2.170) and (3.1)). It can be expanded in Pauli matrices

$$\hat{\rho}_{\vec{k}} = n_{\vec{k}}\hat{I} + \vec{m}_{\vec{k}}\hat{\vec{\sigma}},\tag{7.16}$$

where $\hat{I}$ is the unit 2×2 matrix, and

$$n_{\vec{k}} = \frac{1}{2}\mathrm{Tr}\hat{\rho}_{\vec{k}}\tag{7.17}$$

and

$$\vec{m}_{\vec{k}} = \frac{1}{2}\mathrm{Tr}\left(\hat{\vec{\sigma}}\hat{\rho}_{\vec{k}}\right)\tag{7.18}$$

are charge and pseudospin densities (in the $\vec{k}$ representation). On substituting Eq. (7.16) into Eq. (7.14) we find the separated equations for the charge density

$$\frac{\partial n_{\vec{k}}}{\partial t} = -\frac{e}{\hbar}\left(\vec{E}\cdot\vec{\nabla}_{\vec{k}}\right)n_{\vec{k}},\tag{7.19}$$

and the pseudospin density

$$\frac{\partial\vec{m}_{\vec{k}}}{\partial t} = 2v\left(\vec{k}\cdot\vec{m}_{\vec{k}}\right) - \frac{e}{\hbar}\left(\vec{E}\cdot\vec{\nabla}_{\vec{k}}\right)\vec{m}_{\vec{k}}.\tag{7.20}$$

To calculate the time-dependent current density

$$\vec{j} = \mathrm{Tr}\left(\hat{\vec{j}}\hat{\rho}\right) = 2ev\sum_{\vec{k}}\vec{m}_{\vec{k}},\tag{7.21}$$

we need only Eq. (7.20). It is rigorous (for noninteracting fermions) and can be used to calculate both linear and nonlinear optical properties. The first term on the right-hand side of Eq. (7.20) is nothing other than precession, with a pseudomagnetic "field" proportional to $\vec{k}$ acting on the pseudospin degree of freedom. A similar formalism was used by Anderson (1958) as the most physical way to represent the BCS theory of superconductivity.

To calculate the optical conductivity we will use the first-order perturbation in $\vec{E}$, assuming that it has the form $\vec{E}\exp(-i\omega t)$, and look for the solution of Eq. (7.20) as

$$\vec{m}_{\vec{k}}(t) = \vec{m}_{\vec{k}}^{(0)} + \delta \vec{m}_{\vec{k}} \exp\left(-i\omega t\right), \tag{7.22}$$

where

$$\delta \vec{m}_{\vec{k}} \sim \vec{E}.$$

To calculate $\vec{m}_{\vec{k}}^{(0)}$, we use the unitary transformation

$$\psi_{\vec{k}}1 = \frac{1}{\sqrt{2}}\left(\xi_{\vec{k}1} + \xi_{\vec{k}2}\right),$$

$$\psi_{\vec{k}}2 = \frac{\exp\left(i\varphi_{\vec{k}}\right)}{\sqrt{2}}\left(\xi_{\vec{k}1} - \xi_{\vec{k}2}\right), \tag{7.23}$$

diagonalizing the Hamiltonian $\hat{H}_0$,

$$\hat{H}_0 = \sum_{\vec{k}} \hbar v k \left(\xi_{\vec{k}2}^{+}\xi_{\vec{k}2} - \xi_{\vec{k}1}^{+}\xi_{\vec{k}1}\right). \tag{7.24}$$

So $\xi_{\vec{k}1}$ and $\xi_{\vec{k}2}$ are annihilation operators for holes and electrons, respectively. At equilibrium,

$$\left\langle \xi_{\vec{k}i}^{+}\xi_{\vec{k}i}\right\rangle = f_{\vec{k}i} \tag{7.25}$$

are Fermi distribution functions depending on the energies $\mp \hbar v k$. We obtain

$$\vec{m}_{\vec{k}}^{(0)} = \frac{\vec{k}}{2k}\left(f_{\vec{k}1} - f_{\vec{k}2}\right). \tag{7.26}$$

Eq. (7.20) takes the form

$$\omega \,\delta \vec{m}_{\vec{k}} = 2v\left(\vec{k} \times \delta \vec{m}_{\vec{k}}\right) - \frac{e}{\hbar}\left(\vec{E}\cdot\vec{\nabla}_{\vec{k}}\right)\vec{m}_{\vec{k}}^{(0)}. \tag{7.27}$$

Since the vector (7.26) lies in the xy-plane, the component δm^z is not coupled to the electric field and can be found from Eq. (7.27):

$$\delta m_{\vec{k}}^{z} = \frac{2v}{\omega}\left(k_x \delta m_{\vec{k}}^{y} - k_y \delta m_{\vec{k}}^{x}\right). \tag{7.28}$$

Using Eq. (7.28) to exclude δm^z from the equations for δm^x and δm^y, we find

$$\left(\omega^2 - 4v^2 k_y^2\right)\delta m_{\vec{k}}^{x} + 4v^2 k_x k_y \delta m_{\vec{k}}^{y} = -\frac{ie\omega}{\hbar}E\frac{\partial m_{\vec{k}}^{x(0)}}{\partial k_x},$$

$$4v^2 k_x k_y \delta m_{\vec{k}}^{x} + \left(\omega^2 - 4v^2 k_x^2\right)\delta m_{\vec{k}}^{y} = -\frac{ie\omega}{\hbar}E\frac{\partial m_{\vec{k}}^{y(0)}}{\partial k_x}, \tag{7.29}$$

where we have chosen the direction of the x-axis along the electric field. By solving Eq. (7.29) and calculating the current along the x-axis as

$$j_x = 2ev \sum_{\vec{k}} \delta m_{\vec{k}}^x = \sigma(\omega)E \qquad (7.30)$$

we obtain the following expression for the optical conductivity:

$$\sigma(\omega) = -\frac{8ie^2v^3}{\hbar\omega} \sum_{\vec{k}} \frac{k_y}{\omega^2 - 4v^2k^2} \left(k_y \frac{\partial m_{\vec{k}}^{x(0)}}{\partial k_x} - k_x \frac{\partial m_{\vec{k}}^{x(0)}}{\partial k_x} \right). \qquad (7.31)$$

On substituting Eq. (7.26) into Eq. (7.31) we find

$$\sigma(\omega) = -\frac{4ie^2v^3}{\hbar\omega} \sum_{\vec{k}} \frac{k_y^2}{\omega^2 - 4v^2k^2k} \frac{1}{k} \left(f_{\vec{k}1} - f_{\vec{k}2} \right)$$

$$= -\frac{2ie^2v^3}{\hbar\omega} \sum_{\vec{k}} \frac{k \left(f_{\vec{k}1} - f_{\vec{k}2} \right)}{\omega^2 - 4v^2k^2}. \qquad (7.32)$$

As is usual in calculations of response functions, one should make the replacements $\omega \to \omega + i\delta$ in Eq. (7.32) and $\delta \to +0$ at the end of the calculations (Zubarev, 1974).

To calculate Re $\sigma(\omega)$, one needs to make the replacement

$$\frac{1}{\omega^2 - 4v^2k^2} \to \mathrm{Im} \frac{1}{(\omega + i\delta)^2 - 4v^2k^2} = -\pi i\delta(\omega^2 - 4v^2k^2)$$

$$= -\frac{\pi i\delta(\omega - 2vk)}{4vk}. \qquad (7.33)$$

So,

$$\mathrm{Re}\sigma(\omega) = \frac{\pi e^2 v^2}{2\hbar\omega} \sum_{\vec{k}} \left(f_{\vec{k}1} - f_{\vec{k}2} \right) \delta(\omega - 2vk)$$

$$= \frac{e^2}{16\hbar} \left[f\left(\varepsilon = -\frac{\hbar\omega}{2} \right) - f\left(\varepsilon = \frac{\hbar\omega}{2} \right) \right]. \qquad (7.34)$$

This is the conductivity per valley per spin. On multiplying the result by 4 and setting the temperature to zero one has

$$\mathrm{Re}\sigma(\omega) = \begin{cases} 0, & \omega < 2|\mu|, \\ \dfrac{e^2}{4\hbar}, & \omega > 2|\mu|. \end{cases} \qquad (7.35)$$

This expression corresponds exactly to the absorption coefficient (7.12).

It is important to stress that the universal optical conductivity

$$\sigma_0 = \frac{e^2}{4\hbar} = \frac{\pi e^2}{2h} \qquad (7.36)$$

is of the order of, but not equal to, the static ballistic conductivity

$$\sigma_B = \frac{4e^2}{\pi h} \qquad (7.37)$$

(see Eq. (3.16) and Eq. (3.18)). This is not surprising, since we saw in Chapter 3 that limits $\omega \to 0$, $\mu \to 0$, $T \to 0$, etc. do not necessarily commute with one another, as different ways to regularize the ill-posed expression (3.10).

The imaginary part of the conductivity can be restored from Eq. (7.35) via the Kramers–Kronig relations. The result is (see, e.g., Stauber, Peres, & Geim, 2008)

$$\text{Im}\sigma(\omega) = \frac{\sigma_0}{\pi} \left(\frac{4\mu}{\hbar\omega} - \ln \left| \frac{\hbar\omega + 2\mu}{\hbar\omega - 2\mu} \right| \right). \qquad (7.38)$$

At $\mu \to 0$, Im $\sigma(\omega) \to 0$ for any frequency.

7.3 Many-body corrections to the universal optical conductivity: a phenomenological approach

Experimental data obtained by Nair et al. (2008) agree, to within a few percent, with the theoretical value (7.12) (or, equivalently, (7.35)), which is, actually, a problem. As we will see later, the electron–electron interaction in graphene is not small, and earlier considerations (Fritz et al., 2008; Herbut, Juričič, & Vafek, 2008) predicted a rather strong renormalization of the optical conductivity, of the order of $1/\ln|t/(\hbar\omega)|$. The following first-principles GW (G is the Green function and W is the dynamically screened interaction; Yang et al., 2009) as well as the lattice quantum Monte Carlo (Boyda et al., 2016) calculations show that the many-body corrections to the optical conductivity are either absent or small. A more detailed analytical many-body analysis (Mishchenko, 2008; Sheehy & Schmalian, 2009; de Juan, Grushin, & Vozmediano, 2010, Teber & Kotikov, 2014; Link et al., 2016) leads to the conclusion that, whereas the terms of the order of $1/\ln|t/(\hbar\omega)|$ do not exactly disappear, there is a small numerical factor before them. The situation will be considered in more detail in Chapter 15.

Importantly, the survival of the many-body corrections to the conductivity, albeit with a small numerical prefactor, is a consequence of long-range Coulomb interelectron interaction in graphene. For the case of weak enough short-range interactions the corrections to $\sigma(\omega)$ are absent. Here we present, following Katsnelson (2008), some arguments in support of this statement based on the phenomenological Fermi-liquid theory. Later, the absence of correlation corrections to the optical conductivity of electrons on the honeycomb lattice was proved rigorously for the case of a weak enough, short-range interelectron interaction (Giuliani, Mastropietro, & Porta, 2011). Despite that this is not exactly the case of real graphene, the phenomenological consideration seems to be instructive as a demonstration of power of the density matrix and pseudospin formalism.

The equation of motion for the density matrix can be modified naturally to the kinetic equation for quasiparticles within the framework of Landau Fermi-liquid theory (Landau, 1956; Platzman & Wolf, 1973; Vonsovsky & Katsnelson, 1989). Assuming

$$\hat{\rho} = \hat{\rho}^{(0)} + \delta\hat{\rho} \exp(-i\omega t) \tag{7.39}$$

(cf. Eq. (7.22)), one can write, instead of Eq. (7.14),

$$\hbar\omega \, \delta\hat{\rho}_{\vec{k}} = \hbar v\vec{k}\left[\hat{\vec{\sigma}}, \delta\hat{\rho}_{\vec{k}}\right] - ie\left(\vec{E}\cdot\vec{\nabla}_{\vec{k}}\right)\hat{\rho}_{\vec{k}}^{(0)} + \left[\delta\hat{H}_{\vec{k}}, \hat{\rho}_{\vec{k}}^{(0)}\right], \tag{7.40}$$

where the last term contains the change of the Hamiltonian $\delta\hat{H}$ due to the change of the density matrix. In the spirit of Landau theory it is due to the interaction between quasiparticles characterized by some matrix $\hat{F}$:

$$\delta\hat{H}_{\vec{k}} = \sum_{\vec{k}'} \hat{F}_{\vec{k}\vec{k}'}\delta\hat{\rho}_{\vec{k}'}. \tag{7.41}$$

Eq. (7.41) generalizes the standard Landau theory to the case of a *matrix* distribution function for the quasiparticles.

The (pseudo)spinor structure of the matrix $\hat{F}$ can be found by invoking symmetry considerations. First, it should be rotationally invariant in the two-dimensional space. Second, as was discussed in Chapter 1 (see Eq. (1.42)), the Hamiltonian $\delta\hat{H}$ and, thus, the matrix $\hat{F}$ cannot contain the $\hat{\sigma}_z$ matrix (this follows from the inversion and time-reversal symmetries). Third, it should vanish at $\vec{k}$, $\vec{k}' \rightarrow 0$, together with $\hat{H}_0\left(\vec{k}\right)$. The most general expression satisfying these requirements is

$$\hat{F}_{\vec{k}\vec{k}'} = A\left(\left|\vec{k} - \vec{k}'\right|\right)I\otimes I' + B\left(\left|\vec{k} - \vec{k}'\right|\right)\left(\vec{k}\cdot\vec{\sigma}\right)\otimes\left(\vec{k}'\cdot\hat{\vec{\sigma}}'\right)$$

$$+ C\left(\left|\vec{k} - \vec{k}'\right|\right)\left(\vec{k}\cdot\vec{k}'\right)\left(\hat{\sigma}_x\otimes\hat{\sigma}'_x + \hat{\sigma}_y\otimes\hat{\sigma}'_y\right). \tag{7.42}$$

The long-range Coulomb (Hartree) interaction, singular at $\left|\vec{k} - \vec{k}'\right| \rightarrow 0$ (see Section 8.4), contributes to the function A only, whereas the functions B and C are supposed to be smooth and tend to become constants as $\left|\vec{k} - \vec{k}'\right| \rightarrow 0$.

By substituting Eq. (7.41) and (7.42) into Eq. (7.40) we derive, instead of Eq. (7.29),

$$\omega^2 \delta m_{\vec{k}}^x - 4v^2 k_y^2 \delta \tilde{m}_{\vec{k}}^x + 4v^2 k_x k_y \delta \tilde{m}_{\vec{k}}^y = -\frac{ie\omega}{\hbar} E \frac{\partial m_{\vec{k}}^{x(0)}}{\partial k_x},$$

$$4v^2 k_x k_y \delta \tilde{m}_{\vec{k}}^x + \omega^2 \delta m_{\vec{k}}^y - 4v^2 k_x^2 \delta \tilde{m}_{\vec{k}}^y = -\frac{ie\omega}{\hbar} E \frac{\partial m_{\vec{k}}^{y(0)}}{\partial k_x}, \qquad (7.43)$$

where $\delta \tilde{m} = \delta \vec{m} + \vec{\Delta}$, and the term

$$\vec{\Delta}_{\vec{k}} = \frac{1}{vk} \sum_{\vec{k}'} \left[B_{\vec{k}\vec{k}'} \vec{k} \left(\vec{k}' \delta \tilde{m}_{\vec{k}'} \right) + C_{\vec{k}\vec{k}'} \left(\vec{k} \cdot \vec{k}' \right) \delta \tilde{m}_{\vec{k}'} \right] \qquad (7.44)$$

contains all correlation effects. Also, we have an additional correlation contribution to the current density,

$$j_x^{\text{corr}} = \frac{\delta \hat{H}_{\vec{k}}}{\delta k_x} = \sum_{\vec{k}} \frac{\delta \hat{F}_{\vec{k}\vec{k}'}}{\delta k_x} \delta \hat{\rho}_{\vec{k}}, \qquad (7.45)$$

which can, after some straightforward manipulations, be rewritten as

$$j_x^{\text{corr}} = 8e^2 v^3 \sum_{\vec{k}} \frac{k_y}{\omega^2 - 4\omega^2 k^2} \left(k_y \Delta_{\vec{k}}^x - k_x \Delta_{\vec{k}}^y \right). \qquad (7.46)$$

The remaining work is just direct analysis of the corrections, term by term, which shows that they all vanish by symmetry after the integration over $\vec{k}$ and $\vec{k}'$ (Katsnelson, 2008).

7.4 The magneto-optics of Dirac fermions

Consider now the case of Dirac fermions in a magnetic field. Instead of momentum $\vec{k}$, the eigenstates of the unperturbed problem $|n\rangle$, are characterized by the Landau band index n and the coordinate of the Landau orbit x_0 (see Section 2.2). This does not lead to any difficulties, since the optical conductivity, as well as any response functions, can easily be written in an arbitrary basis. The general formalism has already been presented in Section 2.9 (see Eq. (2.175) and later here). We will use the Hamiltonian (2.177) with the electric field (7.5) and calculate the induced electric current, $\frac{1}{2} \left[\vec{j} \exp(i\omega t) + \vec{j}* \exp(-i\omega t) \right]$, assuming that

$$\vec{j} = \hat{\sigma}(\omega) \vec{E} \qquad (7.47)$$

(in this section, $\vec{j}$ is the *electric* current operator).

Then, using Eq. (2.176) and (2.177), we find

$$\sigma_{\alpha\beta}(\omega) = e \sum_{mn} \frac{f_m - f_n}{E_n - E_m - \hbar(\omega + i\delta)} \langle n|j_\alpha|m\rangle\langle m|r_\beta|n\rangle. \qquad (7.48)$$

We will consider here only the case of finite ω, thus, the term with $m = n$ does not contribute to Eq. (7.48). Keeping in mind Eq. (2.179), we find

$$\langle m|j_\alpha|n\rangle = \frac{ie}{\hbar}\langle m|r_\alpha|n\rangle(E_m - E_n). \qquad (7.49)$$

On substituting Eq. (7.49) into Eq. (7.48), taking into account that

$$\frac{1}{E_n - E_m - \hbar(\omega + i\delta)}\frac{1}{E_m - E_n} = -\frac{1}{\hbar\omega}\left(\frac{1}{E_n - E_m - \hbar(\omega + i\delta)} - \frac{1}{E_m - E_n}\right), \qquad (7.50)$$

we obtain

$$\sigma_{\alpha\beta}(\omega) = \frac{i}{\omega}\left[\Pi_{\alpha\beta}(\omega) - \Pi_{\alpha\beta}(0)\right], \qquad (7.51)$$

where

$$\Pi_{\alpha\beta}(\omega) = \sum_{mn} \frac{f_m - f_n}{E_n - E_m - \hbar(\omega + i\delta)} \langle n|j_\alpha|m\rangle\langle m|j_\beta|n\rangle. \qquad (7.52)$$

In particular, for the quantity Re $\sigma_{xx}(\omega)$, determining the absorption of electromagnetic waves, we have

$$\text{Re } \sigma_{xx}(\omega) = \frac{\pi}{\omega}\sum_{mn}(f_m - f_n)|\langle n|j_x|m\rangle|^2\delta(E_n - E_m - \hbar\omega). \qquad (7.53)$$

For the Dirac electrons $j_x = e\sigma_x$. Without a magnetic field, this immediately gives us the result (7.34). In the presence of a magnetic field, we have to use as the basis functions m and n the solutions of the Landau problem (2.45) and (2.46). They are dependent on the Landau indices and on k_y (see Eq. (2.40) and (2.41)). Obviously, the matrix elements $\langle n|\sigma_x|m\rangle$ are diagonal in k_y. Since the functions $D_n(X)$ are orthogonal, one can see immediately that the allowed transitions are $n \to n \pm 1$ and $n \to -(n \pm 1)$ only and, thus, the expression (7.53) describes absorption peaks at

$$\hbar\omega = |E_n| \pm |E_{n+1}|, \qquad (7.54)$$

or at $\omega = \omega_c\left(\sqrt{p+1} \pm \sqrt{p}\right)$, where $p = 0, 1, 2, \ldots$ The complete expression can be found in Gusynin, Sharapov, and Carbotte (2007a, 2009).

This absorption has been observed experimentally (Sadowski et al., 2006; Jiang et al., 2007a; Witowski et al., 2010). The results are in agreement with Eq. (7.54).

This effect can be used as an alternative method by which to measure the Fermi velocity v in graphene.

Another interesting magneto-optical effect is the polarization rotation of propagating light in the magnetic field, that is, the Faraday effect (Landau & Lifshitz, 1984). The rotation angle is proportional to Re σ_{xy}, which has absorption peaks at the same frequencies (7.54) as Re σ_{xx} (for the theory of the Faraday effect in graphene, see Fialkovsky and Vassilevich [2009]). Near the resonances, the rotation is very large, as was observed experimentally by Crassee et al. (2011). This giant Faraday effect is potentially interesting for applications.

7.5 Optical properties of graphene beyond the Dirac approximation

Now consider the theory of optical conductivity for a honeycomb lattice, beyond the Dirac cone approximation, so that it can be used at $\hbar\omega \geq |t|$ as well (Gusynin, Sharapov, & Carbotte, 2007b; Stauber, Peres, & Geim, 2008). We will start with the expression (2.20) for the Hamiltonian of band electrons in the presence of a vector potential; in the single-band approximation it also works for the time-dependent vector potential $\vec{A}(t)$.

In particular, in the nearest-neighbor approximation the Hamiltonian has the form

$$
\hat{H}\left(\vec{k}\right) = \begin{pmatrix} 0 & tS\left(\vec{k} - \frac{e\vec{A}}{\hbar c}\right) \\ tS^*\left(\frac{\vec{k} - e\vec{A}}{\hbar c}\right) & 0 \end{pmatrix}
\tag{7.55}
$$

(cf. Eq. (1.14) and (1.15)). To calculate the linear response, we need to expand the right-hand side of Eq. (7.55) up to *second* order in $\vec{A}$. Indeed, the electric current operator

$$
\hat{\vec{j}} = c\frac{\delta\hat{H}}{\delta\vec{A}}
\tag{7.56}
$$

has paramagnetic (p) and diamagnetic (d) components

$$
\hat{j}_\alpha = \hat{j}_\alpha^{(\mathrm{p})} + \hat{j}_\alpha^{(\mathrm{d})},
\tag{7.57}
$$

where

$$
\hat{\vec{j}}_\alpha^{(\mathrm{p})} = c\left(\frac{\delta\hat{H}}{\delta\vec{A}_\alpha}\right)_{\vec{A}=0}
\tag{7.58}
$$

and

$$\hat{j}_\alpha^{(d)} = \frac{1}{2}c^2 \sum_\beta \left(\frac{\delta^2 \hat{H}}{\delta \vec{A}_\alpha \delta \vec{A}_\beta} \right)_{\vec{A}=0} A_\beta. \tag{7.59}$$

When calculating the average current density to linear order in $\vec{A}$,

$$j_\alpha = \mathrm{Tr}\left(\hat{j}_\alpha^{(p)} \hat{\rho}' \right) + \mathrm{Tr}\left(\hat{j}_\alpha^{(d)} \hat{\rho}_0 \right), \tag{7.60}$$

both terms contribute to the conductivity. Further calculations are quite straight-forward (Gusynin, Sharapov, & Carbotte, 2007b; Stauber, Peres, & Geim, 2008). Here, we will present only the expressions for Re $\sigma_{xx}(\omega) = $ Re $\sigma(\omega)$:

$$\mathrm{Re}\,\sigma(\omega) = D\delta(\omega) + \frac{\pi t^2 e^2 a^2}{8\hbar^3 \omega A_0} \sum_{\vec{k}} F\left(\vec{k}\right)\left(f_{\vec{k}1} - f_{\vec{k}2}\right)$$

$$\times \left[\delta\left(\omega - \varepsilon\left(\vec{k}\right)\right) - \delta\left(\omega + \varepsilon\left(\vec{k}\right)\right) \right], \tag{7.61}$$

where the first term originates from $j^{(d)}$, the Drude weight is

$$D = -\frac{e^2 a^2}{3\hbar^2 A_0} \sum_{\vec{k}} \varepsilon\left(\vec{k}\right)\left(f_{\vec{k}1} - f_{\vec{k}2}\right), \tag{7.62}$$

$\varepsilon\left(\vec{k}\right) = t\left|S\left(\vec{k}\right)\right|(t>0), f_{\vec{k}1,2}$ are given by Eq. (7.25), $A_0 = 3\sqrt{3}a^2/2$ is the area of the unit cell and

$$F\left(\vec{k}\right) = 18 - 4\left|S\left(\vec{k}\right)\right|^2 + 18\frac{\left[\mathrm{Re}\,S\left(\vec{k}\right)\right]^2 - \left[\mathrm{Im}\,S\left(\vec{k}\right)\right]^2}{\left|S\left(\vec{k}\right)\right|^2}. \tag{7.63}$$

The optical conductivity (7.61) at $\omega \neq 0$ is proportional to the density of states

$$N(E) = \sum_{\vec{k}} \delta\left(|E| - \varepsilon\left(\vec{k}\right)\right) \tag{7.64}$$

(it differs by a factor of 2 from Eq. (1.70)). It can be analytically expressed (Hobson & Nierenberg, 1953) in terms of the elliptic integral

$$K(m) = \int_0^{\pi/2} \frac{d\varphi}{\sqrt{1 - m\,\sin^2\varphi}}, \tag{7.65}$$

namely

$$N(|E|) = \frac{2|E|}{\pi^2 t^2} \begin{cases} \dfrac{1}{\sqrt{\varphi(|E|/t)}} K\left(\dfrac{4|E|/t}{\varphi(|E|/t)}\right), & 0 < |E| < t, \\[3ex] \dfrac{1}{\sqrt{4(|E|/t)}} K\left(\dfrac{\varphi(|E|/t)}{4|E|/t}\right), & t < |E| < 3t, \end{cases} \tag{7.66}$$

where

$$\varphi(x) = (1+x)^2 - \frac{(x^2-1)^2}{4}. \tag{7.67}$$

This function is shown in Fig. 7.2. It has logarithmic divergences at $E = \pm t$ corresponding to Van Hove singularities in the electron density of states.

At $0 < \hbar\omega < t$ the optical conductivity (7.61) coincides with Eq. (7.35). The corrections are (Stauber, Peres, & Geim, 2008)

$$\sigma(\omega) \approx \frac{\sigma_0}{2}\left(\tanh\left(\frac{\hbar\omega + 2\mu}{4T}\right) + \tanh\left(\frac{\hbar\omega - 2\mu}{4T}\right)\right)\left[1 + \frac{(\hbar\omega)^2}{36t^2}\right]. \tag{7.68}$$

The curve for the whole interval is shown in Fig. 7.3 (Yuan, De Raedt, & Katsnelson, 2010a). One can see a singularity at $\hbar\omega = 2t$; however, a moderate disorder (such as 1% of vacancies or resonant impurities) smears it essentially.

For the case of bilayer graphene, we have a Van Hove singularity at low energy, due to trigonal warping and the merging of four Dirac ones to give one paraboloid (see Section 1.4). Also, the gap can be made to open in that case by applying a bias

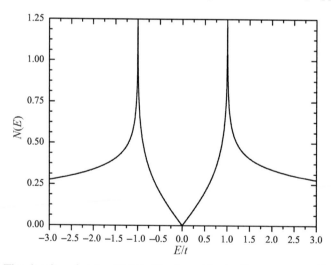

Fig. 7.2 The density of states (7.66). The logarithmic divergences at $E = \pm t$ are Van Hove singularities.

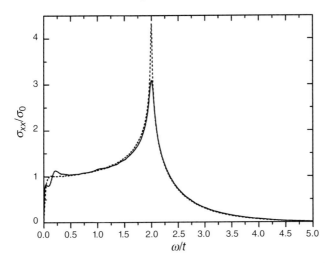

Fig. 7.3 The frequency dependence of Re $\sigma_{xx}(\omega)$ for an ideal honeycomb lattice in the nearest-neighbor approximation (dashed line) and for one with 1% of vacancies, randomly distributed (solid line); σ_0 is given by Eq. (7.36). (Reproduced with permission from Yuan, De Raedt, & Katsnelson, 2010a.)

between the layers. Experimentally, these effects on the infrared optics of bilayer graphene have been studied by Kuzmenko et al. (2009).

7.6 The dielectric function of Dirac fermions

Now we will consider the response function for an *inhomogeneous* external perturbation

$$V_{\text{ext}}(\vec{r}, t) = \sum_{\vec{k}} \Psi_{\vec{k}}^+ \hat{V}_{\vec{q}}^{ext} \Psi_{\vec{k}+\vec{q}} \exp\left(i\vec{q}\vec{r} - i\omega t\right), \tag{7.69}$$

where $\Psi_{\vec{k}}^+ = \left(\psi_{\vec{k}1}^+, \psi_{\vec{k}2}^+\right)$ is the spinor creation operator, $\hat{V}_{\vec{q}}^{\text{ext}}$ is a generic 2×2 matrix, and $\vec{q}$ is the wave vector of the inhomogeneity. We need to pass to electron- and hole-creation operators (7.23). The result is

$$\Psi_{\vec{k}}^+ \hat{V}_{\vec{q}}^{\text{ext}} \Psi_{\vec{k}+\vec{q}} = \Xi_{\vec{k}}^+ \hat{U}_{\vec{q}} \Xi_{\vec{k}+\vec{q}}, \tag{7.70}$$

where $\Xi_{\vec{k}}^+ = \left(\varsigma_{\vec{k}1}^+, \varsigma_{\vec{k},2}^+\right)$ and

$$\hat{U}_{\vec{q}} = \frac{1}{2} \begin{pmatrix} 1 & \exp\left(-i\varphi_{\vec{k}+\vec{q}}\right) \\ 1 & -\exp\left(-i\varphi_{\vec{k}+\vec{q}}\right) \end{pmatrix} \hat{V}_{\vec{q}}^{\text{ext}} \begin{pmatrix} 1 & 1 \\ \exp\left(i\varphi_{\vec{k}}\right) & -\exp\left(i\varphi_{\vec{k}}\right) \end{pmatrix}. \tag{7.71}$$

Then, the perturbation of the density matrix (2.175) is the operator $\hat{\rho}' \exp{(i\vec{q}\vec{r} - i\omega t)}$ with the matrix elements (in the ξ representation)

$$\hat{\rho}'_{\vec{k}+\vec{q},i,\vec{k},j} = \frac{f_{\vec{k},j} - f_{\vec{k}+\vec{q},i}}{E_{\vec{k},j} - E_{\vec{k}+\vec{q},i} - \hbar(\omega + i\delta)} \left(\hat{U}_{\vec{q}}\right)_{ij} \tag{7.72}$$

and the perturbation of the operator

$$\hat{J} = \sum_{\vec{k}\vec{q}} \Psi_{\vec{k}}^{+} \hat{J}_{\vec{q}} \Psi_{\vec{k}+\vec{q}} \equiv \sum_{\vec{k}\vec{q}} \Xi_{\vec{k}}^{+} \hat{J}_{\vec{q}} \Xi_{\vec{k}+\vec{q}} \tag{7.73}$$

is

$$\delta J_{\vec{q}} = \mathrm{Tr}\left(\hat{J}\hat{\rho}'\right) = \sum_{\vec{k}} \frac{f_{\vec{k},j} - f_{\vec{k}+\vec{q},i}}{E_{\vec{k},j} - E_{\vec{k}+\vec{q},i} - \hbar(\omega + i\delta)} \left(\hat{U}_{\vec{q}}\right)_{ij} \left(\hat{J}_{\vec{q}}\right)_{ji}. \tag{7.74}$$

Consider first the case of a scalar potential and the density operator $\hat{J} = \hat{n}$; in that case, both $\hat{V}_{\vec{q}}^{\mathrm{ext}}$ and $\hat{J}_{\vec{q}}$ are proportional to the unit matrix. We obtain

$$\delta n_{\vec{q}\omega} = -\Pi(\vec{q}, \omega)\hat{V}_{\vec{q}\omega}^{\mathrm{ext}}, \tag{7.75}$$

where

$$\Pi(\vec{q}, \omega) = g_s g_v \sum_{\vec{k}} \sum_{s, s'=\pm} \lambda_{ss'}\left(\vec{k}, \vec{q}\right) \frac{f\left[sE\left(\vec{k}\right)\right] - f\left[s'E\left(\vec{k}+\vec{q}\right)\right]}{s'E\left(\vec{k}+\vec{q}\right) - sE\left(\vec{k}\right) + \hbar(\omega + i\delta)} \tag{7.76}$$

is the *polarization operator* $E\left(\vec{k}\right) = \hbar v k$,

$$\lambda_{ss'}\left(\vec{k}, \vec{q}\right) = \frac{1}{2}\left(1 + ss'\frac{k + q\,\cos\,\varphi}{\left|\vec{k} + \vec{q}\right|}\right), \tag{7.77}$$

φ is the angle between $\vec{k}$ and $\vec{q}$, and the factors $g_s = 2$ and $g_v = 2$ take into account spin and valley degeneracy (Ando, 2006; Wunsch et al., 2006; Hwang & Das Sarma, 2007).

Perturbation of the electron density will induce perturbation of the potential

$$V_{\vec{q}\omega}^{ind} = v_C(q)\delta n_{\vec{q}\omega}, \tag{7.78}$$

where

$$v_C(q) = \frac{2\pi e^2}{q\varepsilon_{\mathrm{ext}}} \tag{7.79}$$

is the Fourier component of the Coulomb interaction

$$v_C(r) = \frac{e^2}{r\varepsilon_{ext}} \tag{7.80}$$

in two dimensions and ε_{ext} is the external dielectric constant (e.g., due to screening by a substrate). The total potential perturbation is

$$V_{\vec{q}\omega} = V_{\vec{q}\omega}^{ext} + V_{\vec{q}\omega}^{ind} = \frac{V_{\vec{q}\omega}^{ext}}{\varepsilon(\vec{q},\omega)}. \tag{7.81}$$

The last equality in Eq. (7.81) defines the *dielectric function* $\varepsilon(\vec{q},\omega)$. Within the *random-phase approximation* (RPA) it is assumed that, for a system of *interacting* fermions, the induced density formally has the same expression as for the *non*interacting fermions, (7.75) and (7.76), but with the replacement $V^{ext} \rightarrow V$ in Eq. (7.75). This means that the interaction effects are taken into account via a self-consistent mean field (Vonsovsky & Katsnelson, 1989). As a result,

$$\varepsilon(\vec{q},\omega) = 1 + v_C(q)\Pi(q,\omega). \tag{7.82}$$

If we also take into account the external screening, the total dielectric function is

$$\varepsilon_{tot}(q,\omega) = \varepsilon_{ext}\varepsilon(q,\omega) = \varepsilon_{ext} + \frac{2\pi e^2}{q}\Pi(q,\omega). \tag{7.83}$$

In the case when graphene lies between two subspaces with dielectric constants ε_1 and ε_2, one has (Landau & Lifshitz, 1984)

$$\varepsilon_{ext} = \frac{\varepsilon_1 + \varepsilon_2}{2}. \tag{7.84}$$

For the two most popular substrates, SiO_2 and BN, $\varepsilon_2 \approx 4$, so, assuming $\varepsilon_1 = 1$ (vacuum, or air), one has $\varepsilon_{ext} \approx 2.5$.

Consider first the case of undoped graphene ($\mu = 0$) at zero temperature. Then, only interband transitions ($s = +$ and $s' = -$ or vice versa) contribute to Eq. (7.76) and

$$\Pi_0(q,\omega) = \frac{g_s g_v}{\hbar}\sum_{\vec{k}}\left(1 - \frac{(k + q\cos\varphi)}{|\vec{k} + \vec{q}|}\right)\frac{v\left(k + |\vec{k} + \vec{q}|\right)}{v^2\left(k + |\vec{k} + \vec{q}|\right)^2 - (\omega + i\delta)^2}. \tag{7.85}$$

As the next step, we calculate Im $\Pi_0(q, \omega)$. It contains $\delta\left[v\left(k + \left|\vec{k} + \vec{q}\right| - \omega\right)\right]$, which allows us to calculate the integral (first, in ω and then in k) in a quite elementary manner. The result is

$$\text{Im } \Pi_0(q, \omega) = \frac{g_s g_v}{16\hbar} \frac{q^2}{\sqrt{\omega^2 - v^2 q^2}} \theta(\omega - vq), \tag{7.86}$$

where $\theta(x > 0) = 1$, $\theta(x < 0) = 0$ is the step function. Noticing that the analytic function $1/\sqrt{z + i\delta}$ is purely imaginary at real $z < 0$ and purely real at real $z > 0$, one can do analytic continuation immediately, thus having

$$\text{Re}\Pi_0(q, \omega) = \frac{g_s g_v}{16\hbar} \frac{q^2}{\sqrt{v^2 q^2 - \omega^2}} \theta(vq - \omega). \tag{7.87}$$

On combining Eq. (7.86) and (7.87) we have a very simple answer (Gonzáles, Guinea, & Vozmediano, 1999):

$$\Pi_0(q, \omega) = \frac{g_s g_v}{16\hbar} \frac{q^2}{\sqrt{v^2 q^2 - (\omega + i\delta)^2}}. \tag{7.88}$$

At $\omega = 0$, $\Pi_0(q, \omega) \sim q$, and the dielectric function $\varepsilon(q)$ is actually not dependent on q:

$$\varepsilon = \varepsilon_{\text{ext}} + \frac{\pi e^2}{2\hbar v}. \tag{7.89}$$

For graphene,

$$\alpha = \frac{e^2}{\hbar v} \approx 2.2 \tag{7.90}$$

and the second term on the right-hand side of Eq. (7.89) is about 3.5.

Within the RPA, this result is exact, and high-energy states cannot change the value of $\varepsilon(q = 0)$. Indeed, for arbitrary band structure with the Bloch states $\left|m\vec{k}\right\rangle$, one has (Vonsovsky & Katsnelson, 1989)

$$\Pi(\vec{q}, \omega = 0) = 2 \sum_{mn} \sum_{\vec{k}} \frac{f_{n,\vec{k}} - f_{m,\vec{k}+\vec{q}}}{E_{m,\vec{k}+\vec{q}} - E_{n,\vec{k}}} \left|\left\langle n\vec{k}\middle|m, \vec{k} + \vec{q}\right\rangle\right|^2 \tag{7.91}$$

(the factor of 2 is due to spin degeneracy).

Let us exclude the Dirac point, considering the case when (at $T = 0$) we have completely occupied bands and completely empty bands and some gap in between. Then Eq. (7.91) can be rewritten as

$$\Pi(\vec{q}, \omega = 0) = 4 \sum_{n}^{\text{occ}} \sum_{m}^{\text{empty}} \frac{1}{E_{m, \vec{k}+\vec{q}} - E_{n, \vec{k}}} \left| \left\langle n, \vec{k} | m, \vec{k} + \vec{q} \right\rangle \right|^2, \tag{7.92}$$

which is obviously proportional to q^2 at $q \to 0$. More explicitly, on writing

$$\left| m, \vec{k} + \vec{q} \right\rangle \approx \left(1 + \vec{q} \vec{\nabla}_{\vec{k}} \right) \left| m, \vec{k} \right\rangle \tag{7.93}$$

and using Eq. (2.85), one finds for $\vec{q} \to 0$

$$\Pi(\vec{q}, \omega = 0) = \sum_{\alpha\beta} C_{\alpha\beta} q_\alpha q_\beta, \tag{7.94}$$

where

$$C_{\alpha\beta} = 4 \sum_{n}^{\text{occ}} \sum_{m}^{\text{empty}} \frac{1}{\left(E_{m, \vec{k}} - E_{n, \vec{k}} \right)^3} \left\langle m, \vec{k} \left| \frac{\partial \hat{H}}{\partial k_\alpha} \right| n, \vec{k} \right\rangle \left\langle n, \vec{k} \left| \frac{\partial \hat{H}}{\partial k_\beta} \right| m, \vec{k} \right\rangle \tag{7.95}$$

is some finite tensor. Since $v_c(q) \sim 1/q$, we have, in two dimensions, $\varepsilon(q \to 0, \omega = 0) = 1$ for any gapped state. This means that only the region close to the Dirac point contributes to this quantity. Note that first-principles *GW* calculations do indeed give results quite similar to those obtained by use of Eq. (7.89) (Schilfgaarde & Katsnelson, 2011).

Now consider the case of doped graphene (to be specific, we put $\mu > 0$, i.e., the case of electron doping). The calculations are quite cumbersome but straightforward. The result is (Wunsch et al., 2006; Hwang & Das Sarma, 2007)

$$\Pi(q, \omega) = \Pi_0(q, \omega) + \Pi_1(q, \omega),$$

with

$$\Pi_1(q, \omega) = \frac{g_s g_v \mu}{2\pi\hbar^2 v^2} - \frac{g_s g_v q^2}{16\pi\hbar\sqrt{\omega^2 - v^2 q^2}}$$

$$\times \left\{ G\left(\frac{\hbar\omega + 2\mu}{\hbar v q} \right) - \theta\left(\frac{2\mu - \hbar\omega}{\hbar v q} - 1 \right) \left[G\left(\frac{2\mu - \hbar\omega}{\hbar v q} \right) - i\pi \right] \tag{7.96}$$

$$- \theta\left(\frac{\hbar\omega - 2\mu}{\hbar v q} + 1 \right) G\left(\frac{\hbar\omega - 2\mu}{\hbar v q} \right) \right\},$$

where

$$G(x) = x\sqrt{x^2 - 1} - \ln\left(x + \sqrt{x^2 - 1} \right). \tag{7.97}$$

For generalization of this expression to the case of gapped graphene, see Pyatkovskiy (2009).

Now we will consider different partial cases of this general expression. Keeping in mind the case of graphene, we will put $g_s = g_v = 2$.

7.7 Static screening

We start with the case $\omega = 0$. The result is (Gorbar et al., 2002; Ando, 2006; Wunsch et al., 2006; Hwang & Das Sarma, 2007)

$$
\Pi(q,0) = \frac{2k_F}{\pi\hbar v} \times \begin{cases} 1, & q < 2k_F, \\ 1 - \frac{1}{2}\sqrt{1 - \left(\frac{2k_F}{q}\right)^2} + \frac{q}{4k_F}\cos^{-1}\left(\frac{2k_F}{q}\right), & q > 2k_F. \end{cases}
$$

(7.98)

Interestingly, at $q < 2k_F$, $\Pi(q, 0) = $ constant, due to cancellation of the q dependence in the (formally) μ-dependent contribution

$$
\Pi_+(q,0) = \frac{2k_F}{\pi\hbar v}\left(1 - \frac{\pi q}{8k_F}\right)
$$

(7.99)

and the contribution for the undoped case (see Eq. (7.88)),

$$
\Pi_0(q,0) = \frac{q}{4\hbar v}.
$$

(7.100)

It is instructive to compare Eq. (7.98) with that for a conventional, nonrelativistic two-dimensional electron gas (Stern, 1967):

$$
\Pi_0(q,0) = N(E_F) \times \begin{cases} 1, & q < 2k_F, \\ 1 - \sqrt{1 - \left(\frac{2k_F}{q}\right)^2}, & q > 2k_F. \end{cases}
$$

(7.101)

In both cases, the polarization operator is constant at $q < 2k_F$. At the same time, the behavior at $q > 2k_F$ is essentially different. For the nonrelativistic case $\Pi(q, \omega)$ decays with increasing q, as $1/q^2$ at $q \to \infty$, whereas for the case of massless Dirac fermions $\Pi(q, 0)$ *increases* linearly with increasing q, due to the contribution (7.100). The behavior of expressions (7.98) and (7.101) at $q \to 2k_F$ is also essentially different. Whereas for the nonrelativistic electron gas $\delta\Pi(q,0) \sim \sqrt{q - 2k_F}$, with a divergent derivative, for the case of graphene the singularity is weaker $\delta\Pi(q,0) \sim (q - 2k_F)^{3/2}$.

The result for small q corresponds to the *Thomas–Fermi approximation* (Katsnelson, 2006c; Nomura & MacDonald, 2006). The latter (Lieb, 1981) assumes that the perturbation $V(\vec{r})$ is smooth enough that its effect on the electron density

$$n(\mu) = \int_0^\mu dE \, N(E) \tag{7.102}$$

can be taken into account just by making the replacement $n(\mu) \rightarrow n[\mu - V(\vec{r})]$. This means that the potential just locally shifts the maximum band energy $E_F(\vec{r})$, such that

$$E_F(\vec{r}) + V(\vec{r}) = \mu. \tag{7.103}$$

The self-consistent equation for the total potential, which is similar to Eq. (7.81), reads

$$V(\vec{r}) = V_{\text{ext}}(\vec{r}) + \frac{e^2}{\varepsilon_{\text{ext}}} \int d\vec{r}' \, \frac{n_{\text{int}}(\vec{r}')}{|\vec{r} - \vec{r}'|}, \tag{7.104}$$

where

$$n_{\text{int}}(\vec{r}) = n[\mu - V(\vec{r})] - n(\mu) \tag{7.105}$$

is the induced change of the electron density. Assuming that the perturbation V is small, one can expand (7.105) as follows:

$$n_{\text{int}}(\vec{r}) \approx -\frac{\partial n}{\partial \mu} V(\vec{r}) = -N(E_F) V(\vec{r}), \tag{7.106}$$

where the last identity assumes $T = 0$ (cf. Eq. (2.138)).

On Fourier-transforming Eq. (7.104) and comparing the result with Eq. (7.81) one finds

$$\varepsilon(q, 0) = \varepsilon_{\text{ext}} + \frac{2\pi e^2 N(E_F)}{q} = \varepsilon_{\text{ext}} \left(1 + \frac{\kappa}{q}\right), \tag{7.107}$$

where

$$\kappa = \frac{4e^2 |\mu|}{\varepsilon_{\text{ext}} \hbar^2 v^2} \tag{7.108}$$

is the *inverse Thomas–Fermi screening radius*. This result coincides exactly with Eq. (7.98) and (7.101) at $q < 2k_F$. Thus, for a two-dimensional electron gas, the nonrelativistic and ultrarelativistic versions of Thomas–Fermi theory both give exactly the same result, as does the RPA for static screening with $q < 2k_F$. For a three-dimensional electron gas the situation is different (Vonsovsky & Katsnelson, 1989).

Consider now the real-space effects of static screening. If the external potential $V^{\text{ext}}(r)$ is radially symmetric, with the Fourier component V_q^{ext} depending only on the modulus q, the expression for the total potential is

$$V(r) = \int \frac{d\vec{q}}{(2\pi)^2} \exp\left(i\vec{q}\vec{r}\right) \frac{V_q^{\text{ext}}}{\varepsilon(q,0)} = \int_0^\infty \frac{dq\,q}{2\pi} J_0(qr) \frac{V_q^{\text{ext}}}{\varepsilon(q,0)}. \qquad (7.109)$$

At $r \to \infty$, there are two important contributions to the integral (7.109), from the region of small q (to compensate for large r in the argument of the Bessel function) and from the region $q = 2k_F$, where $\varepsilon(q, 0)$ has a singularity in $\Pi(q, \omega)$ (7.98). In the three-dimensional case, the first contribution decays exponentially at $r \to \infty$, whereas the second oscillates and decays as $\cos{(2k_F r)}/r^3$, being what is called a Friedel oscillation (Vonsovsky & Katsnelson, 1989). In the two-dimensional case, the situation is different since the Thomas–Fermi (small-q) contribution also decays as $1/r^3$ (Katsnelson, 2006c; Wunsch et al., 2006). As a result, the asymptotics of the induced density around the point defect is (Wunsch et al., 2006)

$$n_{\text{ind}}(r) \sim \frac{\alpha + \beta \, \cos{(2k_F r)}}{r^3}, \qquad (7.110)$$

with some parameters α and β dependent on k_F and on the potential.

In a nonrelativistic electron gas in two dimensions, $n_{\text{ind}}(r) \sim \cos(2k_F r)/r^2$ since the singularity in $\Pi(q, \omega)$ at $q \to 2k_F$ is stronger. In graphene, the Thomas–Fermi and Friedel contributions to the induced density around point defects are comparable at $r \to \infty$.

The first-principles *GW* results for the dielectric function $\varepsilon(q, 0)$ of graphene (Schilfgaarde & Katsnelson, 2011) show that the Dirac approximation works for $q \leq 0.05 \,\text{Å}^{-1}$; at $q \approx 0.1 \,\text{Å}^{-1}$ the polarization operator approximately halves in comparison with the value (7.100).

7.8 Plasmons

Let us now consider the opposite limiting case

$$\omega \gg vq. \qquad (7.111)$$

The polarization operator (7.96) in the limit of small q takes the form

$$\Pi(q \to 0, \omega) = \frac{q^2}{2\pi\hbar\omega}\left[\frac{i\pi}{2}\theta(\hbar\omega - 2\mu) - \frac{2\mu}{\hbar\omega} + \frac{1}{2}\ln\left|\frac{\hbar\omega + 2\mu}{\hbar\omega - 2\mu}\right|\right]. \qquad (7.112)$$

At $\hbar\omega > 2\mu$ it has an imaginary part that is at least comparable to the real part, so the equation

$$\varepsilon(q, \omega) = 0, \qquad (7.113)$$

which determines the spectrum of plasma oscillations (Platzman & Wolf, 1973; Vonsovsky & Katsnelson, 1989), has no real solutions. In the opposite limit

$$\hbar\omega \ll 2\mu \qquad (7.114)$$

one has

$$\Pi(q \to 0, \omega) \approx -\frac{\mu q^2}{\pi(\hbar\omega)^2} \qquad (7.115)$$

and the solution of Eq. (7.113) is

$$\omega = \sqrt{\frac{2e^2\mu}{\hbar^3 \varepsilon_{\text{ext}}}} q. \qquad (7.116)$$

At $q \to 0$, the expression (7.116) obviously satisfies the condition (7.114).

The existence of the low-frequency plasmon mode with the dispersion $\omega \sim \sqrt{q}$ is a general property of a two-dimensional electron gas (Ando, Fowler, & Stern, 1982). However, the dependence of the plasmon dispersion relation on the electron density n is different: For graphene, due to Eq. (7.116) $\omega \sim n^{1/4} q^{1/2}$, whereas for the nonrelativistic case $\omega \sim n^{1/2} q^{1/2}$.

Outside the region $qv < \omega < 2\mu$ $\Pi(q, \omega)$ has a large imaginary part and the plasmon is essentially damped. This is a partial case of Landau damping due to a decay into incoherent electron–hole excitations (Vonsovsky & Katsnelson, 1989). It was argued, however, by Gangadharaiah, Farid, and Mishchenko (2008) that higher-order correlation effects, beyond the RPA, can change the situation, leading to a well-defined plasmon mode with $\omega < qv$, even at $\mu = 0$.

Beyond the Dirac approximation, there are two important physical mechanisms that can lead to additional plasmon modes. First, there is Coulomb interaction between electrons from different valleys, resulting in the appearance of intervalley plasmons, with a linear dispersion law $\omega \sim q$ (Tudorovskiy & Mikhailov, 2010). Second, there is a Van Hove singularity in the optical conductivity at $\omega = 2t$ (see Section 7.5), because of which high-energy "optical" plasmons arise (Hill, Mikhailov, & Ziegler, 2009; Stauber, Schliemann, & Peres, 2010; Yuan, Roldán, & Katsnelson, 2011).

Experimental study of plasmons in graphene is currently an intensively developing field (Grigorenko, Polini, & Novoselov, 2012; Woessner et al., 2015; Basov, Fogler, & García de Abajo, 2016; Alonso-González et al., 2017; Lundeberg et al., 2017; Low et al., 2017). There are several important advantages of graphene in comparison with conventional plasmonic materials such as metallic surfaces. First,

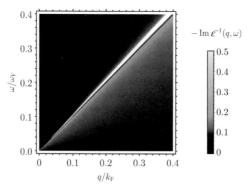

Fig. 7.4 A grayscale plot of the loss function $-\mathrm{Im}\frac{1}{\varepsilon(q,\omega)}$ with k_{F} and $\omega_{\mathrm{F}} = E_{\mathrm{F}}/\hbar$ (Fermi wave vector and Fermi frequency of graphene) as units for the plasmon wave vector and frequency. A sharp acoustic plasmon mode is visible just above the electron–hole continuum line $\omega = vq$ (v is the Fermi velocity in graphene). Calculations are made for the electron density 10^{12} cm^{-2} for graphene and 10^{21} cm^{-3} for metal, Fermi velocity of the metal is taken as $0.35v$. The plasmon velocity for these parameters is $1.04v$.
(Reproduced with permission from Principi et al., 2018).

the electron concentration in graphene is easily tunable by gate voltage which, due to Eq. (7.116), changes plasmon frequency. Second, graphene encapsulated to hexagonal boron nitride (see Chapter 13) has extremely high quality and can be made practically defect-free. Together with a very high intrinsic electron mobility due to a weakness of electron–phonon interaction (Chapter 11), it allows us to excite and observe plasmons with unusually small damping (or unusually high lifetime).

Electrodynamics of graphene on metallic substrate was studied by Principi et al., (2018). In this case, due to a nonlocal metallic screening, acoustic plasmons with linear dispersion are formed (Fig. 7.4). Surprisingly, such plasmons can still have a very low damping which makes them potentially interesting for applications.

7.9 Transverse response functions and diamagnetic susceptibility

Similarly to the previous sections, one can consider the response of electrons in graphene to a vector potential (Principi, Polini, & Vignale, 2009). One just needs to choose $\widehat{V}_q^{\mathrm{ext}} = \hat{\sigma}$ in Eq. (7.70). In general, instead of the polarization operator (7.76), we introduce a set of response functions

$$\Pi_{\alpha\beta}(\vec{q},\omega) = g_s g_v \sum_{\vec{k}} \sum_{s,s'=\pm} \lambda_{ss'}^{\alpha\beta}(\vec{k},\vec{q}) \frac{f\left[sE\left(\vec{k}\right)\right] - f\left[s'E\left(\vec{k}+\vec{q}\right)\right]}{s'E\left(\vec{k}+\vec{q}\right) - sE\left(\vec{k}\right) + \hbar(\omega + i\delta)},$$

$$(7.117)$$

where

$$\lambda^{\alpha\beta}_{ss'}\left(\vec{k},\vec{q}\right) = \left\langle \psi_s\left(\vec{k}\right)|\sigma_\alpha|\psi_{s'}\left(\vec{k}+\vec{q}\right)\right\rangle\left\langle \psi_{s'}\left(\vec{k}+\vec{q}\right)|\sigma_\beta|\psi_s\left(\vec{k}\right)\right\rangle, \tag{7.118}$$

in which $\psi_s\left(\vec{k}\right)$ are electron and hole wave functions (1.30). The density–density response function is, in this notation Π_{00}, where $\sigma_0 = I$. For example,

$$\lambda^{xx}_{ss'}\left(\vec{k},\vec{q}\right) = \frac{1 + ss' \cos\left(\varphi_{\vec{k}} + \varphi_{\vec{k}+\vec{q}}\right)}{2}. \tag{7.119}$$

For the response function, determining the current in the x-direction induced by the vector potential in the x-direction $\left(\hat{j}_x = v\hat{\sigma}_x\right)$

$$j_{\vec{q}}\omega^x = -\frac{e^2v^2}{c}\Pi_{xx}(\vec{q},\omega)A^x_{\vec{q},\omega}. \tag{7.120}$$

When calculating this quantity we are faced with an important problem, showing that sometimes one needs to be very careful when using the Dirac approximation. Let us put $\omega = 0$ and express the vector potential in terms of an external magnetic field $\vec{B} = \vec{\nabla} \times \vec{A} = (0,0,B(x,y))$:

$$B_{\vec{q}} = -\frac{i}{q_y}A^x_q. \tag{7.121}$$

Phenomenologically, the magnetic field induces a magnetization $\vec{M} = (0,0,M(x,y))$ proportional to the magnetic field

$$M_{\vec{q}} = \chi(\vec{q})B_{\vec{q}} \tag{7.122}$$

and the current

$$\vec{j} = c\vec{\nabla} \times \vec{M} \tag{7.123}$$

(Jackson, 1962; Landau & Lifshitz, 1984), or, equivalently,

$$j^x_{\vec{q}} = icq_yM_{\vec{q}}. \tag{7.124}$$

On substituting Eq. (7.121) through Eq. (7.123) into Eq. (7.120) one finds

$$\Pi_{xx}(\vec{q}) = -\frac{q_y^2c^2}{v^2e^2}\chi(\vec{q}) \tag{7.125}$$

and, obviously, $\Pi_{xx}(\vec{q} = 0) = 0$. Physically, this means that, due to the gauge invariance, a *constant* vector potential cannot induce any physical response.

However, on substituting Eq. (7.119) into Eq. (7.117) we have, even at $\mu = 0$, a divergent integral over $\left|\vec{k}\right|$. On introducing by hand a cut-off $\left|\vec{k}\right| \leq k_{max}$, we find the result (Principi, Polini, & Vignale, 2009)

$$\Pi_{xx}(\vec{q}) = -g_s g_v \frac{k_{\max}}{4\pi \hbar v}, \qquad (7.126)$$

which is finite and, moreover, tends to infinity at $k_{\max} \to \infty$. This is a pathological property of our model, reflecting the fact that by introducing the cut-off we break the gauge invariance $\vec{k} \to \vec{k} - e\vec{A}/(\hbar c)$. The contribution (7.126) should just be subtracted from the answer.

By calculating $\Pi_{xx}(q_y, 0)$ at small q_y and using Eq. (7.125), we find the magnetic susceptibility describing the effect of the magnetic field on the orbital motion of electrons:

$$\chi = -\frac{g_s g_v}{24\pi} \frac{e^2 v^2}{c^2} \frac{1}{T} \frac{1}{\cosh^2[\mu/(2T)]} = -\frac{g_s g_v}{6\pi} \frac{e^2 v^2}{c^2} \delta(\mu), \qquad (7.127)$$

where the last equality assumes the limit $T \to 0$. This expression was first obtained by McClure (1956) by differentiation of the thermodynamic potential (2.134) with respect to the magnetic field (see also Sharma, Johnson, & McClure, 1974; Safran & DiSalvo, 1979; Koshino & Ando, 2007, 2010).

The result (7.127) is really unusual. It means that at zero temperature and finite doping the orbital susceptibility of graphene within the Dirac model should be zero! Usually, the contribution of the orbital motion of electrons to the magnetic susceptibility is diamagnetic (Landau–Peierls diamagnetism), but here we have an exact cancellation of intraband and interband contributions; for a general discussion of these contributions, see Wilson (1965). In multilayer graphene and graphite, there is no cancellation but, rather, a strong diamagnetism (Sharma, Johnson, & McClure, 1974; Koshino & Ando, 2010).

As a result, the orbital magnetism of electrons in single-layer doped graphene is completely determined by electron–electron interactions (Principi et al., 2010). Using perturbation theory, one can find that the resulting effect is paramagnetic ($\chi > 0$), with

$$\chi = g_s g_v \frac{e^2 v^2}{c^2} \frac{e^2}{\hbar v \varepsilon_{\text{ext}}} \frac{\Lambda}{E_F}, \qquad (7.128)$$

where Λ is a function of the interaction constant, of the order of 10^{-2} (Principi et al., 2010).

Other nontrivial manifestations of the electron–electron interactions will be considered in the next chapter and in more detail in Chapter 15.

8

The Coulomb problem

8.1 Scattering of Dirac fermions by point charges

Now we come back to the problem of scattering of Dirac electrons by a radially symmetric potential $V(r)$ considered in Section 6.1. The case of a Coulomb potential

$$V(r) = -\frac{Ze^2}{\varepsilon_{ext}r} \equiv -\frac{\hbar v\beta}{r} \tag{8.1}$$

deserves a special consideration for reasons that will be clarified in this chapter. Here ε_{ext} is the dielectric constant due to substrate and other external factors and

$$\beta = \frac{Ze^2}{\varepsilon_{ext}\hbar v} \tag{8.2}$$

is the dimensionless interaction strength (the sign is chosen such that positive β corresponds to attraction). This problem has been considered for the case of two-dimensional massless Dirac equations by Shytov, Katsnelson, and Levitov (2007a, 2007b), Pereira, Nilsson, and Castro Neto (2007), and Novikov (2007). Here we will follow the works by Shytov, Katsnelson, and Levitov.

Instead of using the general expression (5.19), it is convenient to try the solution of the Coulomb problem in the form

$$\Psi(r,\varphi) = \begin{pmatrix} w_+(r) + w_-(r) \\ [w_+(r) - w_-(r)]\exp(i\varphi) \end{pmatrix} r^{s-1/2} \exp\left[i\left(m-\frac{1}{2}\right)\varphi\right]\exp(ikr),$$
$$\tag{8.3}$$

where m is half-integer,

$$m = \pm\frac{1}{2}, \pm\frac{3}{2}, \ldots, \tag{8.4}$$

and the parameters k and s should be found from the behavior of solutions at large and small r, respectively. For the potential (8.1), we find

$$s = \sqrt{m^2 - \beta^2}, \qquad\qquad k = -\frac{E}{\hbar v}, \qquad\qquad (8.5)$$

where E is the energy. On substituting Eq. (8.1) and (8.3) into the Dirac equation (6.1) we find, instead of Eq. (6.2),

$$r\frac{dw_+}{dr} + (s - i\beta + 2ikr)w_+ - mw_- = 0,$$

$$(8.6)$$

$$r\frac{dw_-}{dr} + (s + i\beta)w_- - mw_+ = 0.$$

Note that s can be either real (if $|m| > |\beta|$) or imaginary (if $|m| < |\beta|$); the behaviors of solutions in these two cases are essentially different, as will be discussed later.

Using the second of Eq. (8.6) one can express w_+ in terms of w_- and substitute it into the first equation. Then, after introducing a new independent variable

$$z = -2ikr, \qquad\qquad (8.7)$$

one has a confluent hypergeometric equation, or Kummer's equation (Abramowitz & Stegun, 1964)

$$z\frac{d^2 w_-}{dz^2} + (c - z)\frac{dw_-}{dz} - aw_- = 0, \qquad\qquad (8.8)$$

where

$$c = 2s + 1, \qquad\qquad a = s + i\beta. \qquad\qquad (8.9)$$

Its general solution has the form

$$w_-(z) = A\,_1F_1(a, c; z) + Bz^{1-c}\,_1F_1(a - c + 1, 2 - c; z), \qquad (8.10)$$

where A and B are arbitrary constants and

$$_1F_1(a, c; z) = \frac{\Gamma(c)}{\Gamma(a)} \sum_{n=0}^{\infty} \frac{\Gamma(a + n)z^n}{\Gamma(c + n)n!} \qquad\qquad (8.11)$$

is the confluent hypergeometric function ($_1F_1(a, c; 0) = 1$).

We will start with the case of real s, that is, $|m| > |\beta|$. Then, only the first term in Eq. (8.10) is regular at $r = 0$ and is therefore allowed, thus

$$w_-(z) = A\,_1F_1(s + i\beta, 2s + 1; z). \qquad\qquad (8.12)$$

Using the identity

$$z\frac{d}{dz}\,_1F_1(a, c; z) = a\left[\,_1F_1(a + 1, c; z) - \,_1F_1(a, c; z)\right] \qquad\qquad (8.13)$$

one finds from Eq. (8.6)

$$w_+(z) = A \frac{s + i\beta}{m} {}_1F_1(s + 1 + i\beta, 2s + 1; z). \tag{8.14}$$

Eq. (8.12) and (8.14) give us a formal solution of our problem. Using the asymptotic expression (Abramowitz & Stegun, 1964)

$${}_1F_1(a, c; z) \approx \frac{\Gamma(c)}{\Gamma(c-a)} (-z)^{-a} + \frac{\Gamma(c)}{\Gamma(a)} \exp(z) z^{a-c} \tag{8.15}$$

for $|z| \gg 1$, one finds for $kr \gg 1$

$$w_-(r) = \frac{\lambda \exp\left[-i\beta \ln(2kr)\right]}{(2kr)^s},$$

$$\tag{8.16}$$

$$w_+(r) = \frac{\lambda^* \exp\left[i\beta \ln(2kr)\right] \exp(-2ikr)}{(2kr)^s},$$

where λ is a constant dependent on m and β but not on k. It follows from Eq. (8.16) that w_- and w_+ represent scattered and incident waves, respectively (we have to recall our definition of k (8.5); E is assumed to be positive). Their ratio gives us the scattering phases $\delta_m(k)$ (cf. Eq. (6.11)):

$$\frac{w_-(r)}{w_+(r)} = \exp\left[2i\delta_m(k) + 2ikr\right],$$

$$\tag{8.17}$$

$$\delta_m(k) = -\beta \ln(2kr) + \arg\lambda.$$

The logarithmic dependence in Eq. (8.17) is typical for the phases coming from the $1/r$ Coulomb tail of the potential (Landau & Lifshitz, 1977). Since this contribution does not depend on m, it does not affect the angular dependence of the scattering current, giving just an irrelevant factor $|\exp[-i\beta \ln(2kr)]|^2 = 1$. The relevant scattering phases are $\arg\lambda$. Its explicit dependence on m and β is not important for us; it suffices to know that they are k-independent. From the general expression for the transport cross-section (6.26) one can see immediately that for the Coulomb scattering

$$\sigma_{tr} \sim \frac{1}{k}, \tag{8.18}$$

which gives us for the contribution of Coulomb impurities to the resistivity (cf. Section 6.2):

$$\rho \approx \frac{h}{e^2} \frac{n_{\mathrm{imp}}}{n}. \tag{8.19}$$

This contribution is much larger (by a factor of $(nR^2)^{-1}$) than that of short-range scatterers (Eq. (6.36)) and corresponds, at least qualitatively, to the experimentally observed V-shape of the dependence of the conductivity on the electron concentration (Novoselov et al., 2005a). It is not surprising therefore that charge impurities were initially suggested to be the main factor limiting electron mobility in graphene (Ando, 2006; Nomura & MacDonald, 2006; Adam et al., 2007). The real situation is probably much more complicated and will be discussed in Chapter 11. It is clear, anyway, that long-range scattering potentials deserve special attention in the case of graphene. However, screening effects are important and should be taken into account, as will be discussed later.

Consider now the case $|\beta| > |m|$ where $s = i\gamma$,

$$\gamma = \sqrt{\beta^2 - m^2}. \tag{8.20}$$

Then both terms in Eq. (8.10) are formally allowed:

$$w_-(z) = A_1F_1(i(\gamma + \beta), 1 + 2i\gamma; z) + Bz^{-2i\gamma}{}_1F_1(i(\beta - \gamma), 1 - 2i\gamma; z). \tag{8.21}$$

This means that the Dirac equation with the potential (8.1) for large enough $|\beta|$ is ill-defined. To find a solution, one needs to add some boundary conditions at small but finite r.

For $|kr| \ll 1$,

$$w_-(z) \approx A + B \exp(-\pi\gamma) \exp[-2i\gamma \ln(2kr)]. \tag{8.22}$$

The solution $w_+(z)$ corresponding to Eq. (8.21) is

$$w_+(z) = iA\frac{\gamma + \beta}{m}{}_1F_1(1 + i\gamma + i\beta, 1 + 2i\gamma; z)$$

$$+ iB\frac{\beta - \gamma}{m}z^{-2i\gamma}{}_1F_1(1 + i\beta - i\gamma, 1 + 2i\gamma; z). \tag{8.23}$$

Its asymptotics at $|kr| \ll 1$ is

$$w_+(z) \approx iA\frac{\gamma + \beta}{m} + iB\frac{\beta - \gamma}{m}\exp(-\pi\gamma)\exp[-2i\gamma \ln(2kr)]. \tag{8.24}$$

To be specific, let us use "zigzag" boundary conditions $\psi_2(r) = 0$ at some cutoff radius $r = r_0$, which means (see Eq. (8.3))

$$w_-(r_0) = w_+(r_0). \tag{8.25}$$

By substituting Eq. (8.22) and (8.24) into Eq. (8.25) one can find the ratio B/A and then use it to find the ratio $w_-(r)/w_+(r)$ at $|kr| \gg 1$ and the scattering phases (see Eq. (8.17)). The result is (Shytov, Katsnelson, & Levitov, 2007b)

$$\exp\left[2i\delta_m(k)\right] = \exp\left[\pi i|m|\right]\frac{z + \exp\left[2i\chi(k)\right]}{1 + z^* \exp\left[2i\chi(k)\right]}, \tag{8.26}$$

where

$$z = \frac{\exp(\pi\gamma)}{\eta}\frac{\Gamma(1 + 2i\gamma)}{\Gamma(1 - 2i\gamma)}\frac{\Gamma(1 - i\gamma + i\beta)}{\Gamma(1 + i\gamma + i\beta)} \tag{8.27}$$

and

$$\chi(k) = \gamma\ln(2kr_0) + \arctan\left(\frac{1 + \eta}{1 - \eta}\right), \tag{8.28}$$

with

$$\eta = \sqrt{\frac{\beta - \gamma}{\beta + \gamma}}. \tag{8.29}$$

The factor $\exp\left[2i\chi(k)\right]$ oscillates rapidly at $kr_0 \ll 1$. This conclusion does not depend on a specific choice of the boundary condition (8.25); for a generic boundary condition the first (logarithmic) term in Eq. (8.28) will be the same.

The expressions (8.26) through (8.29) have a very interesting property: They describe the existence of *quasilocalized states* (Shytov, Katsnelson, & Levitov, 2007b). For *localized* states, the wave function is described by a single real exponent, $\exp(-\kappa r)$ $(\kappa > 0)$, at $r \to \infty$, which means the absence of a scattering wave. Considering $\delta_m(k)$ as a function of the complex variable k and taking into account the condition (8.17), one can write the equation for the bound state as

$$\exp[2i\delta_m(k)] = 0 \tag{8.30}$$

for $k < 0$ and

$$\exp[-2i\delta_m(k)] = 0 \tag{8.31}$$

for $k > 0$. To be specific, let us consider the first case, $E > 0$. Then, Eq. (8.30) is equivalent to

$$\exp[2i\chi(k)] = -z, \tag{8.32}$$

which, taking into account Eq. (8.28), reads

$$\ln(2k_n r_0) = -\frac{i}{2\gamma}\ln z - \frac{1}{2\gamma}\ln\arctan\left(\frac{1 + \eta}{1 - \eta}\right) - \frac{\pi n}{\gamma}, \tag{8.33}$$

where n is an integer. For small γ, that is, near the threshold $|\beta| \cong |m|$, Eq. (8.33) describes the series of quasilocalized states corresponding to positive n (for negative n, $kr_0 \gg 1$, which contradicts our choice of the parameter r_0 as a small cut-off). The k values have an imaginary part, due the term $-[i/(2\gamma)]\ln|z|$ in Eq. (8.33). Keeping in mind that

$$\ln|\Gamma(1+i\beta)| = \frac{1}{2}\ln[\Gamma(1+i\beta)\Gamma(1-i\beta)] = \frac{1}{2}\ln\left(\frac{\pi\beta}{\sinh(\pi\beta)}\right), \qquad (8.34)$$

one finds

$$k_n = c\exp\left[-\frac{\pi n}{\gamma} - i\lambda\right], \qquad (8.35)$$

where

$$\lambda = \frac{\pi}{1 - \exp(-2\pi\beta)} \qquad (8.36)$$

and the prefactor c is of the order of r_0^{-1}. The corresponding energies $E_n = -\hbar v k_n$ have an imaginary part, due to the factor λ; however, it is small:

$$-\frac{\mathrm{Im}\, E_n}{\mathrm{Re}\, E_n} = \frac{\pi}{\exp(2\pi\beta) - 1}. \qquad (8.37)$$

The minimal value of β corresponds to $|\beta| = \frac{1}{2}$, and the right-hand side of Eq. (8.37) is about 0.14. This means that the imaginary part is relatively small and the resonances are narrow. The resonances correspond to jumps in the scattering phases and sharp anomalies in the transport scattering cross-section (6.26). The corresponding numerical data are shown in Fig. 8.1 (Shytov, Katsnelson, & Levitov, 2007b). One can see typical Fano resonances (see Section 6.7), as one would expect for quasilocalized states within a continuum spectrum.

These resonances were observed experimentally for the groups of more than three Ca^{2+} ions on graphene, hexagonal boron nitride was used as a substrate (Wang et al., 2013). In this case $\varepsilon_{\mathrm{ext}} \approx 2.5$ (see Eq.(7.84)) and the total screening constant (7.89) is about 8. Thus, the experimental situation corresponds to the effective dimensionless coupling constant $\frac{Ze^2}{\varepsilon\hbar v}$ of the order of unity, which seems to be reasonable. At the same time, the direct quantitative comparison of the theory for the single Coulomb center with the data on the group of four (or more) Coulomb centers seems to be impossible.

8.2 Relativistic collapse for supercritical charges

Our consideration up to now has been rather formal. To understand the physical meaning of the quasilocalized states considered in the previous section we will use a simple semiclassical consideration (Shytov, Katsnelson, & Levitov, 2007b; Shytov et al., 2009). It turns out that these states are related to the phenomenon

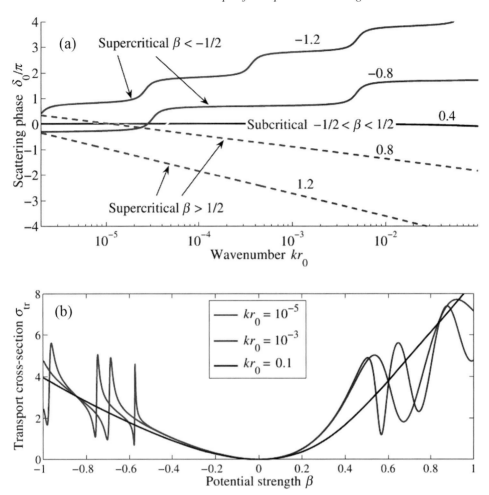

Fig. 8.1 (a) The scattering phase for $m = \frac{1}{2}$ at negative energy $E = -\hbar v k < 0$. The kinks correspond to quasilocalized states trapped by the impurity potential for supercritical β. (b) The transport cross-section as a function of the potential strength; the quasilocalized states are seen as Fano resonances.
(Reproduced with permission from Shytov, Katsnelson, & Levitov, 2007b.)

of *relativistic collapse,* or fall of electrons into the center for superheavy nuclei (Pomeranchuk & Smorodinsky, 1945). This provides us with a new, interesting connection between the physics of graphene and high-energy physics.

To gain some insight into the problem, let us start with a hand-waving derivation of the size of atoms using the Heisenberg uncertainty principle. If an electron is confined within a spatial region of radius R, its typical momentum is of the order of

$$p \approx \frac{\hbar}{R}. \tag{8.38}$$

For nonrelativistic particles with mass m, the kinetic energy is $p^2/(2m)$, and the total energy of the electron, taking into account its attraction to the nucleus, can be estimated as

$$E(R) \approx \frac{\hbar^2}{2mR^2} - \frac{Ze^2}{R}, \tag{8.39}$$

with a minimum at

$$R_0 = \frac{\hbar^2}{mZe^2}, \tag{8.40}$$

which is nothing other than the Bohr radius. For a relativistic particle we have, instead of Eq. (8.39),

$$E(R) \approx \sqrt{\left(\frac{\hbar c}{R}\right)^2 + (mc^2)^2} - \frac{Ze^2}{R}. \tag{8.41}$$

The minimum condition

$$\left(\frac{\partial E}{\partial R}\right)_{R=R_0} = 0$$

gives us the equation

$$1 + \left(\frac{mcR_0}{\hbar}\right)^2 = \left(\frac{\hbar c}{Ze^2}\right)^2, \tag{8.42}$$

which has a solution only for

$$Z < Z_c = \frac{\hbar c}{e^2} = \frac{1}{\alpha} \approx 137. \tag{8.43}$$

For $Z > Z_c$, the energy (8.41) decays monotonically with R, decreasing from $E_\infty = mc^2$ at $R \to \infty$ to $E = -\infty$ at $R = 0$. This means that the electron falls into the center.

Speaking more formally, the Dirac equation for a point charge $Z > Z_c$ is ill-defined and has no unique solutions, without introducing some additional boundary conditions at small R, similarly to what we did in the previous section. The wave function has infinitely many oscillations at $r \to 0$ (cf. Eq. (8.22) and (8.24)), and some of the solutions for the energies (Bjorken & Drell, 1964; Berestetskii, Lifshitz, & Pitaevskii, 1971)

$$E_{n,j} = \frac{mc^2}{\sqrt{1 + \dfrac{(Z\alpha)^2}{\left(n - |j| + \sqrt{j^2 - (Z\alpha)^2}\right)^2}}} \tag{8.44}$$

($n = 0, 1, 2, \ldots, j = \pm1, \pm2, \ldots$) become non-real, which means that the Hamiltonian is not a proper Hermitian operator.

If we draw the positions of the energy levels as a function of $\zeta = Z\alpha$ one can see that at $\zeta = 1$ the energy of the 1s state goes to zero, and the gap between electron and positron states disappears. In this situation, one could expect vacuum reconstruction, with the creation of electron–positron pairs from vacuum (Pomeranchuk & Smorodinsky, 1945; Zel'dovich & Popov, 1972; Greiner, Mueller, & Rafelski, 1985; Grib, Mamaev, & Mostepanenko, 1994); cf. our discussion of the Klein paradox in Section 4.1. The scheme of the energy levels (Zel'dovich & Popov, 1972) is shown in Fig. 8.2.

Taking into account the finite size of atomic nuclei $R = R_n$ and assuming a parabolic potential at $r < R_n$, as should be the case for a uniformly charged sphere, one finds a larger value for the critical radius, $Z_c \approx 170$ (Zel'dovich & Popov, 1972), which is still far beyond the charge of the heaviest known element. One can hope to observe this very interesting effect in collisions of two heavy ions with $Z < Z_c$, but in this case the critical value of total Z is even larger. Therefore, this effect of "relativistic collapse" of superheavy atoms has not been observed, thus far.

In the case of graphene, we have the Fermi velocity $v \approx c/300$, instead of the velocity of light, and the critical value Z_c should be of the order of one, which makes this effect observable (Pereira, Nilsson, & Castro Neto, 2007; Shytov, Katsnelson, & Levitov, 2007a, 2007b; Wang et al., 2013). Actually, some manifestations have been discussed in the previous section, such as the Fano resonances shown in Fig. 8.1 (Shytov, Katsnelson, & Levitov, 2007b). The scanning tunneling microscopy (STM) observation of this feature was claimed by Wang et al. (2013). Strong oscillations of the local density of states for the supercritically charged

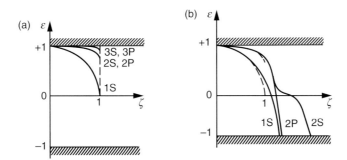

Fig. 8.2 (a) Energy levels of superheavy atoms (in units of mc^2) obtained from the Dirac equation for the Coulomb potential as a function of the coupling constant $\zeta = Z\alpha$. (b) The same, but taking into account the effects of the finite size of atomic nuclei. The critical value of Z is shifted from $Z_c = \alpha^{-1} \approx 137$ to $Z_c \approx 170$ (Zel'dovich & Popov, 1972).

impurities observable in principle by STM could be considered another manifestation (Shytov, Katsnelson, & Levitov, 2007a).

Strictly speaking, the massless case $m = 0$ relevant for graphene deserves
special consideration. We saw in the previous section that relatively narrow
resonances occur in the continuum spectrum. To better understand their origin it
is instructive to consider the problem semiclassically.

For ultrarelativistic particles with the Hamiltonian

$$H(\vec{p}, r) = v|\vec{p}| - \frac{Ze^2}{r}, \tag{8.45}$$

one can introduce the radial momentum p_r and angular momentum $p_\varphi = M$, which
is an integral of motion since the Hamiltonian (8.45) does not depend on φ. One
can find from the energy-conservation condition $H = E$ that

$$p_r^2 = \frac{1}{v^2}\left(E + \frac{Ze^2}{r}\right)^2 - \frac{M^2}{r^2}, \tag{8.46}$$

and the classically allowed regions are determined by the condition $p_r^2 > 0$. If M is
large enough,

$$M > M_c = \frac{Ze^2}{v}, \tag{8.47}$$

the particle can propagate from $r = 0$ to $r = \infty$. At $M < M_c$ the situation is different,
and we have two classically allowed regions, $0 < r < r_1$ and $r > r_2$, separated by a
potential barrier, where

$$r_{1,2} = \frac{Ze^2 \mp Mv}{|E|}. \tag{8.48}$$

If we were to neglect the tunneling through the classically forbidden region, we
could use the semiclassical quantization condition (Bohr–Sommerfeld condition)
for the inner well:

$$\int_0^{r_1} dr p_r = \pi\hbar(n + \mu), \tag{8.49}$$

where $n = 0, 1, 2, \ldots$ and μ is a factor of the order of unity, cf. Section 2.4 (Landau
& Lifshitz, 1977).

One can see, however, that the integral on the left-hand side of Eq. (8.49) is
logarithmically divergent at the lower limits, and a cut-off at $r = r_0 \ll r_1$ should be
introduced. This divergence reflects the fall toward the center discussed earlier.
After that, we will find from the corrected version of Eq. (8.49)

$$E_n \approx -C\frac{\hbar v}{r_0} \exp\left[-\frac{\pi\hbar n}{\sqrt{M_c^2 - M^2}}\right] \tag{8.50}$$

with a factor $C \approx 1$, in very good agreement with the positions of quasilocalized levels found from the exact solution Eq. (8.35).

Owing to the Klein tunneling through the classically forbidden region, the lifetime in the inner well is finite, which leads to the appearance of the imaginary part of the energy,

$$\frac{\Gamma_n}{|E_n|} \approx \exp\left(-\frac{2S}{\hbar}\right), \tag{8.51}$$

$$S = \int_{r_1}^{r_2} dr |p_r(r)| \tag{8.52}$$

(Landau & Lifshitz, 1977). The explicit calculation gives us the answer

$$S = \pi\left(M_c - \sqrt{M_c^2 - M^2}\right). \tag{8.53}$$

At the threshold, $M = M_c$, and this gives us a result that differs from the exact one (8.37) only by the replacement

$$\frac{1}{\exp(2\pi\beta) - 1} \rightarrow \exp(-2\pi\beta). \tag{8.54}$$

The resonances are narrow, and the quasilocalized states are long-lived, because of the numerical smallness $\exp(-\pi) \approx 0.04$, an interesting example of a small *numerical* parameter for a coupling constant of the order of 1!

8.3 Nonlinear screening of charge impurities

Up to now, we have not taken into account electron–electron interactions. However, they are essential in our problem. The Coulomb potential (8.1) induces some redistribution of the electron density $n_{\text{ind}}(\vec{r})$, which will create an additional potential

$$V_{\text{ind}}(\vec{r}) = \frac{e^2}{\varepsilon_{\text{ext}}} \int d\vec{r}' \frac{n_{\text{ind}}(\vec{r}')}{|\vec{r} - \vec{r}'|} + V_{\text{xc}}(\vec{r}), \tag{8.55}$$

where the first term is the Hartree potential and the second is the exchange correlation potential. In the simplest approximation the latter can be neglected, and we restrict ourselves to this approximation. The density-functional approach, taking into account V_{xc} for the case of massless Dirac fermions, was developed by Polini et al. (2008) and Rossi and Das Sarma (2008) (see also Brey & Fertig, 2009; Fogler, 2009; Gibertini et al., 2010).

We will focus on the case of undoped graphene (chemical potential $\mu = 0$). In this situation, the radial dependence of $n_{ind}(r')$ can be written just from dimensional analysis. There is no way to construct any length from the parameters of the potential (Ze^2) and of the electron spectrum (the Fermi velocity v); the only relevant characteristic β, given in (8.2), is dimensionless. At the same time, $n_{ind}(r)$ has a dimensionality of inverse length squared. The most general expression is

$$n_{ind}(r) = A(\beta)\delta(\vec{r}) + \frac{B(\beta)}{r^2},$$
(8.56)

with the dimensionless A and B. The physical roles of these two terms are dramatically different. The term proportional to $A(\beta)$ is nothing other than the renormalization of the point charge:

$$-\frac{Z}{\varepsilon_{ext}} \rightarrow -\frac{Z}{\varepsilon_{ext}} + A(\beta).$$
(8.57)

At the same time, phenomenologically, the answer should be $-Ze^2/\varepsilon$, where ε is the total dielectric constant (7.89), thus

$$A(\beta) = Z\left(\frac{1}{\varepsilon} - \frac{1}{\varepsilon_{ext}}\right).$$
(8.58)

Therefore, the first term on the right-hand side of Eq. (8.56) describes nothing but *linear* screening, that is, the renormalization of the dielectric constant.

The second term gives a logarithmically divergent contribution to the total charge:

$$Q_{ind} = \int d\vec{r}\,' n_{ind}(\vec{r}\,') \approx 2\pi B(\beta) \ln\left(\frac{r_{max}}{r_{min}}\right),$$
(8.59)

where r_{max} and r_{min} are the upper and lower limits of the integration. The obvious choice for r_{min} is the lattice constant a, since at such small distances the Dirac model is not applicable. As for r_{max}, it is of the order of the sample length L. The appearance of such contributions proportional to large $\ln(L/a)$ should have very important consequences.

Let us first consider the case of small Z. The linear-response problem was considered in Chapter 7, and no logarithms appeared. Owing to electron–hole symmetry, Q_{ind} (Z) should be an odd function:

$$Q_{ind} = (-Z) = -Q_{ind}(Z), \tag{8.60}$$

which means that, at small Z, B can be represented as

$$B(Z) = B_3 Z^3 + B_5 Z^5 + \cdots. \tag{8.61}$$

Straightforward calculations show that $B_3 = 0$ (Ostrovsky, Gornyi, & Mirlin, 2006; Biswas, Sachdev, & Son, 2007). Later in this section we will show in a nonperturbative way that $B = 0$ at $Z < Z_c$ (Shytov, Katsnelson, & Levitov, 2007a). To consider the opposite limit of large Z one can use the Thomas–Fermi approximation (Katsnelson, 2006c). For the case of atoms, one can prove that it is asymptotically exact at $Z \to \infty$ (Lieb, 1981). Within this approximation (Landau & Lifshitz, 1977; Lieb, 1981; Vonsovsky & Katsnelson, 1989) the effect of the total potential

$$V(\vec{r}) = -\frac{Ze^2}{r} + V_{ind}(\vec{r}) \tag{8.62}$$

on the electron density dependent on the chemical potential $n(\mu)$, is purely local,

$$n_{ind}(\vec{r}) = n[\mu - V(r)] - n(\mu), \tag{8.63}$$

and the term V_{xc} in Eq. (8.55) can be neglected. The linearized version of this approximation for the doped case was discussed in Section 7.7.

As a result,

$$V_{ind}(\vec{r}) = \frac{e^2}{\varepsilon_{ext}} \int d\vec{r}' \frac{n[\mu - V(\vec{r}')] - n(\mu)}{|\vec{r} - \vec{r}'|}. \tag{8.64}$$

For the case of graphene,

$$n(\mu) = \int_0^\mu dE N(E) = \frac{\mu|\mu|}{\pi \hbar^2 v^2} \tag{8.65}$$

(see Eq. (1.72)).

Let us start with the undoped case ($\mu = 0$). Then, on substituting Eq. (8.62) and (8.65) into Eq. (8.64) and integrating over angles (it is obvious that $V(\vec{r})$ and $n_{ind}(\vec{r})$ depend only on $|\vec{r}| = r$), one finds the integral equation (Katsnelson, 2006c)

$$F(r) = Z - \frac{2q}{\pi} \int_0^\infty \frac{dr'}{r'} \frac{r}{r+r'} K\left(\frac{2\sqrt{rr'}}{r+r'}\right) F(r')|F(r')|, \tag{8.66}$$

where F is related to V by the expression

$$V(\vec{r}) = -\frac{e^2}{\varepsilon_{\text{ext}} r} F(r), \tag{8.67}$$

$K(m)$ is the elliptic integral (7.65) and

$$q = 2\left(\frac{e^2}{\varepsilon_{\text{ext}} \hbar v}\right)^2. \tag{8.68}$$

We will see later that, actually, the integral on the right-hand side of Eq. (8.66) is divergent at $r = 0$; the reason is the inapplicability of the Dirac model at $r \leq a$. Therefore, we need to introduce a cut-off at $r' \approx a$, as was discussed previously. The exact value of a is not relevant, given the logarithmic accuracy.

To proceed further, we make a replacement of variables in Eq. (8.66), $r' = r \exp(t)$, and introduce the notation $\tilde{F}(\ln r) = F(r)$. As a result, Eq. (8.66) takes the form

$$\tilde{F}(x) = Z - q \int_{\ln a}^{x} dt \tilde{F}(t) \left|\tilde{F}(t)\right| - q \int_{-\infty}^{\infty} dt \tilde{F}(x+t) \left|\tilde{F}(x+t)\right| \varphi(t), \tag{8.69}$$

where

$$\varphi(t) = \frac{2}{\pi} \frac{K\left(1/\cosh\left(\frac{t}{2}\right)\right)}{1 + \exp t} - \theta(-t), \tag{8.70}$$

with $\theta(x > 0) = 1$, $\theta(x < 0) = 0$. The function $\varphi(t)$ decays exponentially at $t \to \pm\infty$ and has a logarithmic divergence at $t = 0$ (see Fig. 8.3). For large x, the last term in Eq. (8.69) can be neglected:

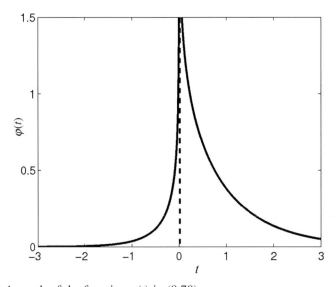

Fig. 8.3 A graph of the function $\varphi(t)$ in (8.70).

$$\tilde{F}(x) = Z - q \int\limits_{\ln a}^{x} dt \tilde{F}(t) \left| \tilde{F}(t) \right|. \tag{8.71}$$

This integral equation is equivalent to the differential one:

$$\frac{d\tilde{F}(x)}{dx} = -q\tilde{F}(x)\left| \tilde{F}(x) \right|, \tag{8.72}$$

with the initial condition $\tilde{F}(0) = Z$. The solution is

$$F(r) = \frac{Z}{1 + |Z|q \ln \left(\frac{r}{a} \right)}, \tag{8.73}$$

which corresponds to a very strong (logarithmic) screening of the effective charge at $r \gg a$:

$$Z_{\text{eff}}(r) = Z + Q_{\text{ind}} \approx \frac{Z_{\text{eff}}(r)}{q \ln \left(\frac{r}{a} \right)}. \tag{8.74}$$

If we were to expand Eq. (8.73) formally in Z, the leading term in Q_{ind} would be

$$Q_{\text{ind}} \approx -Z|Z|q \ln \left(\frac{r_{\max}}{r_{\min}} \right), \tag{8.75}$$

which does not have the form (8.61) (but, of course, satisfies the condition (8.60)). However, as we will see, the expression (8.73) is correct only for $|Z| \gg 1$.

If we took the expression (8.73) literally, it would lead to the conclusion that any charge is completely screened by the vacuum of two-dimensional massless Dirac electrons (Katsnelson, 2006c). The situation is reminiscent of "charge nullification" in quantum electrodynamics (Landau & Pomeranchuk, 1955; Landau, Abrikosov, & Khalatnikov, 1956; Migdal, 1977), which was considered (especially by the Landau school) as a fundamental difficulty of quantum field theory in general. Actually, complete nullification occurs neither in quantum electrodynamics nor in graphene. We will see that in the latter case the screening is stopped at the value $Z = Z_c$ (Shytov, Katsnelson, & Levitov, 2007a).

The simplest way to demonstrate this is to use arguments based on the Friedel sum rule (Friedel, 1952; Vonsovsky & Katsnelson, 1989); its generalization to the case of the Dirac equation has been proposed by Lin (2005, 2006). According to the sum rule, the total induced charge is related to the phase scattering at the Fermi surface

$$Q_{\text{int}} = -\frac{4}{\pi} \sum_{m} \delta_m(k_F), \tag{8.76}$$

where the minus sign corresponds to that in Eq. (8.5) and we introduce the factor of 4 (valley degeneracy multiplied by spin degeneracy), keeping in mind applications to graphene. We are interested in the limit $k_F \to 0$, which, however, requires some careful treatment for the supercritical charges ($|\beta| > \beta_c$), due to the term $\ln(2kr_0)$ in Eq. (8.28). This is, actually, the same logarithmic divergence as in Eq. (8.59), so we will immediately see that the B term in Eq. (8.56) arises naturally at $|\beta| > \beta_c$ (but is equal to zero at $|\beta| > \beta_c$, as has already been mentioned). For the r-dependent term one can estimate, with logarithmic accuracy

$$Q_{\text{int}}(r) \approx -\frac{4}{\pi} \sum_m \delta_m \left(k \sim \frac{1}{r} \right) = -\frac{4}{\pi} \sum_m \gamma_m \ln \left(\frac{r}{a} \right), \tag{8.77}$$

where the sum is taken over all $|m| < |\beta|$.

Thus, we have the following expression for the logarithmically dependent term in Eq. (8.56):

$$B(\beta) = -\frac{2}{\pi^2} \beta \sum_{|m| < |\beta|} \sqrt{\beta^2 - m^2}. \tag{8.78}$$

To proceed further one can use the renormalization group (RG) method, in its simplest form of the "poor man's scaling" (Anderson, 1970). Let us find the dimensionless charge β self-consistently:

$$\beta(r) = \frac{e^2}{\varepsilon_{\text{ext}} \hbar v} (Z + Q_{\text{ind}}(r)) = \beta_0 \left(Z + 2\pi B \ln \left(\frac{r}{a} \right) \right), \tag{8.79}$$

where β_0 is the bare value (8.2). The differential RG equation for the effective coupling constant reads

$$\frac{d\beta}{d \ln r} = 2\pi \beta_0 B(\beta) = -\frac{4e^2 \beta}{\pi \varepsilon_{\text{ext}} \hbar v} \sum_{|m| < |\beta|} \sqrt{\beta^2 - m^2}. \tag{8.80}$$

Eq. (8.80) describes the flow of effective charge from its initial value $\beta(r \approx a) = \beta_0$ to a smaller screened value. The flow stops, however, when $|\beta(r)|$ reaches the critical value $\beta_c = \frac{1}{2}$, since $B(|\beta| < |\beta_c|) = 0$. It happens at a finite screening radius r^* determined by the condition

$$\frac{1}{2\pi |\beta_0|} \int_{\beta_c}^{|\beta|} \frac{d|\beta|}{B(\beta)} = \ln \left(\frac{r^*}{a} \right). \tag{8.81}$$

For the case of $\frac{1}{2} < |\beta_0| < \frac{3}{2}$, for which only one term (with $|m| = \frac{1}{2}$) contributes to $B(\beta)$, the integration can be carried out explicitly:

$$r^* = a \exp\left[\frac{\pi \varepsilon_{\text{ext}} \hbar v}{4e^2} \cosh^{-1}(2\beta_0)\right]. \tag{8.82}$$

This means that the supercritical charge in graphene is surrounded by a cloud of electron–hole pairs (created from the vacuum) of finite radius r^*. For distances $r > r^*$ the supercritical charge looks like the critical one. In our simple theory this critical charge corresponds to $|\beta_c| = \frac{1}{2}$; however, one should keep in mind that a more accurate consideration of electron–electron interactions can renormalize this value. Also note that taking into account the A term (8.58) will lead to the replacement $\varepsilon_{\text{ext}} \rightarrow \varepsilon$ in Eq. (8.80) and (8.82).

Anyway, it is natural to expect that $|\beta_c|$ is of the order of 1. Thus, due to the condition $v \ll c$, the rich and interesting physics of the supercritical charge and vacuum reconstruction, which is hardly reachable for superheavy nuclei, can play an important role in graphene.

To finish this section, let us establish the relations between the Thomas–Fermi approximation and our RG treatment. If we assume that $|Z| \gg 1$ and $|\beta|$ is *much* larger than the critical value, the sum in Eq. (8.78) can be replaced by the integral

$$\sum_{|m| < |\beta|} \sqrt{\beta^2 - m^2} \approx \int_{-|\beta|}^{|\beta|} dm \sqrt{\beta^2 - m^2} = \frac{\pi \beta^2}{2}, \tag{8.83}$$

and Eq. (8.80) coincides with Eq. (8.72), with the solution (8.73). Thus, the Thomas–Fermi approximation works at $Z \rightarrow \infty$, as one would naturally expect.

8.4 Interelectron Coulomb interaction and renormalization of the Fermi velocity

As discussed in Chapter 7, electron–electron interaction in graphene is not weak, the effective coupling constant being of the order of 1. This makes the problem of a many-body description of graphene very complicated. Also, experimental evidence of many-body effects in graphene (except in the quantum Hall regime) remains very poor. For these two reasons, it seems to be a bit early to discuss in detail the correlation effects in graphene. However, one of the predictions, namely, a concentration-dependent renormalization of the Fermi velocity (González, Guinea, & Vozmediano, 1994) is based on relatively simple Hartee–Fock calculations and should be reliable, at least, qualitatively. Very recently, this effect was confirmed experimentally (Elias et al., 2011). It demonstrates the importance of the long-range character of interelectron Coulomb interactions and, therefore, will be considered in this chapter.

The Hamiltonian of the Coulomb interaction reads

$$\hat{H}_{\mathrm{C}} = \frac{e^2}{2} \sum_{\alpha, \beta} \iint d\vec{r} d\vec{r}' \frac{\hat{\psi}_\alpha^+(\vec{r})\hat{\psi}_\alpha(\vec{r})\hat{\psi}_\beta^+(\vec{r}')\hat{\psi}_\beta(\vec{r}')}{|\vec{r} - \vec{r}'|}, \tag{8.84}$$

where $\hat{\psi}_\alpha(\vec{r})$ is the electron annihilation operator at the point $\vec{r}$, α is an intrinsic quantum number (e.g., a set of spin-projection, sublattice, and valley labels). The Hartree–Fock approximation corresponds to the replacement

$$\hat{\psi}_1^+ \hat{\psi}_2 \hat{\psi}_3^+ \hat{\psi}_4 \rightarrow \langle \hat{\psi}_1^+ \hat{\psi}_2 \rangle \hat{\psi}_3^+ \hat{\psi}_4 + \langle \hat{\psi}_1^+ \hat{\psi}_4 \rangle \hat{\psi}_2 \hat{\psi}_3^+ = \rho_{21} \hat{\psi}_3^+ \hat{\psi}_4 + \rho_{41} \hat{\psi}_2 \hat{\psi}_3^+, \tag{8.85}$$

which means a consideration of electron–electron interactions at the mean-field level (Landau & Lifshitz, 1977; Vonsovsky & Katsnelson, 1989). The coupling with

$$\sum_\infty \langle \hat{\psi}_\alpha^+(\vec{r})\hat{\psi}_\alpha(\vec{r}) \rangle = n(\vec{r}) \tag{8.86}$$

corresponds to Hartree (electrostatic) terms and, within the model of a homogeneous electron gas, is exactly compensated for by the interactions with ionic charge density, due to the electroneutrality of the system. The Fock contribution survives:

$$\hat{H}_{\mathrm{F}} = -e^2 \sum_{\alpha, \beta} \int d\vec{r} d\vec{r}' \frac{\langle \hat{\psi}_\alpha^+(\vec{r})\hat{\psi}_\beta(\vec{r}') \rangle \hat{\psi}_\beta^+(\vec{r}')\hat{\psi}_\alpha(\vec{r})}{|\vec{r} - \vec{r}'|}. \tag{8.87}$$

Owing to the translational invariance of the system,

$$\langle \hat{\psi}_\alpha^+(\vec{r})\hat{\psi}_\beta(\vec{r}') \rangle = \sum_{\vec{k}} \rho_{\beta\alpha}(\vec{k}) \exp\left[i\vec{k}(\vec{r} - \vec{r}') \right], \tag{8.88}$$

where

$$\rho_{\beta\alpha}(\vec{k}) = \langle \hat{\psi}_{\vec{k}\alpha}^+ \hat{\psi}_{\vec{k}\beta} \rangle \tag{8.89}$$

(cf. Chapter 7, where we used this single-particle density matrix many times). If we apply this assumption to graphene, this means that we neglect intervalley Coulomb interaction. The corresponding terms contain "*Umklapp* processes" with $\hat{\rho}(\vec{k}, \vec{k} \pm \vec{g})$, where $\vec{g} = \vec{K} - \vec{K}'$ is the vector connecting the valleys. This approximation will be discussed later.

On substituting Eq. (8.88) into Eq. (8.87) we will have an additional term in the single-electron Hamiltonian

$$\hat{H}_{\mathrm{F}} = \sum_{\vec{k}} \sum_{\alpha, \beta} \hat{\psi}_\alpha^+(\vec{k}) h_{\alpha\beta}(\vec{k}) \hat{\psi}_\beta(\vec{k}), \tag{8.90}$$

where

$$h_{\alpha\beta}\left(\vec{k}\right) = -2\pi e^2 \sum_{\vec{k}} \frac{\rho_{\alpha\beta}\left(\vec{k}\right)}{\left|\vec{k} - \vec{k}'\right|}. \tag{8.91}$$

If we consider the electron–electron interaction effects by applying a perturbation theory, the corrections to the energies of electrons and holes are nothing other than the matrix elements of $\hat{h}\left(\vec{k}\right)$ in the corresponding basis. The explicit calculation for the undoped case (which is similar to that in Sections 7.2 and 7.6) gives us the following result:

$$\delta E_{\mathrm{e,h}}\left(\vec{k}\right) = \pm \sum_{\vec{k}'} \frac{2\pi e^2}{\left|\vec{k} - \vec{k}'\right|} \frac{1}{2} \left(1 \pm \frac{\vec{k}\vec{k}'}{kk'}\right). \tag{8.92}$$

The integral in Eq. (8.92) is logarithmically divergent at the upper limit and has to be cut at $k_{\mathrm{c}} \approx 1/a$, due to the inapplicability of the Dirac approximation. It contains the term $\pm\hbar\delta v_{\mathrm{F}}(k)k$, where

$$\delta v_{\mathrm{F}}(k) = \frac{e^2}{4\hbar} \ln\left(\frac{k_{\mathrm{c}}}{k}\right), \tag{8.93}$$

which is logarithmically divergent at $k \to 0$. This means that, strictly speaking, the Dirac cones near the neutrality point are not exactly cones.

For the case of doped graphene, the divergence at $k \to 0$ is cut at $k \approx k_{\mathrm{F}}$, which results in a logarithmic dependence of the Fermi velocity on the electron concentration:

$$\delta v_{\mathrm{F}} = \frac{e^2}{4\hbar} \ln\left(\frac{1}{k_{\mathrm{F}}a}\right). \tag{8.94}$$

If we take into account the screening of the Coulomb interaction by the environment plus virtual electron–hole transitions, the expression (8.94) is replaced by

$$\delta v_{\mathrm{F}} = \frac{e^2}{4\hbar\varepsilon} \ln\left(\frac{1}{k_{\mathrm{F}}a}\right), \tag{8.95}$$

with ε given by Eq. (7.89). This seems to be in agreement with the experimental data published by Elias et al. (2011).

Note that, if we took into account the intervalley Coulomb interaction, the Fourier component $1/\left|\vec{k} - \vec{k}'\right|$ in Eq. (8.92) would be replaced by a constant

$$\frac{1}{\left|\vec{k} - \vec{k}' + \vec{g}\right|} \approx \frac{1}{g}.$$

This interaction does not lead to any singularities and, therefore, can be neglected.

The situation becomes more complicated and interesting if we take into account *dynamical* screening of the Coulomb interaction (see Eq. (7.81), (7.82) and (7.88)). As was shown by González, Guinea, and Vozmediano (1999), this leads to the damping of electron states proportional to $|E|$, in contrast with the typical Fermi-liquid E^2 behavior. This means that graphene in the vicinity of the neutrality point should be a *marginal Fermi liquid,* with ill-defined quasiparticles. Currently, it is not clear how this result will be changed on going beyond the perturbation theory.

The many-body effects in graphene will be considered in much greater detail in Chapter 15.

9

Crystal lattice dynamics, structure, and thermodynamics

9.1 Phonon spectra of graphene

Phonon spectra of two-dimensional and quasi-two-dimensional crystals have some peculiar features that were first analyzed by Lifshitz (1952; see also Belenkii, Salaev, & Suleimanov, 1988; Kosevich, 1999). To explain them we first recall a general description of the phonon spectra in crystals (Kosevich, 1999; Katsnelson & Trefilov, 2002).

Let the coordinates of the nuclei be

$$\vec{R}_{nj} = \vec{R}_{nj}^{(0)} + \vec{u}_{nj}, \tag{9.1}$$

where $\left\{\vec{R}_{nj}^{(0)}\right\}$ form a crystal lattice, n labels elementary cells (or sites of the corresponding Bravais lattice), $j = 1, 2, \ldots, v$ labels the atoms within elementary cell (or sublattices), and $\vec{u}_{nj}$ are displacements. Further, we will use the notation

$$\vec{R}_{nj}^{(0)} = \vec{r}_n + \vec{\rho}_j, \tag{9.2}$$

where $\vec{r}_n$ are translation vectors and $\vec{\rho}_j$ are basis vectors $(\vec{\rho}_1 \equiv 0)$.

The main assumption of the standard theory of crystal lattices is the smallness of average atomic displacements in comparison with the interatomic distance d:

$$\left\langle \vec{u}_{nj}^{\,2} \right\rangle \ll d^2. \tag{9.3}$$

According to Eq. (9.3) one can expand the potential energy $V\left(\left\{\vec{R}_{nj}\right\}\right)$ in terms of atomic displacements and take into account only the lowest second-order term (the linear term obviously vanishes due to mechanical equilibrium conditions):

$$V\left(\left\{\vec{R}_{nj}\right\}\right) = V\left(\left\{\vec{R}_{nj}^{(0)}\right\}\right) + \frac{1}{2}\sum_{\substack{nn' \\ ij \\ \alpha\beta}} A_{ni,n'j}^{\alpha\beta} u_{ni}^{\alpha} u_{n'j}^{\beta}, \tag{9.4}$$

where

$$A^{\alpha\beta}_{ni,n'j} = \left(\frac{\partial^2 V}{\partial u^\alpha_{ni} \partial u^\beta_{n'j}} \right)_{\vec{u}=0} \tag{9.5}$$

is the *force-constant matrix*. Eq. (9.4) defines the *harmonic approximation*. The classical equations of motion for the potential energy (9.4) read

$$M_i \frac{d^2 u^{(\alpha)}_{ni}}{dt^2} = - \sum_{n'j\beta} A^{\alpha\beta}_{ni,n'j} u^\beta_{n'j}. \tag{9.6}$$

By looking for solutions of the form $u^\alpha_{ni}(t) \sim \exp(-i\omega t)$ and using translational symmetry, one can prove that the square eigenfrequencies of the problem $\omega^2 = \omega^2_\zeta(\vec{q})$, are eigenvalues of the *dynamical matrix*

$$D^{\alpha\beta}_{ij}(\vec{q}) = \sum_n \frac{A^{\alpha\beta}_{0i,nj}}{\sqrt{M_i M_j}} \exp(i\vec{q}\vec{r}_n). \tag{9.7}$$

Here $\vec{q}$ is the phonon wave vector running over the Brillouin zone and $\zeta = 1, 2, \ldots,$ $3v$ is the phonon branch label.

After quantization of the classical problem, one can prove that in the harmonic approximation, the Hamiltonian of the system is

$$\hat{H}_0 = \sum_\lambda \hbar\omega_\lambda \left(\hat{b}^+_\lambda \hat{b}_\lambda + \frac{1}{2} \right), \tag{9.8}$$

where $\lambda = (\vec{q}, \zeta)$ are phonon quantum numbers, $\hat{b}^+_\lambda$ and $\hat{b}_\lambda$ are canonical Bose creation and annihilation operators, and the atomic displacement operator is expressed in terms of $\hat{b}^+_\lambda$ and $\hat{b}_\lambda$ as

$$\hat{u}_{nj} = \sum_\lambda \sqrt{\frac{\hbar}{2N_0 M_j \omega_\lambda}} (\hat{b}_\lambda + \hat{b}^+_{-\lambda}) \vec{e}_j(\lambda) \exp(i\vec{q}\vec{r}_n). \tag{9.9}$$

Here N_0 is the number of elementary cells, $(-\lambda) \equiv (-\vec{q}, \zeta)$ and $\vec{e}_j(\lambda)$ are polarization vectors, that is, unit eigenvectors of the dynamical matrix.

There are important restrictions on the force-constant matrix, due to the translational invariance of the problem. If we were to move all nuclei of the crystal by the same displacement vector $\vec{u}$, no force would act on any atom. This means, due to Eq. (9.6), that

$$\sum_{nj} A^{\alpha\beta}_{0i,nj} = 0. \tag{9.10}$$

It follows from the condition (9.10) that in three-dimensional space there are three *acoustic* modes, with $\omega_\xi^2(\vec{q} \to 0) \to 0(\xi = 1, 2, 3)$ and $3(v - 1)$ *optical* modes, with finite $\omega_\xi^2(\vec{q} \to 0)$. The acoustic modes for small $\vec{q}$ correspond to coherent displacements of all atoms in the elementary cell by the same vector $\vec{u}_j \equiv \vec{u}$, whereas optical modes at $\vec{q} = 0$ correspond to the motion of atoms within the elementary cells with the fixed inertia center:

$$\sum_j M_j \vec{u}_j(\vec{q} = 0) = 0. \tag{9.11}$$

Keeping in mind graphene, we will assume further that $M_j = M$ is the mass of the carbon atom. Owing to mirror symmetry in the graphene plane, it is obvious that

$$\hat{A}^{xz} = \hat{A}^{yz} = 0 \tag{9.12}$$

and, thus, the modes with polarization along the z-direction are rigorously separated, within the harmonic approximation, from the modes polarized in the graphene xy-plane. Also, taking into account that the two sublattices A and B are equivalent, one can see that

$$D_{11}^{\alpha\beta} = D_{22}^{\alpha\beta} \tag{9.13}$$

and, due to Eq. (9.7) and (9.10),

$$D_{12}^{\alpha\beta}(\vec{q} = 0) + D_{11}^{\alpha\beta}(\vec{q} = 0) = 0. \tag{9.14}$$

Therefore, there are *six* phonon branches in graphene, namely the following:

(1) The acoustic flexural mode ZA($\vec{u}\|Oz$) with the frequencies

$$\omega_{ZA}^2(\vec{q}) = D_{11}^{zz}(\vec{q}) + D_{12}^{zz}(\vec{q}). \tag{9.15}$$

(2) The optical flexural mode ZA($\vec{u}\|Oz$) with the frequencies

$$\omega_{ZO}^2(\vec{q}) = D_{11}^{zz}(\vec{q}) - D_{12}^{zz}(\vec{q}). \tag{9.16}$$

(3), (4) Two acoustic in-plane modes, with $\omega^2(\vec{q})$ equal to eigenvalues of the 2×2 matrix

$$D_{11}^{\alpha\beta}(\vec{q}) + D_{12}^{\alpha\beta}(\vec{q}) \ (\alpha, \beta = x, y).$$

(5), (6) Two optical in-plane modes, with $\omega^2(\vec{q})$ equal to eigenvalues of the 2×2 matrix

$$D_{11}^{\alpha\beta}(\vec{q}) - D_{12}^{\alpha\beta}(\vec{q}) \;\; (\alpha, \beta = x, y).$$

If the two-dimensional wave vector $\vec{q}$ lies in symmetric directions, branches (3) through (6) can be divided into longitudinal ($\vec{e} \| \vec{q}$) and transverse ($\vec{e} \perp \vec{q}$) modes; for a generic $\vec{q}$ this classification is not possible.

Because of the conditions (9.14), one can assume that for acoustic modes $\omega^2 \sim q^2$ at $\vec{q} \rightarrow 0$, and this is, in general, true. However, for the ZA mode, q^2 terms also disappear, and $\omega_{ZA}^2(q) \sim q^4$ (Lifshitz, 1952). This follows from the *rotational invariance* of the system. Indeed, instead of uniform translation $\vec{u}_n = $ constant, let us use *uniform rotation*

$$\vec{u}_{nj} = \delta\vec{\varphi} \times \vec{R}_{nj}^{(0)}, \tag{9.17}$$

where $\delta\vec{\varphi}$ is the rotation angle. This should also not lead to the appearance of any forces or torques acting on the atoms. If $\delta\vec{\varphi}$ lies in the xy-plane, $\vec{u}_{nj} \| Oz$, additionally to the conditions (9.10), we will have

$$\sum_{nj} A_{0i,nj}^{zz} r_n^\alpha r_n^\beta = 0 \tag{9.18}$$

($\alpha, \beta = x, y$). It follows immediately from Eq. (9.18) and the definition of the dynamical matrix in (9.7) that

$$\frac{\partial^2}{\partial q_\alpha \partial q_\beta} \left[D_{11}^{zz}(\vec{q}) + D_{12}^{zz}(\vec{q}) \right] \Big|_{\vec{q}=0} = 0 \tag{9.19}$$

and, thus the expansion of the right-hand side of Eq. (9.15) starts with terms of the order of q^4; therefore,

$$\omega_{ZA}(q) \sim q^2 \tag{9.20}$$

at $\vec{q} \rightarrow 0$. In the next section we will derive this result by means of phenomenological elasticity theory.

There is no way, until now, to measure phonon dispersion in graphene experimentally, since the number of atoms in graphene flakes is insufficient for inelastic neutron-scattering experiments. It can be calculated using the density-functional method (Mounet & Marzari, 2005) or some semiempirical interatomic potential. The results are quite similar. Later in this chapter we will frequently discuss the results of atomistic simulations obtained using the so-called long-range

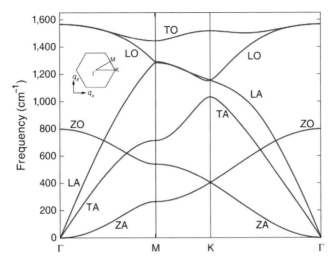

Fig. 9.1 Phonon spectra of graphene.
(Reproduced with permission from Karssemeijer and Fasolino, 2011.)

carbon-bond order potential (LCBOPII) (Los & Fasolino, 2003; Los et al., 2005). Therefore, we show in Fig. 9.1 the phonon spectra calculated within the same model (Karssemeijer & Fasolino, 2011). One can clearly see the six branches of the phonons listed previously.

Let us now consider the case of finite temperatures. In the harmonic approximation, the mean-square atomic displacement is (Kosevich, 1999; Katsnelson & Trefilov, 2002)

$$\left\langle u_{nj}^{\alpha} u_{nj}^{\beta} \right\rangle = \sum_{\lambda} \frac{\hbar}{2N_0 M_j \omega_{\lambda}} \left(e_{nj}^{\alpha} \right)^* \left(e_{nj}^{\beta} \right) \coth \left(\frac{\hbar \omega_{\lambda}}{2T} \right). \qquad (9.21)$$

For in-plane deformations ($\alpha = \beta = x$ or y) at any finite temperature the integral (9.21) is logarithmically divergent due to the contribution of acoustic branches with $\omega \sim q$ at $\vec{q} \rightarrow 0$. This divergence is cut at minimal $q_{min} \sim L^{-1}$ (L is the sample size), thus

$$\left\langle x_{nj}^2 \right\rangle = \left\langle y_{nj}^2 \right\rangle \approx \frac{T}{2\pi M c_s^2} \ln \left(\frac{L}{d} \right), \qquad (9.22)$$

where c_s is the average sound velocity (Peierls, 1934, 1935; Landau, 1937; Landau & Lifshitz, 1980). This led Landau and Peierls to the conclusion that two-dimensional crystals cannot exist. Strictly speaking, this means just the inapplicability of the harmonic approximation, due to violation of the condition (9.3). However, a more rigorous treatment does confirm this conclusion (Mermin, 1968), as a partial case of the *Mermin–Wagner theorem* (Mermin & Wagner, 1966;

Ruelle, 1999). This means that the definition of graphene as a "two-dimensional crystal" requires a detailed and careful discussion, which is one of the main aims of this chapter.

For $\alpha = z$, the situation is even worse, due to the much stronger divergence of ZA phonons Eq. (9.20). One can see from Eq. (9.21) that

$$\langle h_{nj}^2 \rangle \sim T \sum_q \frac{1}{q^4} \sim \frac{T}{E_{at}} L^2, \tag{9.23}$$

where E_{at} is of the order of the cohesive energy. Henceforth we will use the notation $h = u^z$, assuming that $\vec{u} = (u^x, u^y)$ is a two-dimensional vector only.

Before going any further it is important to derive the key results (9.20) and (9.23) from a different point of view.

9.2 The theory of elasticity for thin plates

In this section we present the general equations of the phenomenological elasticity theory, with applications to thin plates (Timoshenko & Woinowsky-Krieger, 1959; Landau & Lifshitz, 1970). This is a necessary preparatory step before we can discuss the unique mechanical properties of graphene (Booth et al., 2008; Lee et al., 2008). Also, it gives us a deeper insight into the properties of flexural phonons.

Let us consider a D-dimensional ($D = 2$ or 3) deformed medium. The particles, which had original coordinates x_α, transformed to the position

$$x'_\alpha = x_\alpha + u_\alpha(\{x_\beta\}). \tag{9.24}$$

The metrics, that is, the distance between infinitesimally distant points, being Pythagorean

$$dl^2 = dx_\alpha dx_\alpha \tag{9.25}$$

(we assume a summation over repeated tensor indices) is changed to

$$dl'^2 = dx'_\alpha dx'_\alpha = \frac{\partial x'_\alpha}{\partial x_\beta} \frac{\partial x'_\alpha}{\partial x_\gamma} dx_\beta dx_\gamma = dl^2 + 2u_{\alpha\beta} dx_\alpha dx_\beta, \tag{9.26}$$

where

$$u_{\alpha\beta} = \frac{1}{2} \left(\frac{\partial u_\alpha}{\partial x_\beta} + \frac{\partial u_\beta}{\partial x_\alpha} + \frac{\partial u_\gamma}{\partial x_\alpha} \frac{\partial u_\gamma}{\partial x_\beta} \right) \tag{9.27}$$

is a so-called *deformation tensor*. It is assumed, in the elasticity theory, that the free energy of a deformed medium is a functional of the deformation tensor $F = F[u_{\alpha\beta}]$.

By definition, the equilibrium state without external forces corresponds to $u_{\alpha\beta} = 0$.

There are two types of external forces resulting in the deformation. First, there are bulk forces acting on each atom of the medium, such as gravitational and electric forces. Their volume density is assumed to be $f_\alpha^{(v)}(\vec{r})$. Second, there are mechanical forces, due to contact with various bodies; they act on the surface only. The hydrostatic pressure P is an example; it leads to the total force

$$\vec{f} = \oint d\vec{S}P, \tag{9.28}$$

where $d\vec{S}$ is the (vector) element of the surface area. In a more general case in which shear forces are also allowed Eq. (9.28) is generalized as

$$f_\alpha = \oint dS_\beta \sigma_{\alpha\beta}, \tag{9.29}$$

where $\sigma_{\alpha\beta}$ is called the *stress tensor*. Using the Gauss theorem, Eq. (9.29) can be represented as an integral over the volume

$$f_\alpha = \int d^D x \frac{\partial \sigma_{\alpha\beta}}{\partial x_\beta}. \tag{9.30}$$

Thus, the condition of local equilibrium can be written as

$$\frac{\partial \sigma_{\alpha\beta}}{\partial x_\beta} + f_\alpha^{(v)} = 0. \tag{9.31}$$

One can prove (Landau & Lifshitz, 1970) that, due to the condition of absence of internal torques, the stress tensor is symmetric:

$$\sigma_{\alpha\beta} = \sigma_{\beta\alpha}. \tag{9.32}$$

Interestingly, this condition is violated in ferromagnetic media, due to gyromagnetic effects (Vlasov & Ishmukhametov, 1964), but we will not consider that case here.

The stress tensor creates deformations that are linear in the stress (Hooke's law). In the approximation of an isotropic elastic medium, the relation is determined by two *Lamé constants*, λ and μ:

$$\sigma_{\alpha\beta} = \lambda \delta_{\alpha\beta} u_{\gamma\gamma} + 2\mu u_{\alpha\beta}. \tag{9.33}$$

It is obvious that for small-enough deformations $|u_{\alpha\beta}| \ll 1$, the renormalization of the local volume is determined by $u_{\gamma\gamma} = \mathrm{Tr}\hat{u}$:

$$\frac{dV'}{dV} = \det\left(\frac{\partial x'_\alpha}{\partial x_\beta}\right) \approx 1 + u_{\gamma\gamma}. \tag{9.34}$$

This component of the deformation tensor is called *dilatation*. The traceless component

$$u'_{\alpha\beta} = u_{\alpha\beta} - \frac{1}{D}\delta_{\alpha\beta}u_{\gamma\gamma}, \tag{9.35}$$

is called *shear deformation*. Hooke's law (9.33) can be rewritten as

$$\sigma_{\alpha\beta} = B\delta_{\alpha\beta}u_{\gamma\gamma} + 2\mu\left(u_{\alpha\beta} - \frac{1}{D}\delta_{\alpha\beta}u_{\gamma\gamma}\right), \tag{9.36}$$

where

$$B = \lambda + \frac{2\mu}{D} \tag{9.37}$$

is the bulk modulus and μ has the meaning of a shear modulus of the system under consideration.

On substituting Eq. (9.33) into Eq. (9.31) we find the equilibrium conditions for the case $f_\alpha^{(v)} = 0$:

$$\frac{\partial}{\partial x_\alpha}\left(\lambda u_{\gamma\gamma}\right) + 2\frac{\partial}{\partial x_\beta}\left(\mu u_{\alpha\beta}\right) = 0. \tag{9.38}$$

Eq. (9.38) corresponds to the extremum of the free energy

$$F = \frac{1}{2}\int d^D x\left[\lambda(u_{\alpha\alpha})^2 + 2\mu u_{\alpha\beta}u_{\alpha\beta}\right]. \tag{9.39}$$

Thermodynamic stability requires

$$B > 0, \quad \mu > 0, \tag{9.40}$$

which is obvious if one considers pure dilatation and pure shear deformation. The inversion of Eq. (9.36) gives us

$$u_{\alpha\beta} = \frac{1}{D^2 B}\delta_{\alpha\beta}\sigma_{\gamma\gamma} + \frac{1}{2\mu}\left(\sigma_{\alpha\beta} - \frac{1}{D}\delta_{\alpha\beta}\sigma_{\gamma\gamma}\right). \tag{9.41}$$

If we apply a uniaxial uniform stress ($\sigma_{xx} = p$ and other components are equal to zero) we can find from Eq. (9.41) that

$$u_{xx} = \frac{p}{Y}, \quad u_{yy} = -\nu u_{xx}, \tag{9.42}$$

where Y is called *Young's modulus* and ν is the *Poisson ratio* determining the change of sizes in directions perpendicular to the stress. For $D = 3$ one has

$$Y = \frac{9B\mu}{3B + \mu},$$

$$v = \frac{1}{2}\frac{3B - 2\mu}{3B + \mu}$$

(9.43)

and, due to Eq. (9.40),

$$-1 < v < \frac{1}{2}.$$

(9.44)

For most solids $v > 0$, which means a constriction of the body in the perpendicular direction. For $D = 2$

$$Y = \frac{4B\mu}{B + \mu},$$

$$v = \frac{B - \mu}{B + \mu}.$$

(9.45)

Now, after recalling these basic definitions of elasticity theory, let us consider the case of a thin plate (its thickness Δ is much smaller than the typical size L, in the x- and y-directions). We start with the case of small deformations, for which the last nonlinear term in the definition (9.27) can be neglected. If we assume that no forces act on the surfaces of the plate, it should be, according to Eq. (9.29), the case that

$$\sigma_{\alpha\beta}n_\beta = 0,$$

(9.46)

where $\vec{n}$ is the unit normal to the surface. For the equation of the surface

$$z = h(x, y)$$

(9.47)

the components of the normal are

$$n_x = -\frac{\partial h}{\partial x}\frac{1}{\sqrt{1 + |\nabla h|^2}},$$

$$n_y = -\frac{\partial h}{\partial y}\frac{1}{\sqrt{1 + |\nabla h|^2}},$$

(9.48)

$$n_z = \frac{1}{\sqrt{1 + |\nabla h|^2}},$$

where

$$\nabla h = \left(\frac{\partial h}{\partial x}, \frac{\partial h}{\partial y}\right)$$

is a two-dimensional gradient (see any textbook on differential geometry, e.g. DoCarmo, 1976; Coxeter, 1989). If $|\nabla h| \ll 1$, the normal is parallel to the z-axis, and Eq. (9.46) reads

$$\sigma_{xz} = \sigma_{yz} = \sigma_{zz} = 0. \tag{9.49}$$

The conditions (9.49) should be satisfied for both surfaces of the plate and, since the plate is thin, should also be valid within the plane. Taking into account Eq. (9.33) and the definitions (9.43), one finds

$$\frac{\partial u_x}{\partial z} = -\frac{\partial u_z}{\partial x}, \quad \frac{\partial u_y}{\partial z} = -\frac{\partial u_z}{\partial y} \tag{9.50}$$

and

$$u_{zz} = -\frac{v}{1-v}\left(\frac{\partial u_x}{\partial x} + \frac{\partial u_y}{\partial y}\right). \tag{9.51}$$

Assuming $u_z = h(x, y)$ to be z-independent within the plane, one finds from Eq. (9.50)

$$u_x = -z\frac{\partial h}{\partial x}, \quad u_y = -z\frac{\partial h}{\partial y}, \tag{9.52}$$

and the components of the deformation tensor are

$$u_{xx} = -z\frac{\partial^2 h}{\partial x^2}, \quad u_{yy} = -z\frac{\partial^2 h}{\partial y^2}, \quad u_{xy} = -z\frac{\partial^2 h}{\partial x \partial y}, \tag{9.53}$$

$$u_{xz} = u_{yz} = 0, \quad u_{zz} = z\left(\frac{\partial^2 h}{\partial x^2} + \frac{\partial^2 h}{\partial y^2}\right)\frac{v}{1-v}.$$

On substituting Eq. (9.53) into Eq. (9.39) and integrating explicitly over $|z| < \Delta/2$ (Δ is the plate thickness) one finds for the energy of bending deformation

$$F_b = \frac{Y\Delta^3}{24(1-v^2)}\int d^2x\left\{(\nabla^2 h)^2 + 2(1-v)\left[\left(\frac{\partial^2 h}{\partial x \partial y}\right)^2 - \frac{\partial^2 h}{\partial x^2}\frac{\partial^2 h}{\partial y^2}\right]\right\}, \tag{9.54}$$

where

$$\nabla^2 = \frac{\partial^2}{\partial x^2} + \frac{\partial^2}{\partial y^2} \tag{9.55}$$

is the two-dimensional Laplacian. The last term in Eq. (9.54)

$$\det\left(\frac{\partial^2 h}{\partial x_i \partial x_j}\right),$$

is proportional to the Gaussian curvature K of the deformed surface (DoCarmo, 1976; Coxeter, 1989); see later for more details. It can be represented as a total derivative:

$$2 \det \left(\frac{\partial^2 f}{\partial x_i \partial x_j} \right) = -\varepsilon_{im} \varepsilon_{jm} \frac{\partial^2}{\partial x_m \partial x_n} \left(\frac{\partial f}{\partial x_i} \frac{\partial f}{\partial x_j} \right) \tag{9.56}$$

($\hat{\varepsilon}$ is the unit antisymmetric 2×2 matrix) and, thus, leads to some integral over the edges of the membrane. It therefore has no effect on the equations of motion. Alternatively, one can refer to the Gauss–Bonnet theorem (DoCarmo, 1976; Coxeter, 1989) that $\int dS\, K$ is a topological invariant that is not changed during smooth deformations. Thus, the bending energy (9.54) can be represented as

$$F_b = \frac{\kappa}{2} \int d^2 x \left(\nabla^2 h \right)^2, \tag{9.57}$$

where

$$\kappa = \frac{Y \Delta^3}{12(1 - v^2)}. \tag{9.58}$$

If we add the kinetic energy

$$T = \frac{1}{2} \int d^2 x \rho \left(\frac{\partial \vec{u}}{\partial t} \right)^2 \approx \frac{1}{2} \int d^2 x \rho \left(\frac{\partial h}{\partial t} \right)^2 \tag{9.59}$$

(ρ is the mass density) and write the Lagrangian $L = T - F_b$ and the corresponding equations of motion

$$\frac{\partial}{\partial t} \left(\rho \frac{\partial h}{\partial t} \right) + \nabla^2 \left(\kappa \nabla^2 h \right)^2 = 0, \tag{9.60}$$

then we find for the frequencies of the bending waves

$$\omega^2 = \frac{\kappa}{\rho} q^4, \tag{9.61}$$

in agreement with Eq. (9.20). The quantity κ is called the *bending rigidity*.

Our consideration up to now has not taken into account the energy of in-plane deformations. To take them into account one needs to add the energy (9.39), where $\alpha, \beta = x, y$. In the definition, one can neglect (9.27) the nonlinear terms

$$\frac{\partial u_x}{\partial x_\alpha} \frac{\partial u_x}{\partial x_\beta} \quad \text{and} \quad \frac{\partial u_y}{\partial x_\alpha} \frac{\partial u_y}{\partial x_\beta},$$

but one should keep the nonlinearities

$$\frac{\partial h}{\partial x_\alpha}\frac{\partial h}{\partial x_\beta}$$

since, as we will see, they can be comparable to $\partial u_\alpha / \partial x_\beta$ (further, $\vec{u} = (u_x, u_y)$ is the *two-dimensional* vector):

$$u_{\alpha\beta} = \frac{1}{2}\left(\frac{\partial u_\alpha}{\partial x_\beta} + \frac{\partial u_\beta}{\partial x_\alpha} + \frac{\partial h}{\partial x_\alpha}\frac{\partial h}{\partial x_\beta}\right). \tag{9.62}$$

The total deformation energy is

$$F = \frac{1}{2}\int d^2x \left\{\kappa\left(\nabla^2 h\right)^2 + \lambda(u_{\alpha\alpha})^2 + 2\mu u_{\alpha\beta}u_{\alpha\beta}\right\}, \tag{9.63}$$

where

$$\lambda = \lambda_3 \Delta, \quad \mu = \mu_3 \Delta \tag{9.64}$$

are the *two-dimensional* Lame constants (henceforth we will write two-dimensional parameters λ, μ without subscripts, and the corresponding three-dimensional parameters with the subscript 3). The equations for equilibrium deformations of the plate can be found by minimization of the functional (9.63), plus interactions with external forces. After rather cumbersome transformations (Landau & Lifshitz, 1970; Timoshenko & Woinowsky-Krieger, 1959) one finds

$$\kappa\nabla^4 h - \Delta\left[\frac{\partial^2\chi}{\partial y^2}\frac{\partial^2 h}{\partial x^2} + \frac{\partial^2\chi}{\partial x^2}\frac{\partial^2 h}{\partial y^2} - 2\frac{\partial^2\chi}{\partial x\partial y}\frac{\partial^2 h}{\partial x\partial y}\right] = P \tag{9.65}$$

$$\nabla^4\chi + Y_3\left[\frac{\partial^2 h}{\partial x^2}\frac{\partial^2 h}{\partial y^2} - \left(\frac{\partial^2 h}{\partial x\partial y}\right)^2\right] = 0, \tag{9.66}$$

where Y_3 is the bulk (three-dimensional) Young modulus, P is the density of external forces (per unit area), and χ is the potential for the stress tensor (*Airy stress function*):

$$\sigma_{xx} = \frac{\partial^2\chi}{\partial y^2}, \quad \sigma_{xy} = -\frac{\partial^2\chi}{\partial x\partial y}, \quad \sigma_{yy} = \frac{\partial^2\chi}{\partial x^2}. \tag{9.67}$$

These equations (the *Föppl–von Karman equations*) are essentially nonlinear, and their solution is, in general, a difficult task. One can, however, estimate the deformation for the situation when $|h| \gg \Delta$, the only one that is relevant for graphene, where Δ is of the order of interatomic distance. The first term in Eq. (9.65) is smaller in this situation than the second one and can be neglected (Landau & Lifshitz, 1970). This means that the bending rigidity κ is irrelevant, and it is in-plane deformation and the corresponding Young modulus that determine the resistance to the external force. Dimensional analysis of Eq. (9.65) and (9.66) gives us a typical value of the deformation:

$$h \sim \left(\frac{L^4 P}{Y_3 \Delta} \right)^{\frac{1}{3}}. \tag{9.68}$$

For example, for a circular plate of radius R with a clamped edge and uniform P, the deformation at the center is (Timoshenko & Woinowsky-Krieger, 1959)

$$h_0 \approx 0.662 R \left(\frac{RP}{Y_3 \Delta} \right)^{\frac{1}{3}}. \tag{9.69}$$

Note that, despite Eq. (9.65), (9.66) do not depend on the Poisson ratio v, the expressions for the deformation tensor and, therefore, the boundary conditions depend on it. The answer (9.69) corresponds to $v = 0.25$. Estimations also show that linear and nonlinear terms in the deformation tensor (9.62) are, in general, of the same order of magnitude.

Graphene is an extremely strong material (the real values of the constants κ, B, and μ will be discussed later in this chapter). Also, being almost defect-free, it can keep a deformation as high as, at least, 10%–15% (Kim et al., 2009). Therefore, according to the classical elasticity theory, for typical flake sizes of the order of 10–100 μm, a flake can bear a weight of the order of billions of times its own weight (Booth et al., 2008). Later we will see that for the single-atomic membrane the Föppl–von Karman equations (9.65) and (9.66) should be reconsidered due to essential role of thermal fluctuations but, qualitatively, the conclusion on the extraordinary strength of graphene membrane remains correct.

There is another way to derive Eq. (9.57), which starts from the model of a membrane as an infinitely thin plate, that is, a single flexible surface (Nelson, Piran, & Weinberg, 2004). It is natural to assume that the energy of a deformed membrane depends on the mutual orientation of normals to the surface at the neighboring points, which determines the orientation of electron orbitals, etc. (Fig. 9.2). If we discretize (e.g., triangulate) the surface, we can write the corresponding free energy as

$$F_b = \tilde{\kappa} \sum_{\langle ij \rangle} \left(1 - \vec{n}_i \vec{n}_j \right), \tag{9.70}$$

where $\tilde{\kappa} > 0$, $\vec{n}_i$ is the normal to the zth triangle and the sum is taken over the neighboring triangles. The bending energy (9.70) is counted from the flat state with all $\vec{n}_i \| Oz$. Since

$$1 - \vec{n}_i \vec{n}_j = \frac{1}{2} \left(\vec{n}_i - \vec{n}_j \right)^2, \tag{9.71}$$

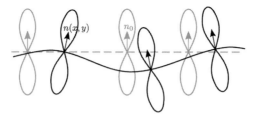

Fig. 9.2 The orientation of normals and the directions of electron orbitals in a fluctuating membrane (black) and in its ground state (gray).

in the continuum limit it will be transformed to the invariant quantity

$$\frac{\partial n_\alpha}{\partial x_\beta} \frac{\partial n_\alpha}{\partial x_\beta},$$

and

$$F_{\rm b} = \frac{\kappa}{2} \int d^2x \frac{\partial n_\alpha}{\partial x_\beta} \frac{\partial n_\alpha}{\partial x_\beta} \tag{9.72}$$

with $\kappa \propto \tilde\kappa$. On substituting Eq. (9.48) into Eq. (9.72) and keeping only the lowest-order terms in $\partial h/\partial x_\alpha$ we have

$$F_{\rm b} = \frac{\kappa}{2} \int d^2x \left[(\nabla^2 h)^2 - 2\det\left(\frac{\partial h}{\partial x_i} \frac{\partial h}{\partial x_j} \right) \right]. \tag{9.73}$$

The last term, which is proportional to the Gaussian curvature, can be skipped for the reasons discussed earlier, and we have the expression (9.57).

One more view of Eq. (9.57) is based on the Helfrich model of liquid membranes (Helfrich, 1973; Jones, 2002). The deformation energy in this model is written in terms of the mean curvature H and Gaussian curvature K of the surface:

$$F = \frac{\kappa}{2} \int dS H^2 + \kappa' \int dS K, \tag{9.74}$$

where, due to the Gauss–Bonnet theorem, the second term is important only for processes during which the topology is changed (e.g., the merging of two vesicles). The first term is also known in mathematics as the Willmore functional; for some recent discussions see Taimanov (2006) and Manyuhina et al. (2010). For a general surface defined by Eq. (9.47) one has (DoCarmo, 1976)

$$dS = dxdy\sqrt{1 + |\nabla h|^2}, \tag{9.75}$$

$$K = \frac{1}{\left[1 + |\nabla h|^2\right]^2} \left[\frac{\partial^2 h}{\partial x^2}\frac{\partial^2 h}{\partial y^2} - \left(\frac{\partial^2 h}{\partial x \partial y}\right)^2\right], \tag{9.76}$$

$$H = \frac{1}{\left[1 + |\nabla h|^2\right]^{3/2}} \left\{\left[1 + \left(\frac{\partial h}{\partial x}\right)^2\right]\frac{\partial^2 h}{\partial y^2} + \left[1 + \left(\frac{\partial h}{\partial y}\right)^2\right]\frac{\partial^2 h}{\partial x^2} - 2\frac{\partial^2 h}{\partial x \partial y}\frac{\partial h}{\partial x}\frac{\partial h}{\partial y}\right\}. \tag{9.77}$$

Keeping only the lowest-order terms in $|\nabla h|$, we have

$$H \approx \nabla^2 h \tag{9.78}$$

and, thus, Eq. (9.74) is equivalent to Eq. (9.57).

9.3 The statistical mechanics of flexible membranes

The expressions (9.62) and (9.63) provide a background for the statistical mechanics of crystalline membranes at finite temperatures (Nelson & Peliti, 1987; Aronovitz & Lubensky, 1988; Abraham & Nelson, 1990; Le Doussal & Radzihovsky, 1992; Nelson, Piran, & Weinberg, 2004). Henceforth we will consider only the classical regime, assuming that $\vec{u}(\vec{r})$ and $h(\vec{r})$ are static fields fluctuating in space. Thus, the partition function is determined by a functional integral

$$Z = \int D\vec{u}(\vec{r})Dh(\vec{r})\exp\left\{-\beta F[\vec{u}(\vec{r}), h(\vec{r})]\right\}, \tag{9.79}$$

where $\beta = T^{-1}$ is the inverse temperature and the free energy F (9.63) plays the role of the Hamiltonian. The nonlinear term in Eq. (9.62) couples the two fields, making the theory highly nontrivial – at least as nontrivial as the famous problem of critical behavior (Wilson & Kogut, 1974; Ma, 1976).

If we neglect this term, the Hamiltonian (9.63) is split into two independent Hamiltonians for the free fields. In the $\vec{q}$ representation, it reads

$$F_0 = \frac{\kappa}{2}\sum_{\vec{q}} q^4 |h_{\vec{q}}|^2 + \frac{1}{2}\sum_{\vec{q}}\left[\mu q^2 |\vec{u}_{\vec{q}}|^2 + (\lambda + \mu)(\vec{q}\cdot\vec{u}_{\vec{q}})^2\right], \tag{9.80}$$

where $h_{\vec{q}}$ and $\vec{u}_{\vec{q}}$ are Fourier components of $h(\vec{r})$ and $\vec{u}(\vec{r})$, respectively. The correlation functions for the free fields can be found immediately using the properties of Gaussian functional integrals (Wilson & Kogut, 1974; Ma, 1976; Faddeev & Slavnov, 1980):

$$G_0(\vec{q}) = \left\langle |h_{\vec{q}}|^2 \right\rangle_0 = \frac{T}{\kappa q^4}, \tag{9.81}$$

$$D_0{}^{\alpha\beta}(\vec{q}) = \left\langle u_{\alpha\vec{q}}{}^* u_{\beta\vec{q}} \right\rangle_0 = P_{\alpha\beta}(\vec{q}) \frac{T}{(\lambda + 2\mu)q^2} + \left[\delta_{\alpha\beta} - P_{\alpha\beta}(\vec{q}) \right] \frac{T}{\mu q^2}, \tag{9.82}$$

where $\langle \ldots \rangle_0$ means averaging with the Hamiltonian F_0 and

$$P_{\alpha\beta}(\vec{q}) = \frac{q_\alpha q_\beta}{q^2} \tag{9.83}$$

is the projection operator on the $\vec{q}$ vector. Note that the normal–normal correlation function is related to $\left\langle |h_{\vec{q}}|^2 \right\rangle$ by

$$\left\langle \delta \vec{n}_{\vec{q}} \delta \vec{n}_{-\vec{q}} \right\rangle = q^2 \left\langle |h_{\vec{q}}|^2 \right\rangle \tag{9.84}$$

as follows from Eq. (9.48), $\delta \vec{n}$ is the deviation of the normal vector from Oz axis. On substituting Eq. (9.81) into Eq. (9.84) we find

$$\left\langle \delta \vec{n}_{\vec{q}} \delta \vec{n}_{-\vec{q}} \right\rangle = \frac{T}{\kappa q^2}. \tag{9.85}$$

However, the approximation (9.81) turns out to be unsatisfactory. It does not describe a flat membrane. Indeed, the membrane is more or less flat if the correlation function

$$\left\langle \vec{n}_0 \vec{n}_{\vec{R}} \right\rangle = \sum_{\vec{q}} \left\langle |\vec{n}_{\vec{q}}|^2 \right\rangle \exp\left(i\vec{q}\vec{R} \right) \tag{9.86}$$

tends to a constant at $R \to \infty$ (normals at large distances have, on average, the same direction). Instead, substitution of Eq. (9.85) into (9.86) leads to a logarithmically divergent integral. Moreover, the mean-square out-of-plane displacement

$$\left\langle h^2 \right\rangle = \sum_{\vec{q}} \left\langle |h_{\vec{q}}|^2 \right\rangle \tag{9.87}$$

after the cut-off at $q_{min} \sim L^{-1}$ gives the result

$$\left\langle h^2 \right\rangle \sim \frac{T}{\kappa} L^2 \tag{9.88}$$

(cf. Eq. (9.23)), which means that the membrane is crumpled (on average, it has all three dimensions of the order of L).

Similarly, the in-plane square deformation

$$\langle \vec{u}^2 \rangle = \sum_{\vec{q}} \left\langle |\vec{u}_{\vec{q}}|^2 \right\rangle \tag{9.89}$$

is logarithmically divergent, as in Eq. (9.22). Thus, we conclude, again, that the statistical mechanics of two-dimensional systems cannot be based on the harmonic approximation, or approximation of free fields.

The nonlinear term

$$\frac{\partial h}{\partial x_\alpha} \frac{\partial h}{\partial x_\beta}$$

in Eq. (9.62) after substitution into Eq. (9.63) results in a coupling of two fields. The integral over $\vec{u}(\vec{r})$ in Eq. (9.79) remains Gaussian and can be calculated rigorously, using the well-known rule (Wilson & Kogut, 1974; Faddeev & Slavnov, 1980)

$$\frac{\int D\vec{u} \exp\left(-\frac{1}{2}\vec{u}\hat{L}\vec{u} - \vec{f}\vec{u} \right)}{\int D\vec{u} \exp\left(-\frac{1}{2}\vec{u}\hat{L}\vec{u} \right)} = \exp\left(\frac{1}{2}\vec{f}\hat{L}^{-1}\vec{f} \right). \tag{9.90}$$

As a result, the partition function (9.79) can be represented as

$$Z = \int Dh(\vec{r}) \exp\left\{ -\beta\Phi[h(\vec{r})] \right\}, \tag{9.91}$$

with the Hamiltonian Φ depending on the out-of-plane deformations only

$$\Phi = \frac{1}{2}\sum_{\vec{q}} \kappa q^4 |h_{\vec{q}}|^2 + \frac{Y}{8}\sum_{\vec{q}\vec{k}\vec{k}'} R\left(\vec{k},\vec{k}',\vec{q}\right)\left(h_{\vec{k}}h_{\vec{q}-\vec{k}} \right)\left(h_{\vec{k}'}h_{-\vec{q}-\vec{k}'} \right), \tag{9.92}$$

where Y is the two-dimensional Young modulus (9.45) and

$$R\left(\vec{k},\vec{k}',\vec{q}\right) = \frac{\left(\vec{q}\times\vec{k}\right)^2\left(\vec{q}\times\vec{k}'\right)^2}{q^4}. \tag{9.93}$$

The term proportional to h^4 in Eq. (9.92) describes anharmonic effects, or self-interaction of the field $h(\vec{r})$, and Y plays the role of the coupling constant.

Thus, we have the problem of interacting fluctuations where the low-q contribution is dominant, which is reminiscent of the problem of a critical point. The difference is that for two-dimensional systems we have such a critical situation at *any* finite temperature.

The correlation function $G(\vec{q}) = \left\langle \left| h_{\vec{q}} \right|^2 \right\rangle$ satisfies the *Dyson equation*

$$G^{-1}(\vec{q}) = G_0^{-1}(\vec{q}) + \Sigma(\vec{q}),\tag{9.94}$$

where $G_0(\vec{q})$ is given by Eq. (9.81) and the *self-energy* $\Sigma(\vec{q})$ can be calculated using perturbation theory in Y via, e.g., Feynman diagrams. We can introduce the *renormalized bending rigidity* $\kappa_R(q)$ by writing

$$G(q) = \frac{T}{\kappa_R(q)q^4},\tag{9.95}$$

and discuss this quantity. The first-order correction gives us (Nelson & Peliti, 1987)

$$\delta\kappa(q) = \kappa_R(q) - \kappa = \frac{TY}{\kappa}\sum_{\vec{k}}\frac{1}{\left|\vec{q}+\vec{k}\right|^4}\left[\frac{\left(\vec{q}\times\vec{k}\right)^2}{q^2k^2}\right]^2.\tag{9.96}$$

On calculating the integral over $\vec{k}$ we find

$$\delta\kappa(q) = \frac{3TY}{16\pi\kappa q^2}.\tag{9.97}$$

At

$$q \le q^* = \sqrt{\frac{3TY}{16\pi\kappa^2}}\tag{9.98}$$

the correction (9.97) is equal to the bare value of κ or larger than κ, and the perturbation theory is obviously not applicable. The value q^* plays the same role as the "Ginzburg criterion" (Ma, 1976; Landau & Lifshitz, 1980) in the theory of critical phenomena: Below q^* the effects of interactions between fluctuations dominate.

The increase of bending rigidity with increasing temperature has a simple physical explanation. It is known, for the case of a corrugated plate, that corrugations of height $h \gg \Delta$ (Δ is the thickness of the plate) should increase its effective rigidity by a factor $(h/\Delta)^2$ (Briassoulis, 1986; Peng, Liew, & Kitipornchai, 2007). Taking into account Eq. (9.88) (with $L \to 1/q$) and $\Delta \approx a$, we will have an estimation like Eq. (9.97)).

Note that in the theory of liquid membranes, where the Hamiltonian is given by Eq. (9.74) and the in-plane deformations $\vec{u}$ are not relevant, there is also a divergent anharmonic correction to $\kappa_R(q)$, due to higher-order (nonlinear) terms in the expression (9.77) for the mean curvature (Peliti & Leibler, 1985):

$$\delta\kappa \approx -\frac{3T}{4\pi}\ln\left(\frac{1}{qa}\right). \tag{9.99}$$

This term has the opposite sign in comparison with that for a crystalline membrane (9.97) and is much smaller than the latter. Thus, the Hamiltonian (9.92) takes into account the *main* nonlinearities, and "liquid" anharmonicities are not relevant for crystalline membranes.

In the next sections we will discuss how to solve this problem and what the real behavior of fluctuations with $q \leq q^*$ is.

9.4 Scaling properties of membranes and intrinsic ripples in graphene

In situations in which one has strongly interacting long-wavelength fluctuations, scaling considerations are extremely useful (Wilson & Kogut, 1974; Ma, 1976; Patashinskii & Pokrovskii, 1979). Let us assume that the behavior of the renormalized bending rigidity at small q is determined by some exponent η:

$$\kappa_R(q) \sim q^{-\eta}, \tag{9.100}$$

which means

$$G(q) = \left\langle \left|\vec{h}_{\vec{q}}\right|^2 \right\rangle = \frac{A}{q^{4-\eta}q_0^\eta}. \tag{9.101}$$

Here we introduce a parameter

$$q_0 = \sqrt{\frac{Y}{\kappa}} \tag{9.102}$$

of the order of a^{-1} to make A dimensionless. One can also assume a renormalization of effective Lamé constants:

$$\lambda_R(q), \mu_R(q) \sim q^{\eta_u}, \tag{9.103}$$

which means

$$D^{\alpha\beta}(\vec{q}) = \left\langle u_{\alpha\vec{q}}{}^* u_{\beta\vec{q}} \right\rangle \sim \frac{1}{q^{2+\eta_u}}. \tag{9.104}$$

Finally, instead of Eq. (9.88) we assume

$$\langle h^2 \rangle \sim L^{2\zeta}. \tag{9.105}$$

The values η, η_u, and ζ are similar to *critical exponents* in the theory of critical phenomena. They are not independent (Aronovitz & Lubensky, 1988).

First, it is easy to express ζ in terms of η. Substituting Eq. (9.101) into Eq. (9.87) and introducing, as usual, a cut-off at $q_{min} \sim L^{-1}$ we have

$$\zeta = 1 - \frac{\eta}{2}. \tag{9.106}$$

If $\eta > 0$, $\zeta < 1$ and the membrane remains flat (in the sense that its effective thickness $\sqrt{\langle h^2 \rangle}$, is much smaller than L at $L \to \infty$). Also, in the correlation function (9.86), due to Eq. (9.84) and (9.101), there is no divergence from the region of small q:

$$\langle \delta\vec{n}_{\vec{q}} \delta\vec{n}_{-\vec{q}} \rangle \sim \frac{1}{q^{2-\eta}} \tag{9.107}$$

is an integrable singularity.

The relation between η_u and η has been derived by Aronovitz and Lubensky (1988) using quite complicated tools, such as the renormalization group and Ward identities in Feynman-diagram technique. Its meaning is, however, rather elementary and related to the requirement that the deformation tensor has the correct structure (9.62) under the renormalization. This means that the correlation functions of $\partial u_\alpha / \partial x_\beta$ and

$$\frac{\partial h}{\partial x_\alpha} \frac{\partial h}{\partial x_\beta}$$

should have the same exponents. The first one follows immediately from Eq. (9.104):

$$\Gamma_1(\vec{q}) = \left\langle \left(\frac{\partial u_\alpha}{\partial x_\beta} \right)_{-\vec{q}} \left(\frac{\partial u_\alpha}{\partial x_\beta} \right)_{\vec{q}} \right\rangle = q^2 D^{\alpha\alpha}(\vec{q}) \sim q^{-\eta_u}. \tag{9.108}$$

For the second one we have a convolution:

$$\Gamma_2(\vec{q}) = \left\langle \left(\frac{\partial h}{\partial x_\alpha} \frac{\partial h}{\partial x_\beta} \right)_{-\vec{q}} \left(\frac{\partial h}{\partial x_\alpha} \frac{\partial h}{\partial x_\beta} \right)_{\vec{q}} \right\rangle$$

$$= \sum_{\vec{k}_1 \vec{k}_2} k_{1\alpha} \left(q_\beta - k_{1\beta} \right) k_{2\alpha} \left(q_\beta + k_{2\beta} \right) \left\langle h_{-\vec{k}_1} h_{-\vec{q}-\vec{k}_1} h_{-\vec{k}_2} h_{\vec{q}+\vec{k}_2} \right\rangle. \tag{9.109}$$

For free fields we have Wick's theorem, and

$$\langle h_1 h_2 h_3 h_4 \rangle = \langle h_1 h_2 \rangle \langle h_3 h_4 \rangle + \langle h_1 h_3 \rangle \langle h_2 h_4 \rangle + \langle h_1 h_4 \rangle \langle h_2 h_3 \rangle. \tag{9.110}$$

For interacting fields this is no longer the case, and we have some irreducible averages (cumulants). It is supposed in the scaling theory that the scaling properties of these cumulants are the same as those for the "reducible" terms (Patashinskii & Pokrovskii, 1979) and, thus, one can use Eq. (9.110) to calculate the exponents. On substituting Eq. (9.110) into Eq. (9.109) one obtains

$$\Gamma_2(\vec{q}) \sim \sum_{\vec{k}} k^2 \left(\vec{q} - \vec{k}\right)^2 G\left(\vec{k}\right) G\left(\vec{k} - \vec{q}\right). \tag{9.111}$$

Finally, on substituting Eq. (9.101) into Eq. (9.111) we have

$$\Gamma_2(\vec{q}) \sim \frac{1}{q^{2-2\eta}}. \tag{9.112}$$

On comparing Eq. (9.112) with Eq. (9.108), we arrive at the result

$$\eta_u = 2 - 2\eta. \tag{9.113}$$

This exponent is positive if $0 < \eta < 1$ (we will see later that this is the case). This means that, due to interactions between out-of-plane and in-plane phonons, the former become harder but the latter become softer.

The temperature dependence of the constant A in Eq. (9.101) can be found from the assumption that q^* in (9.98) is the only relevant wave vector in the theory and that Eq. (9.81) and (9.101) should match at $q \approx q^*$. The result is (Katsnelson, 2010b):

$$A = \alpha \left(\frac{T}{\kappa}\right)^{\zeta}, \tag{9.114}$$

where α is a dimensionless factor of the order of 1.

Before discussing how to calculate the exponent η, it is worth returning to the Mermin–Wagner theorem about the impossibility of long-range crystal order in two-dimensional systems at finite temperatures.

The true manifestation of long-range order is the existence of delta-functional (Bragg) peaks in diffraction experiments; see, e.g., the discussion in Irkhin and Katsnelson (1986). The scattering intensity is proportional to the static structural factor

$$S(\vec{q}) = \sum_{mn'} \sum_{jj'} \left\langle \exp\left[i\vec{q}(\vec{R}_{nj} - \vec{R}_{n'j'})\right]\right\rangle. \tag{9.115}$$

Using Eq. (9.1) and (9.2) the expression (9.115) can be rewritten as

$$S(\vec{q}) = \sum_{nn'} \exp\left[i\vec{q}(\vec{r}_n - \vec{r}_{n'})\right] \sum_{jj'} \exp\left[i\vec{q}\left(\vec{\rho}_j - \vec{\rho}_{j'}\right)\right] W_{nj,n'j'}, \tag{9.116}$$

where

$$W_{nj,n'j'} = \left\langle \exp\left[i\vec{q}(\vec{u}_{nj} - \vec{u}_{n'j'}) \right] \right\rangle. \tag{9.117}$$

Eq. (9.115) and (9.117) are written for the classical case, in which $\vec{u}_{nj}$ are not operators but just classical vectors; for a more detailed discussion of the scattering problem in crystal lattices, see Vonsovsky and Katsnelson (1989) and Katsnelson and Trefilov (2002).

In conventional three-dimensional crystals, one can assume that the displacements $\vec{u}_{nj}$ and $\vec{u}_{n'j'}$ are not correlated, and

$$W_{nj,n'j'} = \left\langle \exp\left(i\vec{q}\vec{u}_{nj}\right) \right\rangle \left\langle \exp\left(-i\vec{q}\vec{u}_{n'j'}\right) \right\rangle \equiv m_j(\vec{q})m_{j'}^*(\vec{q}) \tag{9.118}$$

when $|\vec{r}_n - \vec{r}_{n'}| \to \infty$; here $m_j(\vec{q})$ are Debye–Waller factors that are independent of n due to translational invariance. Therefore, for $\vec{q} = \vec{g}$ (reciprocal lattice vectors), where $\exp(i\vec{q}\vec{r}_n) = 1$, the contribution to $S(\vec{q})$ is proportional to N_0^2, whereas for a generic $\vec{q}$ it is of the order of N_0. The Bragg peaks $\vec{q} = \vec{g}$ are, therefore, sharp; thermal fluctuations decrease their intensity (by the Debye–Waller factor) but do not broaden the peaks. The observation of such sharp Bragg peaks is an experimental manifestation of the existence of long-range crystal order. In the two-dimensional case, the correlation functions of atomic displacements do not vanish at $|\vec{r}_n - \vec{r}_{n'}| \to \infty$. Indeed, in the continuum limit $\vec{u}_{nj} \to (\vec{u}(\vec{r}), h(\vec{r}))$, where $\vec{u}$ is already a two-dimensional vector, and

$$\left\langle [h(\vec{r}) - h(\vec{r}')]^2 \right\rangle = 2\sum_{\vec{q}} \left\langle |h_{\vec{q}}|^2 \right\rangle \{ 1 - \exp\left[i\vec{q}(\vec{r} - \vec{r}') \right] \} \sim |\vec{r} - \vec{r}'|^{2\varsigma}, \tag{9.119}$$

$$\left\langle [\vec{u}(\vec{r}) - \vec{u}(\vec{r}')]^2 \right\rangle = 2\sum_{\vec{q}} \left\langle |\vec{u}_{\vec{q}}|^2 \right\rangle \{ 1 - \exp\left[i\vec{q}(\vec{r} - \vec{r}') \right] \} \sim |\vec{r} - \vec{r}'|^{\eta_u} \tag{9.120}$$

after substitutions of Eq. (9.101) and (9.103) (Abraham & Nelson, 1990). This means that the approximation (9.118) does not work.

To estimate the structural factor near the Bragg peak, $\vec{q} = \vec{g} + \delta\vec{q}$ we can use the identity

$$\left\langle e^c \right\rangle = \exp\left[\frac{1}{2}\left\langle c^2 \right\rangle \right] \tag{9.121}$$

for the correlation function (9.117). Strictly speaking, it follows from Wick's theorem and, therefore, is exact only in the harmonic approximation (Vonsovsky & Katsnelson, 1989) but, as was discussed previously, should give us correct scaling properties. Therefore,

$$W_{nj,n'j'} \sim \exp\left[-\alpha_1 q_\parallel^2 |\vec{r} - \vec{r}'|^{\eta_u} - \alpha_2 (\delta\vec{q}_\perp)^2 |\vec{r} - \vec{r}'|^{2\zeta}\right], \tag{9.122}$$

where $\vec{q}_\parallel$ and $\vec{q}_\perp$ are components of the scattering vector parallel and perpendicular to the crystal plane and we take into account that $\vec{g}_\perp = 0$.

On substituting Eq. (9.122) into Eq. (9.116) one can see that the sum over n' at a given n is convergent, and $S(\vec{q} = \vec{g}) \sim N_0$. Thus, instead of a delta-functional Bragg peak (in the thermodynamic limit) we have a sharp maximum of finite width at $\delta\vec{q} \to 0$ (Abraham & Nelson, 1990). This means that, rigorously speaking, the statement that two-dimensional crystals cannot exist at finite temperatures (Peierls, 1934, 1935; Landau, 1937) is correct. However, the structural factor still can have *very* sharp maxima $\vec{q} = \vec{g}$, and the crystal lattice can be restored from the positions of these maxima. In this (restricted) sense, two-dimensional crystals do exist, and graphene is a prototype example.

It was found experimentally by electron diffraction, namely by transmission electron microscopy, that freely suspended graphene at room temperature is rippled; that is, it exhibits corrugations in the out-of-plane direction (Meyer et al., 2007a, 2007b). The existence of these intrinsic, thermally induced ripples in graphene has been confirmed by atomistic Monte Carlo simulations that use the potential LCBOPII mentioned earlier in Section 9.1 (Fasolino, Los, & Katsnelson, 2007). A typical snapshot is shown in Fig. 9.3. Further detailed studies of the correlation function $G(q)$ by such simulations have been performed for single-layer graphene by Los et al. (2009) and Zakharchenko et al. (2010b) and for bilayer graphene by Zakharchenko et al. (2010a).

According to these simulations, at some intermediate value of q, roughly between 0.1 Å^{-1} and 1 Å^{-1} (for the case of room temperature $T = 300$K), the

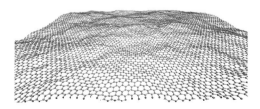

Fig. 9.3 A typical atomic configuration (from atomistic Monte Carlo simulations) for graphene at room temperature.
(Courtesy of A. Fasolino.)

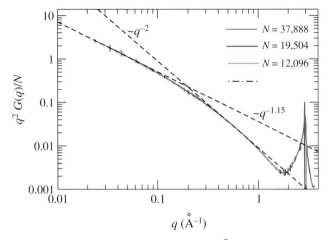

Fig. 9.4 The normal–normal correlation function $q^2\, G(q)$ found from atomistic Monte Carlo (MC) simulations for three samples with indicated number of atoms N.
(Reproduced with permission from Los et al., 2009.)

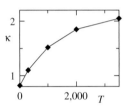

Fig. 9.5 Temperature dependence of the bending rigidity found by fitting $G(q)$ to Eq. (9.81) for $q > q^*$.
(Reproduced with permission from Katsnelson and Fasolino, 2013.)

correlation function $G(q)$ follows the harmonic approximation (9.81) (Fig. 9.4). From the slope of this dependence, one can extract $\kappa \approx 1.1$ eV, which means that graphene at room temperature should be considered a rather hard membrane ($\kappa/T \approx 40$). With the temperature increase, the bending rigidity of graphene grows, as shown in Fig. 9.5. For $q > 1$ Å^{-1} the continuum-medium approximation does not work, and $G(q)$ increases due to closeness to the Bragg peak. At $q \approx q^* \approx 0.2$ Å^{-1} there is a crossover to the behavior described by Eq. (9.101), with

$$\eta \approx 0.85. \tag{9.123}$$

This value is quite close to that predicted by functional renormalization-group analysis of the model (9.92) (Kownacki & Mouhanna, 2009). Thus, both the continuum model and atomistic simulations predict a rather broad, power-law distribution of intrinsic ripples in graphene, without any dominant spatial scale. Ripples in graphene on a substrate will be discussed in Chapter 11, in relation to scattering mechanisms involved in electron transport.

Other evidence for thermally introduced ripples and their effects on thermo-dynamic and mechanical properties will be considered in Sections 9.6 and 9.7.

According to the Monte Carlo simulations, the in-plane and out-of-plane atomic displacements are strongly correlated at $q < q^*$ and uncorrelated for $q > q^*$, in agreement with our general picture (Fig. 9.6).

One needs to make one important comment about the model (9.63) (or, equivalently, (9.92)). In this model of a so-called *phantom membrane*, there is a phase transition at $T \approx \kappa$ to a crumpled phase (Nelson, Piran, & Weinberg, 2004). There are some arguments, however, in favor of the view that this transition is suppressed, and the low-temperature (quasi-) flat phase is stabilized at any temperature if one adds a condition of avoided self-crossing (short-range repulsion forces). It is also assumed that the scaling properties of the (quasi-) flat phase are the same for "phantom" and "real" membranes (Nelson, Piran, & Weinberg, 2004). Anyway, the regime $T \approx \kappa \approx 10^4 \text{K}$ is obviously not reachable for graphene. What happens with graphene with increasing temperature will be discussed in Section 9.6.

To finish this section, let us discuss the case of bilayer graphene. Intrinsic ripples in bilayer graphene were observed experimentally (Meyer et al., 2007b) and studied theoretically (Zakharchenko et al., 2010b). The main difference from the case of single-layer graphene can be seen even at the level of the harmonic approximation for the bilayer membrane. Instead of Eq. (9.57) (or (9.73)) we have

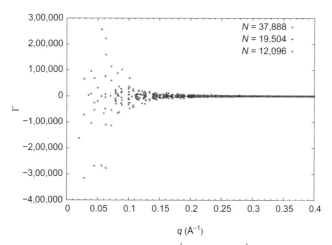

Fig. 9.6 The correlation function $\Gamma_{\vec{q}} = \left\langle (u_x)_{\vec{q}} (h^2)_{-\vec{q}} \right\rangle$ found from atomistic Monte Carlo simulations for three samples with indicated number of atoms N. (Reproduced with permission from Katsnelson and Fasolino, 2013.)

$$F_b = \frac{1}{2} \int d^2x \left[\kappa (\nabla^2 h_1)^2 + \kappa (\nabla^2 h_2)^2 + 2\gamma (\delta h)^2 \right], \tag{9.124}$$

where h_1 and h_2 are out-of-plane deformations in each plane, κ is the bending rigidity per layer

$$\delta h = h_1 - h_2, \tag{9.125}$$

and γ describes a relatively weak van der Waals interaction between the layers. By introducing an average displacement

$$h = \frac{h_1 + h_2}{2} \tag{9.126}$$

one can rewrite Eq. (9.124) as

$$F_b = \frac{1}{2} \int d^2x \left[2\kappa (\nabla^2 h)^2 + \frac{\kappa}{2} (\nabla^2 \delta h)^2 + 2\gamma (\delta h)^2 \right], \tag{9.127}$$

and thus we have, in the harmonic approximation, instead of Eq. (9.81)

$$\left\langle |h_q|^2 \right\rangle = \frac{T}{2\kappa q^4}, \tag{9.128}$$

$$\left\langle |\delta h_q|^2 \right\rangle = \frac{T}{\frac{1}{2}\kappa q^4 + 2\gamma}. \tag{9.129}$$

Atomistic simulations (Zakharchenko et al., 2010a) give, at room temperature, $\gamma \approx 0.025$ eVÅ^4. At

$$q < q_c = \sqrt[4]{\frac{4\gamma}{\kappa}} \tag{9.130}$$

the correlation function (9.129) goes to a constant. In this regime, a bilayer behaves like a single membrane with bending rigidity twice as large as that for a single layer (see Eq. (9.128)). At $q > q_c$ the layers fluctuate more or less independently. The simulations (Zakharchenko et al., 2010a) qualitatively confirm this simple picture; the wavelength of fluctuations at which the crossover happens is about $2\pi/q^* \approx 2$ nm (at room temperature).

This is, however, not the complete story on the bending rigidity of multilayered membranes. Indeed, according to the earlier argument, the effective bending rigidity should grow linearly with the number of layers N. At the same time, phenomenological expression (9.58) shows that $\kappa \propto \Delta^3 \propto N^3$. The crossover between these two regimes was considered by de Andres, Guinea, and Katsnelson (2012). It turns out that the effective bending rigidity of multilayered crystalline membrane is strongly dependent on the wave vector even in harmonic approximation, due to hybridization of in-plane and out-of-plane phonon modes. It interpolates from the N^3 behavior in the limit $q \to 0$ to the linear-in-N for larger wave vectors.

9.5 The self-consistent screening approximation

There are several ways to calculate the exponent η analytically with reasonable accuracy. The simplest approximation is to rewrite Eq. (9.96) in a self-consistent way:

$$\kappa_R(q) = \kappa + TY \sum_{\vec{k}} \frac{1}{\kappa_R\left(\left|\vec{k}+\vec{q}\right|\right)\left|\vec{k}+\vec{q}\right|^4} \left[\frac{\left(\vec{q}\times\vec{k}\right)^2}{q^2 k^2} \right]^2 , \qquad (9.131)$$

assuming that the Young modulus Y is not renormalized (Nelson & Peliti, 1987). On substituting Eq. (9.100) into Eq. (9.131) we find $-\eta = \eta - 2$, or $\eta = 1$.

A more accurate result is given by the *self-consistent screening approximation* (SCSA) (Le Doussal & Radzihovsky, 1992; see also Xing et al., 2003; Gazit, 2009; Zakharchenko et al., 2010b; Roldán et al., 2011; Le Doussal & Radzihovsky, 2018).

The Hamiltonian (9.92) describes the self-interaction of a classical field $h(\vec{r})$ with the momentum-dependent interaction vertex $YR\left(\vec{k},\vec{k}',\vec{q}\right)$. To consider the effects of the interaction one can use a Feynman-diagram technique similar to that used in the theory of critical phenomena (Wilson & Kogut, 1974; Ma, 1976). The basic elements are the Green function $G(\vec{q})$ (solid thick line, in contrast with the solid thin line for $G_0(\vec{q})$ and the interaction vertex (the dashed line), see Fig. 9.7(a). The exact and bare Green functions are related by the Dyson equation (9.94), where, in the lowest order of the perturbation theory, the self-energy $\Sigma\left(\vec{k}\right)$ is given by the diagram shown in Fig. 9.7(b). Its analytic expression corresponds to Eq. (9.96). The next step corresponds to the replacement of $G_0(\vec{q})$ by $G(\vec{q})$ (Fig. 9.7(c)), which corresponds to the Eq. (9.131) and gives $\eta = 1$, as discussed previously.

The SCSA corresponds to the summation of "ladder" diagrams shown in Fig. 9.7(d). This infinite summation is just a geometric progression, with the result

$$1 + \hat{A} + \hat{A}^2 + \cdots = \frac{1}{1-\hat{A}}. \qquad (9.132)$$

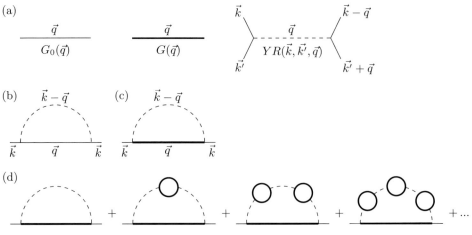

Fig. 9.7 (a) Basic elements of the diagram technique (see the text). (b) The lowest-order perturbation expression for the self-energy corresponding to Eq. (9.96). (c) The self-consistent version of the previous diagram corresponding to Eq. (9.131). (d) The diagram summation equivalent to the SCSA.

The answer is (Le Doussal & Radzihovsky, 1992; Le Doussal & Radzihovsky, 2018)

$$\Sigma(\vec{q}) = \int \frac{d^2\vec{k}}{(2\pi)^2} \frac{Y_{ef}\left(\vec{k}\right)}{T} \left[\frac{\left(\vec{q} \times \vec{k}\right)^2}{k^2}\right]^2 G\left(\vec{k} - \vec{q}\right), \qquad (9.133)$$

where

$$Y_{ef}\left(\vec{k}\right) = \frac{Y}{1 + \dfrac{Y}{2T} I\left(\vec{k}\right)}, \qquad (9.134)$$

$$I\left(\vec{k}\right) = \int \frac{d^2\vec{p}}{(2\pi)^2} \left(\frac{\left(\vec{k} \times \vec{p}\right)^2}{k^2}\right)^2 G(\vec{p})G\left(\vec{k} - \vec{p}\right). \qquad (9.135)$$

Eq. (9.134) and (9.135) describe renormalization of the Young modulus as a result of summation of the infinite series of diagrams according to Eq. (9.132).

Of course, the summation shown in Fig. 9.7(d) is not exact. This approximation was introduced by Bray (1974) in the context of the theory of critical phenomena for an n-component order parameter. It can be justified rigorously if $n \gg 1$. In our problem, the number of components of the field $h(\vec{r})$ is $n = 1$; therefore, the applicability of the SCSA is not clear. The reasonable agreement with the Monte

Carlo simulations (Zakharchenko et al., 2010b) and an explicit analysis of the higher-order diagrams (Gazit, 2009) justify it as a reasonable, relatively simple, approximation in the theory of fluctuating membranes.

Let us consider Eq. (9.133) through (9.135) in the limit of small q, assuming that $\Sigma(q) \gg G_0^{-1}(q)$ and using Eq. (9.101) for the Green function. Thus,

$$I(\vec{k}) = \frac{A^2}{q_0^{2\eta}} \int \frac{d^2\vec{p}}{(2\pi)^2} \left(\frac{\left(\vec{k} \times \vec{p}\right)^2}{k^2} \right)^2 \frac{1}{p^{4-\eta}\left|\vec{k} - \vec{p}\right|^{4-\eta}} = \frac{A^2}{q_0^{2\eta} k^{2-2\eta}} I_1(\eta), \qquad (9.136)$$

where

$$I_1(\eta) = \int \frac{d^2\vec{x}}{(2\pi)^2} \frac{(\vec{x} \times \vec{x}_0)^4}{x^{4-\eta}|\vec{x} - \vec{x}_0|^{4-\eta}} \qquad (9.137)$$

and $\vec{x}_0 = (1, 0)$. The expression (9.136) diverges at $k \to 0$ and, therefore, one can neglect 1 in the denominator of Eq. (9.134), assuming

$$Y_{ef}\left(\vec{k}\right) \approx \frac{2T}{I(\vec{k})} = \frac{2Tq_0^{2\eta}}{A^2} \frac{k^{2-2\eta}}{I_1(\eta)}. \qquad (9.138)$$

On substituting Eq. (9.138) into Eq. (9.133) we have

$$\frac{q^{4-\eta}q_0^{\eta}}{A} = \frac{2q_0^{\eta}q^{4-\eta}}{AI_1(\eta)} I_2(\eta) \qquad (9.139)$$

where

$$I_2(\eta) = \int \frac{d^2x}{(2\pi)^2} \frac{x^{2-2\eta}(\vec{x} \times \vec{x}_0)^4}{|\vec{x} - \vec{x}_0|^{4-\eta}}. \qquad (9.140)$$

Eq. (9.139) is satisfied at arbitrary A, and η can be found from the equation

$$I_1(\eta) = 2I_2(\eta). \qquad (9.141)$$

The integrals I_1 and I_2 can be expressed via a Γ-function and calculated explicitly (Le Doussal & Radzihovsky, 1992; Gazit, 2009; Le Doussal & Radzihovsky, 2018). The answer is

$$\eta = \frac{4}{1 + \sqrt{15}} \approx 0.821, \qquad (9.142)$$

which is not far from the more accurate value $\eta \approx 0.85$ discussed in the previous section (Eq. (9.123)).

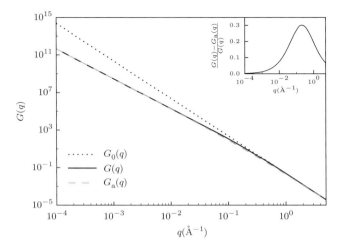

Fig. 9.8 The correlation function $G(q)$ calculated in self-consistent screening approximation for $T = 300$K and parameters characteristic of graphene. One can see that the interpolation formula (9.143) has a pretty high accuracy. (Courtesy of A. Mauri.)

To find $G(q)$ for the whole range of q one needs to solve Eq. (9.133) through (9.135) numerically. The results shown in Fig. 9.8 are in reasonable agreement with the Monte Carlo simulations.

Keeping in mind possible applications, it is worth mentioning that $G(q)$ for all q can be approximated as an interpolation between the high-q limit (9.81) and the low-q limit (9.101) and (9.114):

$$G_a^{-1}(q) = \frac{\kappa q^4}{T} + \left(\frac{\kappa}{T}\right)^{1-\frac{\eta}{2}} \frac{q_0{}^\eta q^{4-\eta}}{\alpha} \qquad (9.143)$$

for some numerical factor α. This fitting is also shown in Fig. 9.8.

The SCSA can also be used to consider the effects of an external stress $\sigma_{\alpha\beta}^{\text{ext}}$ on the properties of membranes (Roldán et al., 2011). The former can be described as an additional term in Eq. (9.63):

$$F = \frac{1}{2}\int d^2x \left\{\kappa\left(\nabla^2 h\right)^2 + \lambda(u_{\alpha\alpha})^2 + 2\mu u_{\alpha\beta} u_{\alpha\beta} + \sigma_{\alpha\beta}^{\text{ext}} u_{\alpha\beta}\right\}, \qquad (9.144)$$

where

$$\sigma_{\alpha\beta}^{\text{ext}} = \lambda\delta_{\alpha\beta} u_{\gamma\gamma}^{\text{ext}} + 2\mu u_{\alpha\beta}^{\text{ext}} \qquad (9.145)$$

can be expressed in terms of an *external* strain tensor $u_{\alpha\beta}^{\text{ext}}$. By substituting Eq. (9.62) into Eq. (9.144) one can see that, in the harmonic approximation, the bare Green function (9.81) is modified as follows:

$$G_0(\vec{q}) = \frac{T}{q^2 \left(\kappa q^2 + \lambda u_{\alpha\alpha}{}^{ext} + 2\mu u_{\alpha\beta}{}^{ext} \frac{q_\alpha q_\beta}{|\vec{q}|^2} \right)}. \tag{9.146}$$

Assuming for simplicity the case of isotropic external deformation,

$$u_{\alpha\beta}^{ext} = u \delta_{\alpha\beta}, \tag{9.147}$$

we have

$$G_0(\vec{q}) = \frac{T}{q^2 [\kappa q^2 + 2(\lambda + \mu)u]} \tag{9.148}$$

where we consider only the case of expansion ($u > 0$); the effect of compression on the membrane is actually very complicated (Moldovan & Golubovic, 1999; Sharon et al., 2002; Cerda & Mahadevan, 2003; Brau et al., 2011). One can see that flexural fluctuations are suppressed by the strain at

$$q < q_u = q_0 u^{1/2} \tag{9.149}$$

(see Eq. (9.102)). If $q_u \geq q^*$, that is,

$$u \geq 0.1 \frac{T}{\kappa} \tag{9.150}$$

(see Eq. (9.98)), the anharmonic effects are assumed to be strongly suppressed, and the harmonic approximation (9.148) should work up to $q \to 0$. This conclusion will be important for our discussion of the transport properties of freely suspended graphene flakes in Chapter 11.

9.6 Thermodynamic and other thermal properties of graphene

The existence of the soft acoustic flexural (ZA) mode (9.15) and the related tendency to intrinsic ripple formation is crucial to the thermodynamic properties of graphene, first of all, to its thermal expansion.

In the quasiharmonic approximation, the lattice thermodynamic properties are assumed to be described by harmonic expressions but with phonon frequencies ω_λ, dependent on the lattice constant. In this approximation, the thermal expansion coefficient

$$\alpha_p = \frac{1}{\Omega} \left(\frac{\partial \Omega}{\partial T} \right)_p \tag{9.151}$$

(where Ω is the volume for three-dimensional crystals and area for two-dimensional ones; p is the pressure) is given by the Grüneisen law (Vonsovsky & Katsnelson, 1989; Katsnelson & Trefilov, 2002)

$$\alpha_p = \frac{\gamma C_V(T)}{\Omega B_T}, \tag{9.152}$$

where B_T is the isothermal bulk modulus

$$C_V(T) = \sum_{\lambda} C_{\lambda}, \tag{9.153}$$

where

$$C_{\lambda} = \left(\frac{\hbar\omega_{\lambda}}{T}\right)^2 \frac{\exp\left(\frac{\hbar\omega_{\lambda}}{T}\right)}{\left[\exp\left(\frac{\hbar\omega_{\lambda}}{T}\right) - 1\right]^2}, \tag{9.154}$$

is the constant-volume heat capacity in the harmonic approximation, and

$$\gamma = \frac{\sum_{\gamma} \gamma_{\lambda} C_{\lambda}}{\sum_{\gamma} C_{\lambda}} \tag{9.155}$$

is the *macroscopic Grüneisen parameter*, where

$$\gamma_{\lambda} = -\frac{\partial \ln \omega_{\lambda}}{\partial \ln \Omega} \tag{9.156}$$

are *microscopic Grüneisen parameters*.

Graphite is known to have a negative thermal expansion coefficient up to 700 K (Steward, Cook, & Kellert, 1960). This behavior has been explained, in terms of the Grüneisen law, by Mounet and Marzari (2005) via density-functional calculations of ω_{λ} and γ_{λ}. It turns out that the Grüneisen parameters γ_{λ} are negative, both in graphene and in graphite, for ZA phonons over the whole Brillouin zone. The same results follow from atomistic simulations with the LCBOPII potential (Karssemeijer & Fasolino, 2011), see Fig. 9.9. The theory explained the change in sign of α_p at $T \approx 700$ K for the case of graphite and predicted that $\alpha_p < 0$ at all temperatures for the case of graphene. Negative thermal expansion of graphene at room temperature and slightly above has been confirmed experimentally by Bao et al. (2009). The linear thermal expansion coefficient at these temperatures was about -10^{-5} K^{-1}, a very large negative value. According to the quasiharmonic theory of Mounet and Marzari (2005), it was supposed to be more or less constant up to temperatures of the order of at least 2,000 K.

However, straightforward Monte Carlo atomistic simulation with the LCBOPII potential, not assuming the quasiharmonic approximation (Zakharchenko,

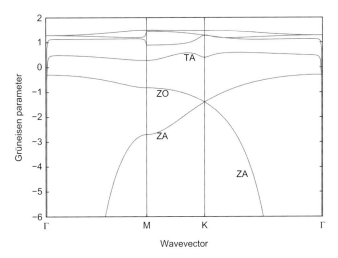

Fig. 9.9 Grüneisen parameters calculated in graphene with the potential LCBOBII.
(Reproduced with permission from Katsnelson & Fasolino, 2013.)

Katsnelson, & Fasolino, 2009), gave an essentially different result (see Fig. 9.10). One can see that, according to this calculation, α_p is supposed to change sign at $T \approx 700\text{--}900$ K. Later, it was confirmed experimentally that α_p, while remaining negative, decreases in modulus with increasing temperature up to 400 K (Yoon, Son, & Cheong, 2011). This temperature dependence of $\alpha_p(T)$, beyond the quasi-harmonic approximation, is a true anharmonic effect.

Similar calculations for the case of bilayer graphene have been performed by Zakharchenko et al. (2010a). The results (Fig. 9.10) show that the change of sign of da/dT happens at lower temperatures than for the case of single-layer graphene and that in this sense bilayer graphene should be similar to graphite. The thermal expansion perpendicular to the graphene plane turns out to be positive, $dc/dT > 0$.

The Lamé constants λ and μ have also been found from atomistic simulations (Zakharchenko, Katsnelson, & Fasolino, 2009). The room-temperature values of the elastic constants are

$$\mu \approx 10 \; \text{eVÅ}^{-2}, \quad B \approx 12 \; \text{eVÅ}^{-2}, \quad \nu \approx 0.12. \tag{9.157}$$

The calculated Young modulus (9.45) lies within the error bars of the experimental value $Y \approx 340 \pm 50 \; \text{Nm}^{-1}$ (Lee et al., 2008). Note that, per atomic layer, it is an order of magnitude higher than that of steel. However, this is correct only when we are talking about the bare values of the elastic constants. As one can see from Eq. (9.103) and will be discussed in detail in the next section, the renormalization effects are very important. Strictly speaking, at finite temperatures and in the limit

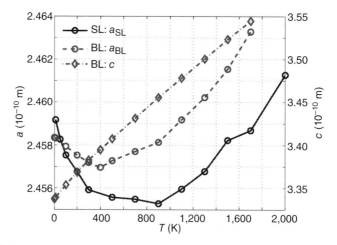

Fig. 9.10 Temperature dependences of the lattice constant a for single-layer (SL) and bilayer (BL) graphene and of the interlayer distance c in interlayer graphene. (Reproduced with permission from Zakharchenko et al., 2010a.)

of infinite sample size ($L \rightarrow \infty$, $q \rightarrow 0$), the effective moduli of single-atom crystalline membrane tend to zero.

Now consider the thermal expansion of graphene at low temperatures. Negative Grüneisen parameter for the ZA phonons can be derived analytically (de Andres, Guinea, & Katsnelson, 2012). Taking into account Eq. (9.144) through (9.148) one can see that in the presence of uniform expansion u, instead of Eq. (9.61), the square frequency of ZA modes reads

$$\omega^2 = \frac{\kappa q^4 + Buq^2}{\rho}. \tag{9.158}$$

Substituting Eq. (9.158) into Eq. (9.156) one obtains

$$\gamma_{\vec{q}} = -\frac{B}{2\kappa q^2}. \tag{9.159}$$

This expression is negative and, importantly, divergent at $q \rightarrow 0$. The integral over the wave vectors in Eq. (9.152) is divergent at the lowest limit. To regularize it, one needs to take into account the renormalization of the bending rigidity as described in the previous section (see, e.g., Eq. (9.143)). With the logarithmic accuracy, the result is

$$\alpha_p \approx -\frac{1}{4\pi\kappa} \int_{q^*}^{q_T} \frac{dq}{q} = -\frac{1}{8\pi\kappa} \ln \frac{T}{\hbar\omega^*} \approx -\frac{1}{16\pi\kappa} \ln \frac{\kappa^3 \rho}{\hbar^2 Y^2}, \tag{9.160}$$

where q^* is the crossover wave vector (9.98),

$$\hbar\omega^* = \hbar\sqrt{\frac{\kappa}{\rho}q^{*2}} = \frac{3T}{16\pi}\frac{Y}{\sqrt{\kappa^3\rho}}, \tag{9.161}$$

the corresponding photon energy and q_T is determined by the condition

$$\hbar\sqrt{\frac{\kappa}{\rho}q_T^2} = T. \tag{9.162}$$

Importantly,

$$\hbar\omega^* \sim \sqrt{\frac{m}{M}}T \ll T \tag{9.163}$$

and therefore $q^* \ll q_T$ (m and M are electron and atomic masses, respectively). The expression (9.160) gives a very accurate estimate of the experimental thermal expansion coefficient of the order of -10^{-5} K^{-1}.

Due to thermodynamic identity (Landau & Lifshitz, 1980)

$$\left(\frac{\partial\Omega}{\partial T}\right)_p = -\left(\frac{\partial S}{\partial p}\right)_T \tag{9.164}$$

and the third law of thermodynamics (the entropy $S \to 0$ at $T \to 0$), one can expect that the thermal expansion coefficient should vanish at zero temperature. The expression (9.160) does not satisfy this requirement. Usually the third law of thermodynamics is protected by quantum statistics of relevant elementary excitations (in our case, phonons) but according to divergent Grüneisen parameter (9.159), the main contribution to the thermal expansion coefficient follows from the classical region (Eq. (9.163)); taking into account anharmonic effects does not help (de Andres, Guinea, & Katsnelson, 2012). To solve the problem and to find the behavior of the thermal expansion at low temperatures, one needs to have the theory of *quantum* membranes, which is still in its infancy. Burmistrov et al. (2016) has shown that at $T \to 0$ the thermal expansion coefficient vanishes very slowly

$$\alpha_p \sim |\ln T|^{-a} \tag{9.165}$$

$(0 < a < 1)$ but the issue certainly deserves more investigation, both experimentally and theoretically.

Let us again come back to the case of high temperatures and consider another high-temperature anharmonic effect, the growth of the heat capacity with the temperature beyond the Dulong–Petit value $3R$

$$C_V(T) = 3R\left(1 + \frac{T}{E_0}\right) \tag{9.166}$$

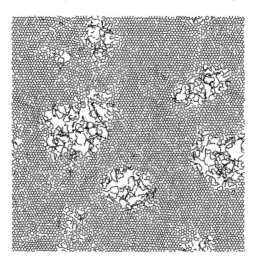

Fig. 9.11 A typical atomic configuration of graphene at $T = 5,000$ K from atomistic Monte Carlo simulations.
(Courtesy of K. Zakharchenko.)

(Katsnelson & Trefilov, 2002). The atomistic simulations (Zakharchenko, Katsnelson & Fasolino, 2009) confirm this behavior, with $E_0 \approx 1.3$ eV.

At high enough temperature, graphene is destroyed. This process was studied by atomistic Monte Carlo simulations in Zakharchenko et al. (2011) and Los et al. (2015). In these simulations the temperature of destruction of graphene was estimated as 4,500 K, which makes graphene probably the most refractory material (it is 210 K higher than the melting temperature of bulk graphite). The word "destruction" is used instead of "melting" to stress that it is a rather peculiar process, leading to the formation of carbon chains, with these chains being strongly entangled and forming something like a polymer melt, rather than a simple liquid (Fig. 9.11).

Probably the most interesting thermal property of graphene, in view of potential applications, is its extraordinarily high thermal conductivity (Balandin et al., 2008; Ghosh et al., 2010; Balandin, 2011). Usually, solids with high thermal conductivity are metals, and the thermal conductivity is determined by conduction electrons, whereas the phonon contribution is negligible (for a general theory of phonon thermal conductivity, see Ziman, 2001). Carbon materials (diamond, nanotubes, and graphene) are exceptional. Their thermal conductivity, being of phonon origin, can be higher than for any metal (for a review, see Balandin, 2011). The very general reason is the high phonon group velocity, due to the very strong chemical bonding and the relatively low mass of the carbon nucleus. Currently, graphene has the largest thermal conductivity among all known materials (Balandin et al., 2008).

The theory of this thermal conductivity was considered in Ghosh et al. (2010). It is a complicated phenomenon, which is not yet fully understood (in particular, the role of flexural phonons needs to be clarified). Its practical importance is related to the problem of heat removal in the electronics industry.

9.7 Mechanical properties of graphene

It follows already from Eq. (9.100) to (9.103) that neither Hooke's law (9.33) nor Föppl–von Karman equations (9.65) through (9.67) are valid for atomically thin membranes at finite temperatures. Indeed, q-dependence of elastic constants means spatial dispersion, that is, nonlocality of relations between stress and deformation (similar to nonlocality of relations between electric induction and electric field in electrodynamics of continuous media (Landau & Lifshitz, 1984). Also, the condition (9.149) means that the thermal fluctuations can be switched off by external deformation, which assumes essential nonlinearity of relations between stress and deformation. Experimental data show that the effective bending rigidity of graphene can be modified in orders of magnitude via the change of sample size L (Blees et al., 2015), and the effective Young modulus can be modified in several times by relatively small disorder (López-Polín et al., 2015) or by external deformation and by change of temperature (Nicholl et al., 2015). We still have no a complete theory that can replace the Föppl–von Karman theory for ultrathin membranes; there are just first attempts in this direction (Košmrlj & Nelson, 2016; Los, Fasolino, & Katsnelson, 2016; Bowick et al., 2017; Los, Fasolino, & Katsnelson, 2017). They include some simple scaling considerations and atomistic simulations. Here, we will review the corresponding results.

First of all, let us introduce characteristic lengths, which determine regions of different mechanical behavior (Košmrlj & Nelson, 2016). In the absence of extrenal stress, there are two characteristic scales: the sample size L and the thermal length

$$L_{th} = \frac{\pi}{q^*} = \sqrt{\frac{16\pi^3\kappa^2}{3TY}} \tag{9.167}$$

(see Eq. (9.98)). Since the revelant wave vectors at the deformation of a plate of the size L are of the order of π/L one can assume that for the case $L < L_{th}$ the in-plane and out-of-plane phonons are decoupled (see Fig. (9.6)) and the elastic constants are more or less equal to their bare values (for the case of graphene, (9.157)). For the opposite case $L > L_{th}$ the effective bending rigidity should be enhanced (see Eq. (9.100)) and the effective Lamé constants, and therefore Young modulus, are suppressed (see Eq. (9.103)).

In the presence of external stress, the flexural fluctuations and anharmonic effects are suppressed (see Eq. (9.149)). Looking at Eq. (9.146) and (9.148) and replacing their bending rigidity by its renormalized value (9.100) one can see that the change of the regime happens when

$$\kappa_R(q)q^4 \approx \sigma q^2, \tag{9.168}$$

or $q \approx \pi/L_\sigma$ where

$$L_\sigma = \left(\frac{\kappa}{(L_{th})^\eta \sigma} \right)^{1/(2-\eta)} = L_{th} \left(\frac{3TY}{16\pi^3 \sigma \kappa} \right)^{1/(2-\eta)}, \tag{9.169}$$

which represents the other characteristic length scale in the system (Roldán et al., 2011; Košmrlj & Nelson, 2016). Detailed scaling analysis of various regimes (Košmrlj & Nelson, 2016) leads to the following answer for the renormalized elastic constants:

$$\frac{\kappa_R(L)}{\kappa} \sim \begin{cases} 1, & L < L_{th} \\ (L/L_{th})^\eta, & L_{th} < L < L_\sigma \\ (L_\sigma/L_{th})^\eta \ln(L/L_\sigma), & L_\sigma < L \end{cases}, \tag{9.170}$$

$$\frac{\lambda_R(L)}{\lambda}, \frac{\mu_R(L)}{\mu}, \frac{Y_R(L)}{Y} \sim \begin{cases} 1, & L < L_{th} \\ (L/L_{th})^{-\eta_u}, & L_{th} < L < L_\sigma \\ (L_\sigma/L_{th})^{-\eta_u}, & L_\sigma < L \end{cases}. \tag{9.171}$$

Scaling behavior of mechanical parameters of graphene was also studied by atomistic simulations with the potential LCBOBII (Los, Fasolino, & Katsnelson, 2016; Los, Fasolino, & Katsnelson, 2017). First of all, the results do confirm the renormalization of the effective Lamé constants in agreement with Eq. (9.103) and (9.113), as one can see from Fig. 9.12.

Under the stress the dependence of the effective Young modulus on the sample size at a given temperature (or, equivalently, on the temperature at a given sample size) is suppressed, see Fig. 9. 13.

The renormalization of the Poisson ratio for thermally fluctuating membranes was discussed in detail by Burmistrov et al. (2018a, 2018b).

Los, Fasolino, and Katsnelson (2017) has estimated, from the combination of atomistic simulations for small sample size and scaling considerations presented earlier, the critical load under which the graphene membrane is destroyed. In the regime of the destruction, the stress is strong enough to suppress the thermal fluctuations, and the results are not very different from those which can be estimated based on the Föppl–von Karman equations. In particular, the quantitative conclusion on extraordinary mechanical stress of graphene is confirmed. For

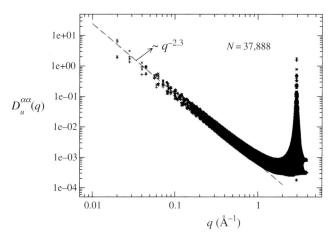

Fig. 9.12 Correlation functions $D_u^{\alpha\alpha}(\alpha = x, y)$ of in-plane displacement fields u_α for graphene at room temperature, the number of atoms in the crystallite is 37,888. The scaling exponent is consistent with Eq. (9.103) and (9.113), $\eta_u = 2 - 2 \times 0.85 = 0.3$ (dashed line).
(Reproduced with permission from Los, Fasolino, & Katsnelson, 2016.)

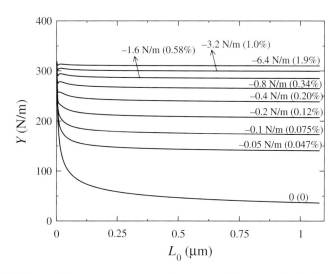

Fig. 9.13 Effective Young modulus as a function of the sample size for graphene at room temperature for the indicated values of the negative stress (the sample is expanded). L_0 is the sample size at zero stress. The value in the bracket is the corresponding elongation under the stress.
(Reproduced with permission from Los, Fasolino, & Katsnelson, 2016.)

graphene drum of 1 m in diameter at room temperature, it can keep the load up to 2.65 kg if all of this load is concentrated in the center and up to 8 kg if it is uniformly distributed over the whole drum – not bad for the single-atom-thick support!

9.8 Raman spectra of graphene

The main experimental tool allowing us to study phonon spectra throughout the Brillouin zone is inelastic neutron scattering (Vonsovsky & Katsnelson, 1989; Katsnelson & Trefilov, 2002). Unfortunately, this method is not applicable (up to now) to graphene because it requires rather massive samples. Optical tools such as infrared and Raman spectroscopy only provide us with information about phonons at some special points of the Brillouin zone. However, even this information is of crucial importance. Raman spectroscopy is one of the main techniques used in graphene physics (Ferrari et al., 2006; for a review see Malard et al., 2009). Here, we discuss some basic ideas about Raman spectra of graphene.

The Raman effect is inelastic light scattering; "inelastic" means that the frequency of the scattered light, ω', is not equal to that of the incident light, ω (Landsberg & Mandelstam, 1928; Raman, 1928; Raman & Krishnan, 1928). Its quantum explanation is based on the Kramers–Heisenberg formula for the light-scattering cross-section (Berestetskii, Lifshitz, & Pitaevskii, 1971)

$$\frac{d\sigma}{do'} = \frac{\omega\omega'^3}{\hbar^2 c^4} \left| \sum_n \left\{ \frac{(\vec{d}_{fn}\vec{e}'^*)(\vec{d}_{ni}\vec{e})}{\omega_{ni} - \omega - i\delta} + \frac{(\vec{d}_{fn}\vec{e})(\vec{d}_{ni}\vec{e}'^*)}{\omega_{ni} + \omega - i\delta} \right\} \right|^2_{\delta \to +0}, \tag{9.172}$$

where do' is the element of solid angle of scattering light, $\vec{e}$ and $\vec{e}'$ are photon polarization vectors for incident and scattered light, respectively, $|f\rangle$ and $|i\rangle$ are the final and initial states of the scattering system, respectively, $|n\rangle$ is its intermediate state, $\vec{d}_{mn}$ are matrix elements of the electric dipole momentum operator

$$\omega_{ni} = \frac{E_n - E_i}{\hbar}, \tag{9.173}$$

and, due to the energy-conservation law,

$$\omega' = \omega + \frac{E_i - E_f}{\hbar}. \tag{9.174}$$

The general expression (9.172) can be applied both to elastic ($\omega = \omega'$) and to inelastic ($\omega \neq \omega'$) cases; we will be interested here in the latter.

The electric dipole moment can be represented as a sum of contributions from electrons and nuclei (phonons):

$$\vec{d} = \vec{d}^{(e)} + \vec{d}^{(ph)}. \tag{9.175}$$

Correspondingly, we have the electron Raman effect when the state $|n\rangle$ corresponds to some electron excitation in the system and the phonon Raman effect when $|n\rangle$ differs from $|i\rangle$ by the creation or annihilation of a phonon with frequency ω_λ. In the latter case

$$\omega' = \omega \pm \omega_\lambda, \tag{9.176}$$

the $+$ and $-$ signs correspond to annihilation and creation of the phonon, respectively. Keeping in mind that for visual light the wave vector of a photon is much smaller than the inverse interatomic distance $1/a$ and bearing in mind the momentum-conservation law, one can conclude that in crystals only phonons at the Γ point ($\vec{q} = 0$) can normally be probed to leading order of perturbation by the Raman effect. As we will see, this is not the case for graphene.

There are selection rules determining whether a given optical phonon can be Raman-active (that is, it contributes to the Raman scattering) or infrared-active (that is, it contributes to absorption of the photon), or both. In general, such analysis requires the use of group theory (Heine, 1960).

For the case of graphene, at the Γ point there are the infrared-active ZO mode and a doubly degenerate Raman-active optical mode, with deformations lying in the plane (see Fig. 9.1). The latter corresponds to the so-called E$_g$ (g for *gerade*) representation of the point group of the honeycomb lattice. The atomic displacements for this mode are shown in Fig. 9.14 (the mode is doubly degenerate since there are two equivalent, mutually perpendicular, directions of the displacements). Therefore, one could expect a single line with the frequency $\omega_\lambda = |\omega' - \omega|$ equal to that of $\omega_{\mathrm{LO}}(\vec{q} = 0) \approx 1580$ cm^{-1}. Indeed, this line was observed long ago in graphite (Tuinstra & Koenig, 1970). It is called usually the *G peak*. However, the

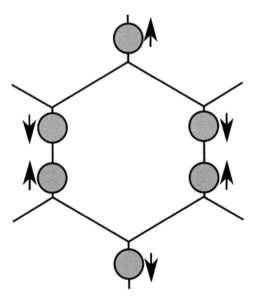

Fig. 9.14 Atomic displacements for a Raman-active optical phonon at the Γ point.

Raman spectra of graphite are characterized by the second sharp and intensive feature in Fig. 9.16 (Nemanich & Solin, 1977, 1979), which is usually called the *2D peak* in the literature on graphene. (In the literature on nanotubes and in the review by Malard et al. (2009), it is called the *G'peak*.) It was interpreted from the very beginning as *a two-phonon peak*. (A detailed theory has been proposed by Thomsen & Reich, 2000; Maultzsch, Reich, & Thomsen, 2004). The basic idea is that in this case, the intermediate state $|n\rangle$ in Eq. (9.159) is a combined electron–phonon excitation.

The basic physics originates from the existence of two valleys, K and K′; the vector $\vec{q}$ connecting K and K′ is equivalent to the vector TK (Fig. 9.15(a)). Therefore, the process is allowed when (i) an incident photon initiates a transition from hole to electron bands at the K point, the electron energy being E_0, (ii) the excited electron is transferred from K to K′, emitting a phonon with $\vec{q} = \vec{K}$ and frequency ω_0; (iii) it is transferred back to K′, emitting a second phonon, with $\vec{q} = -\vec{K}$ and the frequency ω_0; and (iv) the scattered photon is emitted from the state with $E_n = E_0 - 2\hbar\omega_0$ (Fig. 9.15(b)). In this case $\omega' = \omega - 2\omega_0$. This is a higher-order process in the *electron–phonon* coupling; however, this does not give any additional smallness since the process is resonant: The electron bands in K and K′ are identical, and we know that, for the case of perturbation of *degenerate* energy levels, the effect of the perturbation has no smallness (Landau & Lifshitz, 1977). In the electron–photon interaction this is a second-order process, as is a normal Raman effect; therefore, its probability can be comparable to that of single-phonon Raman scattering.

(a)

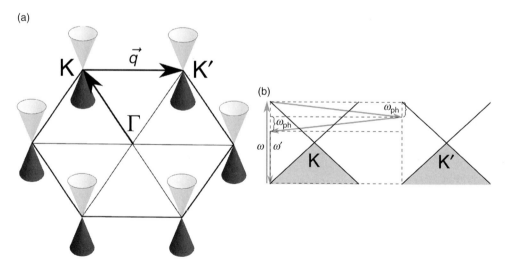

(b)

Fig. 9.15 The origin of the 2D Raman peak. (a) The scheme of momentum conservation. (b) The scheme of the energy transfer (see the text).

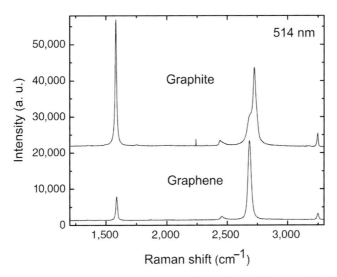

Fig. 9.16 The Raman spectra of graphite and graphene. The wavelength of the incident light is 514 nm.
(Reproduced with permission from Ferrari et al., 2006.)

Actually, there are several types of phonons at the K point (see Fig. 9.1); both electrons and phonons have dispersion, so the 2D peak at $\approx 2,700$ cm^{-1} is not a single line, but a band (see the high-frequency peak in Fig. 9.16). Detailed study of its shape provides information about phonon dispersion near the K point (Mafra et al., 2007). A theoretical analysis of the electron–phonon coupling, which is responsible for the 2D peak for various modes, has been done by Jiang et al. (2005), within a tight-binding model, and by Park et al. (2008) using density-functional calculations. The electron–phonon coupling is essentially different for different modes. Also, effects of destructive interference between contributions to the double resonance should be taken into account (Maultzsch, Reich, & Thomsen, 2004). As a result of all these factors, the main contribution originates from TO phonons along the K–M direction (Mafra et al., 2007). There is also a satellite line (at smaller frequencies), which originates from the processes with one TO phonon and one LA phonon involved (Mafra et al., 2007).

There is a noticeable shift in position of Raman peaks between graphene and graphite (Ferrari et al., 2006), see Fig. 9.16. Moreover, one can easily distinguish single-layer, bilayer, ..., N-layer graphene (up to $N \approx 5$) by Raman spectroscopy, which makes it a very suitable tool for the identification of graphene.

If some defects are present, one of the phonon-induced scattering processes responsible for the 2D peak can be replaced by *elastic* scattering by the defects (the *D peak*, with the frequency $|\omega' - \omega| \approx \omega_0$. "Resonant" impurities that change locally the sp^2 state of carbon atoms to sp^3, such as hydrogen, fluorine and C–C chemical bonds (see Section 6.6), give the main contribution to the origin of this peak, and its intensity can be used to estimate the concentration of such locally modified sp^3 centers in graphene (Elias et al., 2009; Nair et al., 2010; Ni et al., 2010).

We hope these examples suffice to illustrate the importance of Raman spectroscopy in graphene physics and chemistry.

To summarize, in this chapter we have considered some of the peculiarities of the structural state, mechanical properties, dynamics, and thermodynamics of graphene. The consequences of these peculiarities for the electronic properties of graphene will be considered in the next two chapters.

10

Gauge fields and strain engineering

10.1 Strain-induced pseudomagnetic fields

We saw in the previous chapter that graphene at finite temperatures is unavoidably corrugated. As a result, in any real atomic configuration the three bonds of each atom with its neighbors are no longer equivalent; see a snapshot from Monte Carlo simulations by Fasolino, Los, and Katsnelson (2007) in Fig. 10.1. Apart from atomically sharp inhomogeneities, there is a large-scale, macroscopic nonequivalence, which survives in a continuum-medium description of graphene, and is described in terms of the deformation tensor $u_{\alpha\beta}$.

Let us assume that the hopping parameters t_1, t_2, and t_3 are different throughout the whole sample and repeat the tight-binding derivation of the Dirac Hamiltonian (Section 1.2). As a result, instead of Eq. (1.22) we find the following effective Hamiltonian near the K point (Kane & Mele, 1997; Suzuura & Ando, 2002; Sasaki, Kawazoe, & Saito, 2005; Katsnelson & Novoselov, 2007):

$$\hat{H} = \vec{\sigma}\left(-i\hbar v\vec{\nabla} - \vec{A}\right), \qquad (10.1)$$

where

$$A_x = \frac{\sqrt{3}}{2}(t_1 - t_2),$$
$$A_y = \frac{1}{2}(2t_3 - t_1 - t_2) \qquad (10.2)$$

play the role of components of the vector potential. Thus, the difference in t_1, t_2, and t_3 shifts the Dirac conical point in some random direction. It does not produce a mass term proportional to σ_z, since the sublattices remain equivalent. The field $\vec{A}$ is a typical *gauge field*, similar to the vector potential in electrodynamics. It was first discussed in the context of electron–phonon interaction in carbon nanotubes (Kane & Mele, 1997; Suzuura & Ando, 2002; Sasaki, Kawazoe, & Saito, 2005)

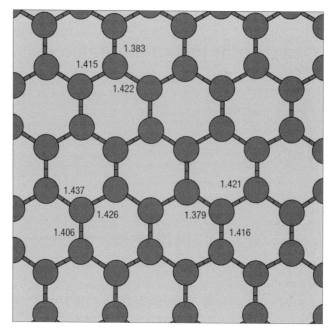

Fig. 10.1 A snapshot of a typical atomic configuration in atomistic Monte Carlo simulations of graphene at $T = 300$ K; the number indicates the bond length (Å). (Reproduced with permission from Fasolino, Los, & Katsnelson, 2007.)

and then introduced in the physics of graphene by Morozov et al. (2006) and Morpurgo and Guinea (2006) as a mechanism suppressing weak (anti)localization. Note that the vector potential $\vec{A}$ in Eq. (10.1) has the dimension of energy; in conventional units, it should be written as $(ev/c)\vec{A}$.

In the weakly deformed lattice, assuming that the atomic displacements $\vec{u}$ are small in comparison with the interatomic distance a, the length of the nearest-neighbor vectors $\vec{\rho}_i$ will be changed by the quantity

$$\delta a_i = \sqrt{(\vec{\rho}_i + \vec{u}_i - \vec{u}_0)^2} - a \approx \frac{\vec{\rho}_i(\vec{u}_i - \vec{u}_0)}{a}, \tag{10.3}$$

where $\vec{u}_i$ and $\vec{u}_0$ are displacement vectors for the corresponding atoms and we take into account that $|\vec{\rho}_i| = a$. As a result, the new hopping integrals will be

$$t_i \approx t - \frac{\beta t}{a^2}\vec{\rho}_i(\vec{u}_i - \vec{u}_0), \tag{10.4}$$

where

$$\beta = -\frac{\partial \ln t}{\partial \ln a} \tag{10.5}$$

is the *electron Grüneisen parameter* describing the dependence of the nearest-neighbor hopping integral on the interatomic distance. This value lies in the interval $\beta \approx 2 - 3$ (Heeger et al., 1988; Vozmediano, Katsnelson, & Guinea, 2010). In the continuum limit (elasticity theory)

$$(\vec{u}_i - \vec{u}_0) \sim (\vec{\rho}_i \nabla) \vec{u}(\vec{r}) \tag{10.6}$$

and, thus

$$A_x = -2c\beta t u_{xy}.$$
$$A_y = -c\beta t (u_{xx} - u_{yy}). \tag{10.7}$$

(Suzuura & Ando, 2002; Mañes, 2007), where c is a numerical factor depending on the detailed model of chemical bonding. In particular, one should take into account that the nearest-neighbor hopping parameter depends not only on the interatomic distance but also on the angles. Keeping in mind an uncertainty in the value of β, we will put $c = 1$ from now on.

Thus, the two components of the vector potential are proportional to the two *shear* components of the deformation tensor. On general symmetry grounds, strains should also lead to a scalar potential proportional to dilatation (Suzuura & Ando, 2002; Mañes, 2007):

$$V(\vec{r}) = g (u_{xx} + u_{yy}). \tag{10.8}$$

It originates from a redistribution of electron density under the deformation. A naïve estimation would be to assume that it should be of the order of the bandwidth, $g \approx 20\text{eV}$ (Ono & Sugihara, 1966; Sugihara, 1983; Suzuura & Ando, 2002). Recent density-functional calculations for single-layer graphene give a much smaller value, $g \approx 4\text{eV}$ (Choi, Jhi, & Son, 2010). However, these two values are not actually contradictory, since the density functional takes into account the effect of electron screening, which should lead to a replacement $g \rightarrow g/\varepsilon$. Taking into account that for undoped single-layer graphene $\varepsilon \approx 4.5$ (see Eq. (7.89)), we see that *screened* $g \approx 4\text{eV}$ corresponds to unscreened $g_0 \approx 18$ eV. This value seems to be in agreement with experimental data on electron mobility in freely suspended graphene (Castro et al., 2010b); for more details, see Chapter 11.

Within the framework of the Dirac approximation, a uniform strain cannot open a gap in the spectrum, but just leads to a shift of conical points. However, if the strain is very strong and t_1, t_2, and t_3 are *essentially* different, the gap can be opened. As was shown by Hasegawa et al. (2006), there is no gap if the "triangular inequalities"

$$\left| t_{l_1} - t_{l_2} \right| \leq \left| t_{l_3} \right| \leq \left| t_{l_1} + t_{l_2} \right| \tag{10.9}$$

are satisfied, where (l_1, l_2, l_3) is a permutation of $(1, 2, 3)$. This issue was later studied in more detail within the framework of the tight-binding model (Pereira, Castro Neto, & Peres, 2009; Cocco, Cadelano, & Colombo, 2010; Pellegrino, Angilella, & Pucci, 2010). According to the last of these papers, the minimum shear deformation that leads to the gap opening is about 16%. This is, in principle, possible in graphene without its destruction (Lee et al., 2008). Henceforth we restrict ourselves to the case of smaller deformations, for which the linear approximation (10.4) is applicable. We can see in this chapter that this already provides very rich and interesting physics, with the prospect of important applications.

If the strain is not uniform, the vector potential (10.7) creates, in general, a *pseudomagnetic field* (in normal units)

$$\frac{evB}{c} = \frac{\partial A_y}{\partial x} - \frac{\partial A_x}{\partial y}.$$ (10.10)

It is important to stress that the pseudomagnetic field acting on electrons from the valley K' is exactly opposite to that acting on electrons from the valley K:

$$B_{\mathrm{K}} = - B_{\mathrm{K}'}.$$ (10.11)

This follows from explicit calculations and is obvious from the time-reversal symmetry: Deformations cannot break it for the honeycomb lattice as a whole. However, if we have only smooth deformations and no scattering processes between the valleys, the electrons in a nonuniformly strained graphene will behave as if the time-reversal symmetry were broken (Morozov et al., 2006; Morpurgo & Guinea, 2006). This has very important consequences for the physics of the quantum Hall effect, weak localization, etc., as will be discussed in later this chapter.

10.2 Pseudomagnetic fields of frozen ripples

As the first example, we consider the pseudomagnetic field created by a frozen ripple. This means that we substitute Eq. (9.62) for the deformation tensor into Eq. (10.7) and (10.10) and take into account only the last term

$$u_{\alpha\beta} = \frac{1}{2} \frac{\partial h}{\partial x_\alpha} \frac{\partial h}{\partial x_\beta}.$$ (10.12)

The effects of in-plane relaxation will be taken into account in the next section.

Thus, the amplitude of the pseudomagnetic field can be estimated as

$$\bar{B} \approx \frac{\hbar c}{e} \frac{h^2}{aR^3},$$ (10.13)

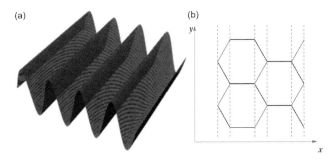

Fig. 10.2 (a) A sketch of the sinusoidal ripple. (b) Atomic rows of a honeycomb lattice. (Reproduced with permission from Guinea, Katsnelson & Vozmediano, 2008.)

where h is the typical height of the ripple and R is its radius (Morozov et al., 2006). This field can be as large as 1 T, for typical sizes of the ripples observed in exfoliated graphene (Morozov et al., 2006).

To perform some quantitave analysis, let us start with the case of the simple sinusoidal deformation shown in Fig. 10.2 (Guinea, Katsnelson, & Vozmediano, 2008). We will assume a modulation along the x-axis $t_{ij} \equiv t_{ij}(x)$. Thus, the problem is effectively one-dimensional, and k_y remains a good quantum number. One can consider hopping parameters between the rows (see Fig. 10.2, right panel) that are equal to t (for horizontal bonds) and

$$2t \cos \left(k_y \frac{\sqrt{3}a}{2} \right)$$

for other bonds.

If we assume a modulation of the hopping parameters,

$$t(x) = t + \delta t(x), \tag{10.14}$$

then the two hoppings are renormalized as

$$t \to t(x),$$
$$2t \cos \varphi \to \sqrt{t^2(x) \cos^2 \varphi + [\delta t(x)]^2 \sin^2 \varphi}, \tag{10.15}$$

where $\varphi = k_y \frac{\sqrt{3}a}{2}$. Let us assume

$$\delta t(x) = \delta t \sin \left(\frac{2\pi x}{l} \right), \tag{10.16}$$

where l is the period of modulation. The electron spectrum has been calculated numerically for a strip with periodic boundary conditions; the results are shown in Fig. 10.3 (Guinea, Katsnelson, & Vozmediano, 2008).

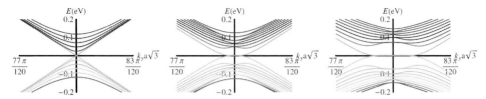

Fig. 10.3. Low-energy states induced by the ripple shown in Fig. 10.2. The average hopping is 3 eV. The width of the ripple is 1,200 $a1200a = 168$ nm. The modulations of the hopping $\delta t/t$ are 0, 0.02, and 0.04 (from left to right). (Reproduced with permission from Guinea, Katsnelson, & Vozmediano, 2008.)

The most important result is the appearance of a dispersionless zero-energy mode, its phase volume grows with increasing $\delta t/t$. This is related to the topologically protected zero-energy Landau level in an *inhomogeneous* magnetic field for the Dirac equation (Section 2.3). There are also some features that are reminiscent of other Landau levels, but they are essentially dispersive, which changes the situation dramatically from the case of a real magnetic field (but see Section 10.4). The real magnetic field B can be included in the calculations via the replacement

$$k_y \rightarrow k_y + \frac{eB}{\hbar c}x. \tag{10.17}$$

The results are shown in Fig. 10.4 (Guinea, Katsnelson, & Vozmediano, 2008). Two important features of these results should be mentioned. First, the combination of the pseudomagnetic field due to rippling and a real magnetic field leads to a broadening of all Landau levels *except* the zero-energy one; this is a consequence of the topological protection of the zero-energy Landau level.

Second, due to Eq. (10.11) for the pseudomagnetic field, the effective total fields acting on electrons from the valleys K and K′ are different, which results in a *valley polarization*. One can clearly see in Fig. 10.4 that the phase space of the dispersionless zero-energy level for the valley K′ is larger than that for the valley K.

The first of these conclusions seem to be relevant for the interpretation of some of the peculiarities of the quantum Hall effect in graphene (Giesbers et al., 2007). The activation gaps for the quantum Hall plateau at $v = 2$ and $v = 6$ have been extracted from the temperature dependences of the resistivity $\rho_{xx}(T)$. Their dependences on the magnetic field are presented in Fig. 10.5. In the ideal case, they would follow $\sqrt{B}$ dependences (see Eq. (2.30) and (2.31)). However, due to disorder there are deviations from this law, and the stronger the disorder the higher the magnetic field at which the $\sqrt{B}$ law is restored. One can see that, for $v = 2$, for which the zero-energy Landau level is involved, it happens much earlier than it does for $v = 6$. This was explained by Giesbers et al. (2007) by postulating that

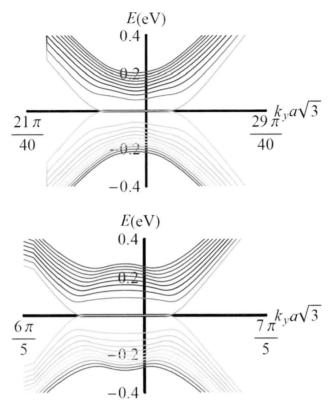

Fig. 10.4 The same as in Fig. 10.3 ($\delta t/1 = 0.04$) but with a magnetic field of $B =$ 10 T. Upper panel, K value; lower panel, K' value.
(Reproduced with permission from Guinea, Katsnelson, & Vozmediano, 2008.)

random pseudomagnetic fields created by ripples (Morozov et al., 2006) contribute essentially to the broadening of all Landau levels *except* the zero-energy one, due to its topological protection (Novoselov et al., 2005a; Katsnelson, 2007a). The same situation should also occur for the case of bilayer graphene (Katsnelson & Prokhorova, 2008).

The electronic structure of the frozen sinusoidal ripple has been studied by Wehling et al. (2008a) by carrying out density-functional calculations.

These calculations confirm the qualitative predictions of the tight-binding model concerning the existence of zero-energy states. A schematic view of the ripple is shown in Fig. 10.6, and the results for the width of the dispersionless zero-energy mode are illustrated in Fig. 10.7. This qualitative agreement is not trivial, since the tight-binding model takes into account neither next-nearest-neighbor hopping nor the *electrostatic* potential (10.8). The reason why the latter is not relevant here will be clear later (see Section 10.6).

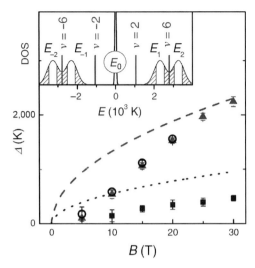

Fig. 10.5 Energy gaps 2Δ between two Landau levels extracted from the temperature dependence of the resistivity p_{xx} as a function of the magnetic field for $v = +2$ (full triangles), $v = -2$ (open circles) and $v = +6$ (full squares). The dashed and dotted lines are the theoretically expected energy gaps for sharp Landau levels. The inset shows schematically the density of states for a sharp zero-energy Landau level and broadened higher Landau levels for electrons and holes at $B = 30$ T. Extended states are represented by the white areas, localized states by the dashed areas.

(Reproduced with permission from Giesbers et al., 2007.)

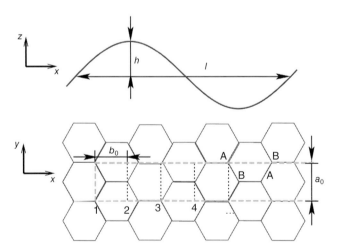

Fig. 10.6 Schematic top and side views of the ripple used in the electronic structure calculations by Wehling et al. (2008a).

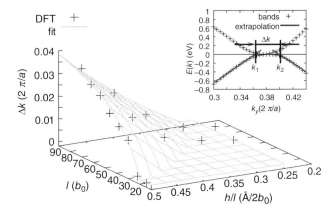

Fig. 10.7 Pseudo-Landau-level extension obtained from the density-functional calculations (DFT) by Wehling et al. (2008a). The definition of Δk is clear from the inset; the parameters h and l are defined in Fig. 10.6. Crosses show the fit to the expression $a\Delta k/(2\pi) = A_1(h/l)^2 - A_2(h/l)$ with some constants A_i. (Courtesy of T. Wehling.)

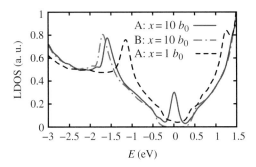

Fig. 10.8 The local density of states (LDOS) inside the cells at $x = 1b_0$ (low pseudomagnetic field) and $x = 10b_0$ (high field region). For the low field region, the LDOS is the same in both sublattices (only sublattice A is plotted here, dashed line), whereas in the high field region the LDOS in the sublattice A (solid line) and B (dash-dotted line) differ significantly, only the first one indicates the formation of zero-energy Landau level.
(Reproduced with permission from Wehling et al., 2008a.)

The calculations by Wehling et al. (2008a) demonstrate a complete sublattice polarization for the zero-energy pseudo-Landau states Fig. 10.8. This follows from Eq. (10.11): In contrast with the usual quantum Hall effect (Sections 2.2 and 2.3), the solutions for both valleys belong to the same sublattice.

It was shown by Wehling et al. (2008a) that, if in-plane relaxation of atoms is allowed, the dispersionless zero-energy mode disappears for the geometry under consideration. The reason for this behavior will be discussed in the next section.

10.3 Pseudomagnetic fields of ripples: the effect of in-plane relaxation

Let us assume a fixed distribution of out-of-plane deformation $h(x, y)$. If in-plane relaxation is allowed, the in-plane deformations u_x and u_y should be found from the minimum of the total energy (9.63) and excluded (Guinea, Horovitz &, Le Doussal, 2008; Wehling et al., 2008a). It is convenient to use the complex-number notation $z = x + iy$, $z^* = x - iy$,

$$\partial = \frac{\partial}{\partial z} = \frac{1}{2}\left(\frac{\partial}{\partial x} - i\frac{\partial}{\partial y}\right),$$

$$\partial^* = \frac{\partial}{\partial z^*} = \frac{1}{2}\left(\frac{\partial}{\partial x} + i\frac{\partial}{\partial y}\right), \tag{10.18}$$

$$\nabla^2 = 4\partial\partial^*$$

and

$$u(z, z^*) = u_x - iu_y,$$

$$A(z, z^*) = A_x - iA_y. \tag{10.19}$$

We will express the deformation tensor via A using Eq. (10.7). As a result, the free energy (9.63) can be rewritten as (Wehling et al., 2008a)

$$F = \int d^2z\Bigg\{ 8\kappa(\partial\partial^* h)^2 + (\lambda + \mu)\left[\frac{1}{2}(\partial^* u + \partial^* u) + \partial h \partial^* h\right]$$

$$+ \mu\left[\partial u + (\partial h)^2\right]\left[\partial^* u^* + (\partial^* h)^2\right]\Bigg\}$$

$$= \int d^2z\Bigg\{ 8\kappa(\partial\partial^* h)^2 + \frac{\mu a^2}{\beta^2 t^2}AA^* + (\lambda + \mu)\left[\frac{a}{2\beta t}\frac{1}{\partial\partial^*}\left(\partial^{*2}A + \partial^2 A^*\right) + \frac{1}{\partial\partial^*}R[h]^2\right]\Bigg\} \tag{10.20}$$

where

$$R[h] = \partial^2 h \partial^{*2} h - (\partial\partial^* h)^2 = \frac{\partial(\partial h, \partial^* h)}{\partial(z, z^*)} \tag{10.21}$$

is proportional to the Gaussian curvature of the surface, Eq. (9.76). On minimizing Eq. (10.20) for a given $h(z, z^*)$ one finds

$$A = -\beta t a^2 \frac{(\lambda + \mu)}{(\lambda + 2\mu)} \frac{\partial^2}{\partial\partial^*} R, \tag{10.22}$$

$$\frac{evB}{c} = i\beta t a^2 \frac{(\lambda + \mu)}{(\lambda + 2\mu)} \frac{\partial^3 - \partial^{*3}}{(\partial\partial^*)^2} R. \tag{10.23}$$

One can see from Eq. (10.22) and (10.23) that for the case of a free membrane both the pseudomagnetic field and the vector potential vanish identically if h depends only on one Cartesian coordinate, which means $R \equiv 0$. This is not so, as we will see in Section 10.5, if the membrane is under strain, in which case an additional term should be added to Eq. (10.20).

This explains the disappearance of zero-energy states created by a frozen sinusoidal ripple under relaxation mentioned at the end of the previous section. If we induce the field

$$f_{\alpha\beta}(\vec{r}) = \frac{\partial h}{\partial x_\alpha} \frac{\partial h}{\partial x_\beta} \tag{10.24}$$

and its Fourier component

$$f_{\alpha\beta}(\vec{k}) = -\sum_{\vec{k}_1} k_{1\alpha}(k_\beta - k_{1\beta}) h_{\vec{k}_1} h_{\vec{k}-\vec{k}_1} \tag{10.25}$$

then the symbolic expression (10.23) can be represented in an explicit form (Guinea, Horovitz, & Le Doussal, 2008)

$$\frac{eB(\vec{k})}{hc} = i k_y \frac{3k_x^2 - k_y^2}{k^4} \frac{\beta}{a} \frac{\lambda + \mu}{\lambda + 2\mu} \left[k_y^2 f_{xx}(\vec{k}) + k_x^2 f_{yy}(\vec{k}) - 2k_x k_y f_{xy}(\vec{k}) \right]. \tag{10.26}$$

This gives us a formal solution of the problem.

Importantly, Eq. (10.23) and (10.26) reflect the trigonal symmetry of the problem: If we have an isotropic ripple, $h = h(r)$, and thus $R = R(r)$, the pseudomagnetic field will have an angular dependence

$$B(r, \varphi) = \sin(3\varphi) B_0(r), \tag{10.27}$$

where ϕ is the polar angle (Wehling et al., 2008a).

In the next chapter, when discussing electron scattering by the ripples, we will be interested in the correlation functions of vector and scalar potentials created by the intrinsic ripples. They are proportional to

$$F_{\alpha\beta,\gamma\delta}(\vec{q}) = \langle f_{\alpha\beta}(\vec{q}) f_{\gamma\delta}(-\vec{q}) \rangle$$

$$= \sum_{\vec{q}_1 \vec{q}_2} q_{1\alpha}\left(q_{1\beta} - q_\beta\right) q_{2\gamma}\left(q_{2\gamma} + q_\delta\right) \langle h_{\vec{q}_1} h_{\vec{q}-\vec{q}_1} h_{-\vec{q}_2} h_{-\vec{q}-\vec{q}_2} \rangle. \tag{10.28}$$

To estimate the correlation function on the right-hand side of Eq. (10.28), one can use Wick's theorem (9.110) and the results of Section 9.4. The answer is (Katsnelson, 2010b)

$$
F(\vec{q}) \sim \begin{cases} \left(\dfrac{T}{\kappa}\right)^2 \dfrac{\ln q/q^*}{q^2}, & q > q^*, \\[2ex] \left(\dfrac{T}{\kappa}\right)^{2-\eta} \dfrac{1}{q_0^{2\eta} q^{2-2\eta}}, & q < q^*, \end{cases}
\tag{10.29}
$$

where q^* is the crossover wave vector (9.98). This means that the correlation function of the vector potential is singular at $q \to 0$. At the same time, the correlation function

$$
\left\langle \left| B_{\vec{q}} \right|^2 \right\rangle \sim q^2 \left\langle \left| A_{\vec{q}} \right|^2 \right\rangle
\tag{10.30}
$$

tends to zero at $q \to 0$. Similarly to Eq. (10.27) in real space, it has the angular dependence $\sin^2(3\varphi_{\vec{q}})$, where $\varphi_{\vec{q}}$ is the polar angle of the vector $\vec{q}$ (Guinea, Horovitz, & Le Doussal, 2008).

10.4 The zero-field quantum Hall effect by strain engineering

In the previous sections we discussed the gauge fields created by ripples, which are almost unavoidable in graphene. However, one can use Eq. (10.7) and (10.10) to intentionally create a magnetic field with the desired properties to manipulate the electronic structure of graphene via "strain engineering." First of all, let us consider an opportunity to create a uniform, or almost uniform, pseudomagnetic field and thus realize Landau quantization and the quantum Hall regime without a real magnetic field (Guinea, Katsnelson, & Geim, 2010; Guinea et al., 2010).

Let us consider the simplest case of plane geometry, where $h = 0$ and the strain tensor is created by the w-field only. Within linear two-dimensional elasticity theory the general solution for the strain tensor can be written in terms of two arbitrary analytic functions $g(z)$ and $k(z)$, namely

$$
\sigma_{xx} = \frac{\partial^2 f}{\partial y^2}, \quad \sigma_{yy} = \frac{\partial^2 f}{\partial x^2}, \quad \sigma_{xy} = -\frac{\partial^2 f}{\partial x \partial y},
\tag{10.31}
$$

where

$$
f(x, y) = \mathrm{Re}\, [z^* g(z) + k(z)]
\tag{10.32}
$$

(Landau & Lifshitz, 1970; Vozmediano, Katsnelson, & Guinea, 2010). For a purely shear deformation, $\sigma_{xx} = -\sigma_{yy}$, which means that $g(z) = 0$. Thus, the components of the vector potential which are expressed in terms of stress as

$$A_x = -\frac{2c\beta t}{\mu}\sigma_{xy},$$

$$A_y = -\frac{c\beta t}{\mu}\left(\sigma_{xx} - \sigma_{yy}\right) \tag{10.33}$$

are proportional to the real and imaginary parts of $d^2 k(z)/dz^2$, respectively, and

$$B \sim \text{Im}\frac{d^3 k(z)}{dz^3}. \tag{10.34}$$

A pure shear deformation that leads to a uniform pseudomagnetic field is

$$k(z) = Az^3 \tag{10.35}$$

(A is a constant). The general deformation (including dilatation), which leads to a uniform pseudomagnetic field, is determined by the function

$$f(z) = Az^3 + Bz^*z^2 \tag{10.36}$$

(A and B are constants). It corresponds to the strain tensor linearly dependent on coordinates

$$u_{\alpha\beta} = \frac{\bar{u}}{L}x_\alpha e_\beta, \tag{10.37}$$

where u is a typical stress, L is the sample size, and $\vec{e}$ is an arbitrary unit vector. The effective pseudomagnetic field is associated with a magnetic length

$$\frac{1}{l_B^2} = \frac{eB}{\hbar c} \approx \frac{\beta\bar{u}}{aL}. \tag{10.38}$$

For $\bar{u} = 10^{-2}$ and $L \approx 10$ μm we obtain $l_B \approx 0.2$ μm, which corresponds to a magnetic field of about 0.3 T. Actually, much higher deformations and, thus, much higher pseudomagnetic fields can be created in graphene.

In reality, the stress can only be applied normally to the boundary of a sample. Numerical solutions of the equations of the theory of elasticity show that it is not difficult to have a *quasiuniform* pseudomagnetic field in a quite general situation; what is really important is to keep the trigonal symmetry of the stress (Guinea, Katsnelson, & Geim, 2010; Guinea et al., 2010). One can also show that the presence of dilatation and, thus, of an electrostatic potential (10.8) does not affect the results (Guinea et al.,

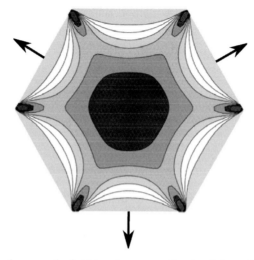

Fig. 10.9 A pseudomagnetic field in a hexagon of a size 1.4 μm that is strained by the forces applied to three sides. The maximum strain of 20% creates an effective field of about 10 T at the hexagon's center. The counters correspond to 8, 6, 4, 2, 0, –2 T, from inside to outside.
(Reproduced with permission from Guinea, Katsnelson, & Geim, 2010.)

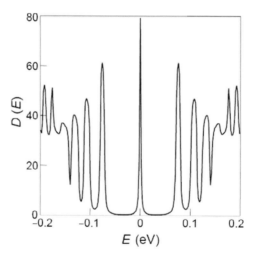

Fig. 10.10 The average density of states in the central region of diameter 0.5 μm for the hexagon shown in Fig. 10.9.
(Reproduced with permission from Guinea, Katsnelson, & Geim, 2010.)

2010). As an example, we present here the results obtained by Guinea, Katsnelson, and Geim (2010) for a hexagonal flake with external forces applied to three edges (Fig. 10.9). One can see that the value of the pseudomagnetic field in the central part of the flake is of uniform to high accuracy. As a result, the density of states, averaged over the central region, clearly exhibits pronounced Landau levels (Fig. 10.10).

It was suggested by Guinea, Katsnelson, and Geim (2010) that one should use electron Raman scattering to observe the Landau levels created by strain. Soon after that, this effect was observed by scanning tunneling microscopy for graphene bubbles on a platinum surface (Levy et al., 2010). It is significant that these bubbles have a shape with trigonal symmetry. The value of the pseudomagnetic field created by spontaneous deformation in these bubbles was estimated by Levy et al. (2010) to be approximately 300 T, which is much higher than any real magnetic field attainable to date. Later, Georgi et al. (2017) have estimated pseudomagnetic fields in their samples as 1,000 T. Importantly, they were able to measure the density of states separately in sublattices A and B and confirmed a strong pseudospin polarization predicted theoretically (see Fig. 10.8).

Owing to the condition (10.11), the system as a whole remains time-reversal-invariant, and, due to the Onsager relations (Zubarev, 1974), one should have $\sigma_{xy} = 0$ (here σ is the conductivity, not the stress!). In terms of edge states (Section 5.8) this results from the existence of two counter-propagating edge states, from values K and K', without total charge transfer. This situation can be described as a "valley quantum Hall effect" analogous to the spin quantum Hall effect (Kane & Mele, 2005a, 2005b). Inhomogeneities at the edges will lead to a scattering between the valleys; however, one can show that, due to the smallness of the parameter a/l_B, the mixture of the counter-propagating edge states can be very small (Guinea, Katsnelson, & Geim, 2010).

10.5 The pseudo-Aharonov–Bohm effect and transport gap in suspended graphene

As the next example, we consider the pseudomagnetic field arising in a freely suspended graphene membrane (Fogler, Guinea, & Katsnelson, 2008). If it is charged with the electron density n, the electrostatic pressure acts on the membrane (Jackson, 1962)

$$p = \frac{2\pi e^2}{\varepsilon} n^2, \tag{10.39}$$

where ε is the dielectric constant. Under this pressure, the membrane will be bent (Fig. 10.11), with the equation of equilibrium

$$\kappa \frac{d^4 h(x)}{dx^4} - \tau \frac{d^2 h}{dx^2} = p, \tag{10.40}$$

where τ is the external strain (this follows from the minimization of the total energy (9.144) in the presence of an external strain $\sigma_{xx}^{\text{ext}} = \tau$, and u_{xx} is given by Eq. (10.12). If we assume that the membrane is supported at $x = \pm L/2$ then the

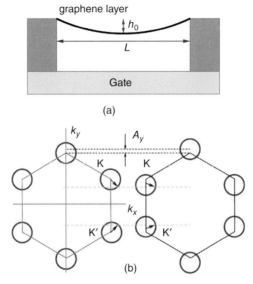

Fig. 10.11 (a) A sketch of the model of a suspended graphene membrane under consideration (see the text). (b) Fermi-circle positions in the Brillouin zone in the leads (left) and in the suspended region (right).
(Reproduced with permission from Fogler, Guinea, & Katsnelson, 2008.)

solution of Eq. (10.40) satisfying the boundary conditions is (Timoshenko & Woinowsky-Krieger, 1959)

$$h(x) = \frac{pL^4}{16u^4\kappa}\left[\frac{\cosh\left(2ux/L\right)}{\cosh u} - 1\right] + \frac{pL^2\left(L^2/4 - x^2\right)}{8u^2\kappa}, \tag{10.41}$$

$$u^2 = \frac{\tau l^2}{4\kappa}. \tag{10.42}$$

The strain has to be found self-consistently, as

$$\tau = \tau_0 + Y\int_{-L/2}^{L/2} dx u_{xx} = \tau_0 + \frac{Y}{2L}\int_{-L/2}^{L/2} dx\left(\frac{dh}{dx}\right)^2, \tag{10.43}$$

where τ_0 is an external strain of nonelectrostatic origin. First we will assume, for simplicity, that $\tau_0 = 0$ and

$$n \gg \sqrt{\frac{\varepsilon\kappa}{e^2 L^3}}, \tag{10.44}$$

which gives us $u \gg 1$. In this regime only the last term on the right-hand side of Eq. (10.49) survives, and the profile $h(x)$ is represented by a simple parabola:

$$h(x) = h_0 \left(1 - \frac{4x^2}{L^2} \right), \tag{10.45}$$

where

$$h_0 = \left(\frac{3\pi}{64} \frac{e^2}{\varepsilon Y} n^2 L^4 \right)^{1/3} \tag{10.46}$$

and

$$\tau = \frac{pL^2}{8h_0} = \frac{\pi e^2 n^2 L}{4\varepsilon h_0}. \tag{10.47}$$

The deformation u_{xx} creates the vector potential. Its effect is largest if the zigzag direction is along the y-axis, thus

$$A_x = 0, \quad A_y = \pm \frac{\beta t}{a} u_{xx}, \tag{10.48}$$

where the signs $\pm$ correspond to the valleys K and K′, respectively. Thus, the conical points will be shifted, inside the membrane, in the y-direction (see Fig. 10.11(b)). If this shift is larger than the Fermi wave vector k_F, that is,

$$k_F < \frac{|A_y|}{\hbar v}, \tag{10.49}$$

the matching of wave functions in the leads and in the membrane becomes impossible, then the transport though the membrane will be totally suppressed; that is, the transport gap will be open. Here we assume, for simplicity, that the concentrations of charge carriers for the leads and membrane are the same.

To proceed further, let us replace the deformation u_{xx} in Eq. (10.48) by its average value

$$u_{xx} = \frac{\tau}{Y}. \tag{10.50}$$

Thus, taking into account Eq. (10.46) through (10.48), we have an estimation

$$\frac{|A_y|}{\hbar v} \approx \left(\frac{e^2}{Y\varepsilon} \right)^{2/3} n^{4/3} L^{-1/3} \approx \frac{a^2 n^{4/3} L^{-1/3}}{\varepsilon^{2/3}}. \tag{10.51}$$

Keeping in mind that $k_F \sim n^{1/2}$, we see that, if all of the strain is purely electrostatic, the condition (10.57) is not satisfied, and the gap never opens. However, it can be open (and will certainly be open, if n is small enough) if $\tau_0 \neq 0$ in Eq. (10.43). This gap opening is an effect of the vector potential itself, not of the pseudomagnetic field, and it takes place even if the vector potential is constant: $A_y = $ constant, $B = 0$. Therefore, it can be considered to be an analog of the Aharonov–Bohm effect for pseudomagnetic fields.

The scattering problem can be solved exactly if one assumes, for simplicity, $A_y = $ constant. The calculations are absolutely similar to those in Chapters 3 and 4. We assume (as has already been mentioned) that we have the same value of k_F in the leads and in the membrane. This means that the y-component of the wave vector in the leads is

$$k_y = k_F \sin \varphi \tag{10.52}$$

(φ is the incidence angle), and within the membrane it is replaced by

$$k_y \rightarrow k_y - q \equiv k_y - \frac{A_y}{hv}. \tag{10.53}$$

The transmission coefficient is (Fogler, Guinea, & Katsnelson, 2008)

$$T(k_y) = \frac{k_1^2 k_2^2}{k_1^2 k_2^2 + k_F^2 q^2 \sin^2(k_2 L)}, \tag{10.54}$$

where

$$k_1 = \sqrt{k_F^2 - k_y^2} = k_F \cos \varphi,$$

$$k_2 = \sqrt{k_F^2 - (k_y - q)^2}. \tag{10.55}$$

The total conductance can be calculated, using the Landauer formula, as

$$G = \frac{4e^2}{h} W \int\limits_{-k_F}^{k_F} \frac{dk_y}{2\pi} T(k_y), \tag{10.56}$$

where W is the width of the membrane.

Pereira and Castro Neto (2009) have suggested that one could use this effect for *strain engineering:* By applying some external strain distribution to graphene, one can create a desirable distribution of the vector potential and thereby manipulate the electronic transport through graphene. This type of strain engineering is different from that considered in the previous section, since no real gaps due to Landau quantization are required, *transport* gaps due to the "pseudo-Aharonov–Bohm effect" suffice.

Low et al. (2012) and Jiang et al. (2013) considered *electron pumping* through the device shown in Fig. 10.11a. The pumping arises in the systems with slow (adiabatic) periodic modulation of the parameters, and the quantized electric charge pumped through the system is expressed via some topological (or geometric) characteristics reminiscent of the Hall conductivity quantization, see Section 2.9 (Thouless, 1983; Brouwer, 1998; Makhlin & Mirlin, 2001). Modulation of voltage applied to the suspended graphene sheet and of its in-plane deformation results in pumping of electrons, one by one, through the device (Low et al., 2012). Dependent on crystallographic orientation of the suspended graphene stripe, one can also reach full valley polarization of the current when all electrons from K valley move to the right and all electrons from K′ valley move to the left, or vice versa (Jiang et al., 2013).

10.6 Gap opening by combination of strain and electric field

Let us now consider the case of *coexistence* of pseudomagnetic fields and electrostatic potential. We will assume that all these perturbations are smooth, and therefore the intervalley scattering can be neglected. Thus, the Hamiltonian of the system is

$$\hat{H} = \hat{H}_0 + \hat{H}_A + \hat{H}_V, \tag{10.57}$$

where

$$\hat{H} = -i\hbar v \hat{\vec{\sigma}} \vec{\nabla},$$

$$\hat{H}_A = -\hat{\sigma}_x A_x(\vec{r}) - \hat{\sigma}_y A_y(\vec{r}),$$

$$\hat{H}_V = V(\vec{r}). \tag{10.58}$$

We will assume that both perturbations are weak and use the perturbation theory for the Green function:

$$\hat{G} = \frac{1}{E - \hat{H} + i\delta} \tag{10.59}$$

(cf. Sections 4.2 and 6.4). We can formally write the answer via the Dyson equation,

$$\hat{G} = \frac{1}{E - \hat{H}_0 - \hat{\Sigma}(E)}, \tag{10.60}$$

where $\hat{\Sigma}(E)$ is the self-energy operator, which can be written as a perturbation series

$$\hat{\Sigma}(E) = \hat{H}_A + \hat{H}_V + (\hat{H}_A + \hat{H}_V)\hat{G}_0(\hat{H}_A + \hat{H}_V) + \cdots, \tag{10.61}$$

where $\hat{G}_0$ is the Green function of the Hamiltonian $\hat{H}_0$, Eq. (4.35) and (4.36). Both $\hat{G}_0$ and $\hat{H}_A$ contain terms proportional to $\hat{\sigma}_x$ and $\hat{\sigma}_y$, and their product can generate $\hat{\sigma}_z$, that is, the mass term:

$$\hat{\sigma}_x \hat{\sigma}_y = -\hat{\sigma}_y \hat{\sigma}_x = i\hat{\sigma}_z. \tag{10.62}$$

In the lowest order, such terms originate from the term linear in $\hat{H}_A$ and linear in $\hat{G}_0$. It is also linear in $\hat{H}_V$. This cross-term has the form

$$\hat{\Sigma}_0(E) = \hat{H}_V \frac{1}{E - \hat{H}_0 + i\delta} \hat{H}_A + \hat{H}_A \frac{1}{E - \hat{H}_0 + i\delta} \hat{H}_V$$

$$= \hat{H}_V (E - \hat{H}_0) \frac{1}{(E + i\delta)^2 - \hat{H}_0^2} \hat{H}_A$$

$$+ \hat{H}_A (E - \hat{H}_0) \frac{1}{(E + i\delta)^2 - \hat{H}_0^2} \hat{H}_V. \tag{10.63}$$

Perturbatively, the correction to the effective Hamiltonian is $\hat{\Sigma}\left(\vec{k}, \vec{k}, E\right)$ (the self-energy depends on two wave vectors since the Hamiltonian (10.57) is not translationally invariant, but we need only terms diagonal in k). The second-order correction containing the mass term is

$$\hat{\Sigma}^{(2)}\left(\vec{k}, \vec{k}, E\right) = \sum_{\vec{k}} \hat{W}_{\vec{k}-\vec{k}'} \hat{G}_0\left(\vec{k}', E\right) \hat{W}_{\vec{k}'-\vec{k}}, \tag{10.64}$$

where

$$\hat{W}_{\vec{q}} = V_{\vec{q}} + \hat{\vec{\sigma}} \vec{A}_{\vec{q}} \tag{10.65}$$

and $\hat{G}_0$ is given by Eq. (4.36). We are interested in the gap opening at the neutrality point and thus should put $E = 0$. By substituting Eq. (4.36) into Eq. (10.64) we find

$$\hat{\Sigma}^{(2)}\left(\vec{k}, \vec{k}, 0\right) = -\frac{1}{\hbar v} \sum_{\vec{q}} \hat{W}_{\vec{k}-\vec{q}} \frac{\vec{q}\hat{\vec{\sigma}}}{q^2} \hat{W}_{\vec{q}-\vec{k}}. \tag{10.66}$$

Since

$$\hat{W}\hat{\sigma}_\alpha - V\hat{\sigma}_\alpha + A^\alpha + i\varepsilon_{\beta\alpha\gamma} A^\beta \hat{\sigma}_\gamma \tag{10.67}$$

the expression (10.66) contains the gap term $\Delta \hat{\sigma}_z$, where

$$\Delta_{\vec{k}} = \frac{2}{\hbar v} \sum_{\vec{q}} \frac{\mathrm{Im}\left[V_{\vec{k}-\vec{q}}\left(q_x A^x_{\vec{q}-\vec{k}} - q_y A^x_{\vec{q}-\vec{k}}\right)\right]}{q^2}. \tag{10.68}$$

At $\vec{k} = 0$, it can be expressed in terms of the Fourier component of the pseudo-magnetic field,

$$B_{\vec{k}} = k_x A_{\vec{k}}^y - k_y A_{\vec{k}}^x, \tag{10.69}$$

namely

$$\Delta_{\vec{k}=0} = \frac{2}{\hbar v} \sum_{\vec{q}} \frac{\mathrm{Im}\left[V_{-\vec{q}} B_{\vec{q}}\right]}{q^2} \tag{10.70}$$

(Low, Guinea, & Katsnelson, 2011).

Before discussing this expression, we derive an important result for $\partial \hat{\Sigma}/\partial E$. It follows from Eq. (10.66) and (4.36) that

$$\left.\frac{\partial \Sigma^{(2)}\left(\vec{k}, \vec{k}, E\right)}{\partial E}\right|_{E=0} = -\sum_{\vec{q}} \frac{\hat{W}_{\vec{k}-\vec{q}} \hat{W}_{\vec{q}-\vec{k}}}{(\hbar v q)^2}. \tag{10.71}$$

The integral (10.71) contains an infrared divergence at $q \to 0$, which should be cut, at some $q_{\min}$. The result is

$$\left.\frac{\partial \Sigma^{(2)}\left(\vec{k}, \vec{k}, E\right)}{\partial E}\right|_{E=0} \approx -\sum_{\vec{q}} \frac{|\ln(q_{\min}a)|}{2\pi(\hbar v)^2} \hat{W}_{\vec{k}} \hat{W}_{-\vec{k}}. \tag{10.72}$$

This divergence is very important for the theory of electron transport in graphene, as will be discussed in the next chapter.

It follows from Eq. (10.70) that the gap is determined by correlations between the electrostatic potential and the pseudomagnetic field. Let us characterize these correlations by a parameter

$$C = \lim_{\vec{k} \to 0} (BV)_{\vec{k}}, \tag{10.73}$$

which has the dimension of energy. It is roughly given by the value of the electrostatic potential times the number of flux quanta of the pseudomagnetic field over the region where the field and the electrostatic potential are correlated. The gap can be estimated, according to Eq. (10.70), as

$$\Delta \approx C \mid \ln(q_{\min}a) \mid. \tag{10.74}$$

The minimal value of q in pure graphene is determined by the gap itself,

$$q_{min} \approx \frac{\Delta}{\hbar v}, \tag{10.75}$$

so Eq. (10.74) is an equation for Δ. In dirty samples, the cut-off is determined by disorder.

Since the ripples create both an electrostatic potential and a vector potential, it is natural to ask whether this effect can result in gap opening or not. To check this, we will use the expression for the deformation tensor created by ripples with the in-plane relaxation taken into account (Guinea, Horovitz, & Le Doussal, 2008):

$$u_{\alpha\beta}\left(\vec{k}\right) = \left[\frac{\left|\vec{k}\right|^2}{2} \delta_{\alpha\beta} - \frac{\lambda + \mu}{\lambda + 2\mu} k_\alpha k_\beta \right] \frac{k_x^2 f_{yy}\left(\vec{k}\right) + k_y^2 f_{xx}\left(\vec{k}\right) - 2k_x k_y f_{xy}\left(\vec{k}\right)}{\left|\vec{k}\right|^4} \tag{10.76}$$

(cf. Eq. (10.26) for the magnetic field). On substituting Eq. (10.76), (10.7), and (10.8) into Eq. (10.70) we obtain (Low, Guinea, & Katsnelson, 2011)

$$\Delta = g \frac{\beta \, \mu(\lambda + \mu)}{a \, (\lambda + 2\mu)^2} \sum_{\vec{k}} \frac{\left| k_x^2 f_{yy}\left(\vec{k}\right) + k_y^2 f_{xx}\left(\vec{k}\right) - 2k_x k_y f_{xy}\left(\vec{k}\right) \right|^2}{\left|\vec{k}\right|^4} \cos\left(3\varphi_{\vec{k}}\right), \tag{10.77}$$

where $\varphi_{\vec{k}}$ is the polar angle of the vector $\vec{k}$. This expression is zero since on making the replacement $\vec{k} \to -\vec{k}$ the cosine changes sign $\left(\phi_{-\vec{k}} = \pi + \varphi_{\vec{k}}\right)$ and $f_{\alpha\beta}\left(-\vec{k}\right) = f_{\alpha\beta}^*\left(\vec{k}\right)$ (since the expression (10.24) is real). This means that, while the scalar and vector potentials originate from the same deformations, the gap is not open. To achieve gap opening one needs to apply an inhomogeneous electrostatic potential *together* with strains. Some specific devices of this kind were considered by Low, Guinea, and Katsnelson (2011). Under some quite realistic assumptions about parameters of the devices, a gap of the order of 0.1 eV can reasonably be expected. In general, this direction in strain engineering looks quite promising.

In this chapter we have considered only the simplest gauge field, that is, a pseudomagnetic one, which can be created by smooth deformations. Topological defects in graphene such as dislocations and disclinations can create *non-Abelian* gauge fields acting on two valleys. This issue and more formal aspects of gauge fields in graphene are reviewed by Vozmediano, Katsnelson, and Guinea (2010).

11

Scattering mechanisms and transport properties

11.1 The semiclassical Boltzmann equation and limits of its applicability

The conventional theory of electronic transport in metals and semiconductors (Ziman, 2001) is based on the *Boltzmann equation* (or kinetic equation) for the *distribution function* $f\left(\vec{k}, \vec{r}, t\right)$, which is nothing other than a probability density in the single-electron phase space (instead of the canonical variables $\vec{p}$ and $\vec{r}$, we will use $\vec{k}$ and $\vec{r}$, $\vec{k} = \vec{p}/\hbar$). It has the form (Lifshitz, Azbel, & Kaganov, 1973; Abrikosov, 1988; Vonsovsky & Katsnelson, 1989; Ziman, 2001)

$$\frac{\partial f}{\partial t} + \dot{\vec{k}} \nabla_{\vec{k}} f + \dot{\vec{r}} \nabla_{\vec{r}} f = I_{\vec{k}}[f], \tag{11.1}$$

where $\dot{\vec{k}}$ and $\dot{\vec{r}}$ are determined by the canonical equations of motion

$$\hbar \dot{\vec{k}} = e\left(\vec{E} + \frac{1}{c} \vec{v}_{\vec{k}} \times \vec{B}\right), \tag{11.2}$$

$$\dot{\vec{r}} = \vec{v}_{\vec{k}} = \frac{1}{\hbar} \frac{\partial \varepsilon\left(\vec{k}\right)}{\partial \vec{k}}, \tag{11.3}$$

where $\varepsilon(\vec{k})$ is the band dispersion and $\vec{E}$ and $\vec{B}$ are the electric and magnetic fields. The right-hand side of Eq. (11.1) is called the *collision integral.* If we neglect electron–electron scattering processes and assume that there is only elastic scattering by some external (with respect to the electron subsystem) sources, the collision integral takes the form

$$I_{\vec{k}}[f] = \sum_{\vec{k}'} w\left(\vec{k}, \vec{k}'\right) \left[f_{\vec{k}'}\left(1 - f_{\vec{k}}\right) - f_{\vec{k}}\left(1 - f_{\vec{k}'}\right)\right] = \sum_{\vec{k}'} w\left(\vec{k}, \vec{k}'\right)\left(f_{\vec{k}'} - f_{\vec{k}}\right), \tag{11.4}$$

where $w(\vec{k}, \vec{k}')$ is the quantum-mechanical scattering probability and the factors $(1 - f)$ in Eq. (11.4) take into account the Pauli principle forbidding scattering into occupied states. One can see, however, that these factors are not essential. If the scattering Hamiltonian has the form

$$\hat{H}' = \sum_{\vec{k}\,\vec{k}} V_{\vec{k}\,\vec{k}'} \hat{c}_{\vec{k}}^{+} \hat{c}_{\vec{k}'} \tag{11.5}$$

and V is a static potential (quenched disorder) then, in the Born approximation, according to "Fermi's golden rule,"

$$w(\vec{k}, \vec{k}') = \frac{2\pi}{\hbar} \left\langle \left| V_{\vec{k}\,\vec{k}'} \right|^2 \right\rangle \delta \left(\varepsilon_{\vec{k}} - \varepsilon_{\vec{k}'} \right) \tag{11.6}$$

(angular brackets denote the average over the states of the scatterers). Note that in this approximation $w(\vec{k}, \vec{k}') = w(\vec{k}', \vec{k})$, which is already taken into account in Eq. (11.4). For simplicity, we omit spin indices and do not take into account summation over them; otherwise, the right-hand side of Eq. (11.4) should be multiplied by 2, the spin degeneracy factor.

Here, we will consider only a linear response, assuming that the external electric field $\vec{E}$ is small enough. Then,

$$f_{\vec{k}}\left(\vec{r}, t\right) = f_0 \left(\varepsilon_{\vec{k}}\right) + \delta f_{\vec{k}}\left(\vec{r}, t\right), \tag{11.7}$$

where $f_0(\varepsilon)$ is the Fermi–Dirac distribution function, and we need to take into account only linear terms in Eq. (11.1). Then, the collision integral is

$$I_{\vec{k}}[f] = \sum_{\vec{k}'} w\left(\vec{k}, \vec{k}'\right) \left(\delta f_{\vec{k}'} - \delta f_{\vec{k}} \right). \tag{11.8}$$

The current and the perturbation of the electron charge density can be calculated as

$$\vec{j}\left(\vec{r}, t\right) = e \sum_{\vec{k}} \vec{v}_{\vec{k}} \delta f_{\vec{k}}, \tag{11.9}$$

$$\delta\rho\left(\vec{r}, t\right) = e \sum_{\vec{k}} \delta f_{\vec{k}}. \tag{11.10}$$

The rigorous quantum-mechanical derivation of the Boltzmann equation from fundamental physical laws, that is, from the Schrödinger equation, is a very complicated problem. It is part of the general problem of the derivation of statistical physics and of macroscopic irreversibility (the Boltzmann equation is irreversible; that is, it has no time-reversal symmetry, whereas the Schrödinger equation does have time-reversal symmetry); see, e.g., Zubarev (1974), Ishihara (1971), and Balescu (1975). For the particular case of elastic scattering with randomly distributed impurities

$$V\left(\vec{r}\right) = \sum_i u\left(\vec{r} - \vec{R}_i\right) \tag{11.11}$$

($\vec{R}_i$ are their positions), the problem was solved by Kohn and Luttinger (1957). The idea was as follows. First, the Schrödinger equation is equivalent to Eq. (2.173) for the density matrix

$$\rho_{\vec{k},\vec{k}'} = \left\langle \hat{c}_{\vec{k}'}^{+} \hat{c}_{\vec{k}} \right\rangle \tag{11.12}$$

(cf. Eq. (2.170)). For the case of a spatially uniform system,

$$f_{\vec{k}} = \rho_{\vec{k},\vec{k}}. \tag{11.13}$$

One can prove that, if V is weak enough, the off-diagonal terms of the density matrix (11.12) are small in comparison with the diagonal ones, with the latter satisfying the Boltzmann equation (11.1), (11.4), and (11.6). Assuming a random distribution of the impurities, one has

$$\left\langle \left| V_{\vec{k}\vec{k}'} \right|^2 \right\rangle = n_{\text{imp}} \left| u_{\vec{k}-\vec{k}'} \right|^2, \tag{11.14}$$

where n_{imp} is the impurity concentration. Luttinger and Kohn (1958) proved that if n_{imp} is small, one can repeat the whole derivation without assuming the smallness of potential u, and Eq. (11.1), (11.4), and (11.6) remain correct, but with replacement of the potential $\hat{u}$ by the single-site $\hat{T}$-matrix:

$$\left\langle \left| V_{\vec{k}\vec{k}'} \right|^2 \right\rangle = n_{\text{imp}} \left| T_{\vec{k}\vec{k}'} \left(E = \varepsilon_{\vec{k}} \right) \right|^2. \tag{11.15}$$

This result has already been mentioned and was used in Chapter 6; see e.g., Eq. (6.22) through (6.26).

If neither the potential nor the concentration of the defects is small, the Boltzmann equation is, in general, incorrect. For example, it does not take into account the effects of Anderson localization, which are crucially important for *strongly* disordered systems (Mott, 1974; Mott & Davis, 1979; Shklovskii & Efros, 1984; Lifshitz, Gredeskul, & Pastur, 1988).

Some general and powerful tools with which to derive kinetic equations, such as Kadanoff–Baym nonequilibrium Green functions and the Keldysh diagram technique for their calculation (Kadanoff & Baym, 1962; Keldysh, 1964;

Rammer & Smith, 1986; Wagner, 1991; Kamenev & Levchenko, 2009; Kamenev, 2011; Stefanucci & van Leeuwen, 2013) and the nonequilibrium statistical operator (NSO) method and similar approaches (Zubarev, 1974; Kalashnikov & Auslender, 1979; Akhiezer & Peletminskii, 1981; Luzzi, Vasconcellos, & Ramos, 2000; Kuzemsky, 2005) were developed thereafter. They are all based on the idea of a *coarse-grained* description. If the disorder is weak (due to either weakness of the scattering potential or smallness of the concentration of defects) the off-diagonal elements of the density matrix have a very fast dynamics in comparison with that of the diagonal ones and can be eliminated. On time scales much larger than typical electron times (e.g., $\hbar/|t|$, where t is the hopping integral) the dynamics of the whole system can be described by a small number of degrees of freedom (we have N_0 diagonal elements (11.13) and N_0^2 elements of the total density matrix (11.12)). If there are no small parameters in the problem under consideration, the coarse-grained approach cannot be justified and one needs other methods (see, e.g., Efetov, 1997; Evers & Mirlin, 2008).

Earlier we discussed the case of a spatially uniform system. If we have inhomogeneities on an atomic scale and no small parameters, the kinetic equation does not work. For the case of smooth enough inhomogeneities, the Boltzmann equation (11.1) can be justified for the *Wigner distribution function*

$$f_{\vec{k}}\left(\vec{r},t\right) = \int d\vec{\xi}\ \exp\left(-i\vec{k}\vec{\xi}\right)\rho\left(\vec{r}+\frac{\vec{\xi}}{2},\vec{r}-\frac{\vec{\xi}}{2};t\right), \qquad (11.16)$$

where $\rho\left(\vec{r},\vec{r}'\right) = \left\langle \hat{\psi}^+\left(\vec{r}'\right)\hat{\psi}\left(\vec{r}\right)\right\rangle$ is the density matrix in the coordinate representation (Kadanoff & Baym, 1962). Henceforth we will not consider the inhomogeneous case. We also restrict ourselves to the case of dc transport with a time-independent $\vec{E}$. Therefore, the terms with $\partial/\partial t$ and $\nabla_{\vec{r}}$ in Eq. (11.1) can be neglected.

For the case of graphene, the applicability of the Boltzmann equation is not obvious. In the standard theory of electron transport in solids, the current operator commutes with the unperturbed Hamiltonian $\hat{H}_0$, thus we start with states that have simultaneously well-defined values of energy and well-defined values of momentum. The perturbation $\hat{H}'$ does not commute with the current operator, leading to scattering between these states. For the Dirac Hamiltonian (3.1), the current operator (3.2) does *not* commute with it (*Zitterbewegung*, see Chapter 3). At the same time, for the case of a scalar potential

$$\hat{H}' = \sum_{\vec{k}\vec{k}'} \hat{\psi}^+_{\vec{k}} V_{\vec{k}\vec{k}'} \hat{\psi}_{\vec{k}'}, \qquad (11.17)$$

with V proportional to the unit matrix in pseudospin space, the current operator commutes with $\hat{H}'$. It is not at all clear how important this huge formal difference can be. Also, it is not clear when interband scattering processes can be neglected; thus, at least, instead of the scalar quantity (11.13), one needs to consider the matrix (7.15) in pseudospin space. If we have atomically sharp scattering, the valley index should also be taken into account, but we will not consider that case here. The *matrix* Boltzmann equation for the case of graphene has been derived by Auslender and Katsnelson (2007) (see also Kailasvuori & Lüffe, 2010; Trushin et al., 2010). They used the NSO approach; Kailasvuori and Lüffe (2010) used the Keldysh diagram technique and discussed the relation between these two approaches.

The corresponding derivations are rather complicated and cumbersome, but the physical results are quite clear. Therefore, we will only present the general idea and the answers here.

First, let us diagonalize the Dirac Hamiltonian by the transformation (7.23) to the form (7.24). The scattering operator (11.5) takes the form

$$\hat{H}' = \sum_{\vec{k}\,\vec{k}'} \hat{\Xi}^+_{\vec{k}} V'_{\vec{k}\,\vec{k}'} \hat{\Xi}^+_{\vec{k}'},\tag{11.18}$$

where $\hat{\Xi}^+_{\vec{k}} = \left(\hat{\varsigma}^+_{\vec{k}1}, \hat{\varsigma}^+_{\vec{k}2}\right)$ and

$$V_{\vec{k}\,\vec{k}'} \rightarrow V'_{\vec{k}\,\vec{k}'} = \frac{1}{2} V_{\vec{k}\,\vec{k}'} \begin{pmatrix} 1 + \exp\left[i\left(\varphi_{\vec{k}'} - \varphi_{\vec{k}}\right)\right] & 1 - \exp\left[i\left(\varphi_{\vec{k}'} - \varphi_{\vec{k}}\right)\right] \\ 1 - \exp\left[i\left(\varphi_{\vec{k}} - \varphi_{\vec{k}'}\right)\right] & 1 + \exp\left[i\left(\varphi_{\vec{k}'} - \varphi_{\vec{k}}\right)\right] \end{pmatrix}.\tag{11.19}$$

It contains both diagonal and nondiagonal elements. In the NSO method, one first needs to postulate the set of "coarse-grained" variables for which a closed set of equations of motion is assumed to exist. In our case, this is the 2×2 density matrix $\left\langle \hat{\Xi}^+_{\vec{k}} \hat{\Xi}_{\vec{k}} \right\rangle$ or, equivalently,

$$D_{\vec{k}} = \left\langle \hat{\varsigma}^+_{\vec{k}1} \hat{\varsigma}_{\vec{k}1} \right\rangle + \left\langle \hat{\varsigma}^+_{\vec{k}2} \hat{\varsigma}_{\vec{k}2} \right\rangle - 1,$$

$$N_{\vec{k}} = \left\langle \hat{\varsigma}^+_{\vec{k}1} \hat{\varsigma}_{\vec{k}1} \right\rangle + 1 - \left\langle \hat{\varsigma}^+_{\vec{k}2} \hat{\varsigma}_{\vec{k}2} \right\rangle,\tag{11.20}$$

$$g_{\vec{k}} = \left\langle \hat{\varsigma}^+_{\vec{k}1} \hat{\varsigma}_{\vec{k}2} \right\rangle = \left\langle \hat{\varsigma}^+_{\vec{k}2} \hat{\varsigma}_{\vec{k}1} \right\rangle^*.$$

Note that the function $g_{\vec{k}}$ is complex. The generalized Boltzmann equation to second order in V reads (Auslender & Katsnelson, 2007)

$$\frac{\partial D_{\vec{k}}}{\partial t} + \frac{eE}{\hbar}\frac{\partial D_{\vec{k}}}{\partial k_x} = -\frac{2\pi}{\hbar}\sum_{\vec{q}}\left|V_{\vec{k},\vec{q}}\right|^2 \cos^2\left(\frac{\varphi_{\vec{k}} - \varphi_{\vec{q}}}{2}\right)\delta\left(\varepsilon_{\vec{k}} - \varepsilon_{\vec{q}}\right)\left(D_{\vec{k}} - D_{\vec{q}}\right),$$

(11.21)

$$\frac{\partial N_{\vec{k}}}{\partial t} + eE\frac{\partial N_{\vec{k}}}{\partial k_x} - \frac{2eE\sin\phi_{\vec{k}}}{\hbar k}\,\mathrm{Img}_{\vec{k}}$$

$$= \frac{2\pi}{\hbar}\sum_{\vec{q}}\left|V_{\vec{k},\vec{q}}\right|^2\left\{\frac{1}{\pi}\sin\left(\varphi_{\vec{k}} - \varphi_{\vec{q}}\right)\mathrm{Reg}_{\vec{q}}\left(\frac{1}{\varepsilon_{\vec{q}} + \varepsilon_{\vec{k}}} + \frac{1}{\varepsilon_{\vec{q}} - \varepsilon_{\vec{k}}}\right)\right.$$

$$\left. - \left[\cos^2\left(\frac{\varphi_{\vec{k}} - \varphi_{\vec{q}}}{2}\right)\left(N_{\vec{k}} - N_{\vec{q}}\right) + \sin\left(\varphi_{\vec{k}} - \varphi_{\vec{q}}\right)\mathrm{Img}_{\vec{q}}\right]\right\}\delta\left(\varepsilon_{\vec{k}} - \varepsilon_{\vec{q}}\right),$$

(11.22)

$$\frac{\partial g_{\vec{k}}}{\partial t} - 2ivkg_{\vec{k}} + \frac{eE}{\hbar}\frac{\partial g_{\vec{k}}}{\partial k_x} + \frac{iE}{2\hbar k}\left(N_{\vec{k}} - 1\right)\sin\varphi_{\vec{k}}$$

$$= -\frac{\pi}{\hbar}\sum_{\vec{q}}\left|V_{\vec{k},\vec{q}}\right|^2\left\{-\frac{i}{2}\sin\left(\varphi_{\vec{k}} - \varphi_{\vec{q}}\right)D_{\vec{q}}\left[\delta\left(\varepsilon_{\vec{k}} - \varepsilon_{\vec{q}}\right) + \frac{i}{\pi}\frac{1}{\varepsilon_{\vec{k}} - \varepsilon_{\vec{q}}}\right]\right.$$

$$+ 2\cos^2\left(\frac{\varphi_{\vec{k}} - \varphi_{\vec{q}}}{2}\right)\left[\left(g_{\vec{k}} - g_{\vec{q}}\right)\delta\left(\varepsilon_{\vec{k}} - \varepsilon_{\vec{q}}\right) + \frac{i}{\pi}\frac{g_{\vec{k}} + g_{\vec{q}}}{\varepsilon_{\vec{k}} - \varepsilon_{\vec{q}}}\right]$$

$$\left. + \frac{1}{2\pi}\frac{N_{\vec{q}}}{\varepsilon_{\vec{k}} + \varepsilon_{\vec{q}}}\sin\left(\varphi_{\vec{k}} - \varphi_{\vec{q}}\right) - \frac{2i}{\pi}\frac{g_{\vec{k}} + g_{\vec{q}}^*}{\varepsilon_{\vec{k}} + \varepsilon_{\vec{q}}}\sin^2\left(\frac{\varphi_{\vec{k}} - \varphi_{\vec{q}}}{2}\right)\right\},$$

(11.23)

where $\varepsilon_{\vec{k}} = \hbar vk$ and the electric field E is supposed to be directed along the x-axis. The current is expressed in terms of these functions as

$$j_x = ev\sum_{\vec{q}}\left(N_{\vec{q}}\cos\varphi_{\vec{q}} + 2\sin\varphi_{\vec{q}}\mathrm{Im}\,g_{\vec{q}}\right).$$

(11.24)

The Eq. (11.21) is decoupled from Eq. (11.22) and (11.23) and is formally equivalent to the usual Boltzmann equation (11.1), (11.4), and (11.6), but the other two equations have an essentially different structure. The most important difference is that the "collision integral" now contains not only "dissipative" terms with $\delta(\varepsilon_{\vec{k}} - \varepsilon_{\vec{q}})$ but also "reactive" terms with $1/(\varepsilon_{\vec{k}} \pm \varepsilon_{\vec{q}})$. These terms are associated with virtual interband transitions, that is, with *Zitterbewegung* (see Chapter 3). As a result, the linearized kinetic equations are singular, and their solutions contain logarithmic divergences at small enough chemical potential μ and temperature T. For the case of the contact potential $V_{\vec{k},\vec{q}} = \text{constant}$, these integral equations can be solved exactly (Auslender & Katsnelson, 2007).

First, let us neglect off-diagonal terms, that is, $g_{\vec{k}}$. Then we will have the standard Boltzmann equation for the Dirac fermions and the corresponding expression for the resistivity (6.23) with the inverse Drude mean-free-path time (Shon & Ando, 1998)

$$\frac{1}{\tau_{\vec{k}}} = \frac{\pi}{\hbar} \sum_{\vec{q}} \left| V_{\vec{k},\vec{q}} \right|^2 \sin^2 \left(\varphi_{\vec{k}} - \varphi_{\vec{q}} \right) \delta \left(\varepsilon_{\vec{k}} - \varepsilon_{\vec{q}} \right)$$

$$= \frac{\pi \varepsilon_{\vec{k}}}{(2\pi\hbar v)^2} n_{\text{imp}} \int_0^{2\pi} d\varphi \left| u \left(2k \sin \frac{\varphi}{2} \right) \right|^2 \sin^2 \varphi \tag{11.25}$$

where $u(q)$ is the Fourier component of $u(r)$ from Eq. (11.11).

If we now find the off-diagonal terms of the density matrix $g_{\vec{k}}$, by iterations we will see that they have a smallness in the parameter

$$\lambda = \frac{\hbar}{|\varepsilon_F| \tau(k_F)} \sim \frac{e^2 \rho_B}{h}, \tag{11.26}$$

where ρ_B is the resistivity (6.23) and (11.25) calculated by applying the ordinary semiclassical Boltzmann equation. If we go closer to the neutrality point the off-diagonal terms are divergent. For the case of the contact potential, the exact solution of the integral equations mentioned previously gives a typical energy scale (Auslender & Katsnelson, 2007)

$$\varepsilon_K = W \exp \left(-\frac{\pi h}{e^2 \rho_B} \right), \tag{11.27}$$

where W is a cut-off energy of the order of the bandwidth. The conventional Boltzmann equation is valid if

$$|\varepsilon_F|, \, T \gg \varepsilon_K. \tag{11.28}$$

The subscript K in Eq. (11.27) refers to Kondo, due to a formal similarity between the energy scale discussed here and the Kondo effect in the scattering of electrons in metals by a magnetic impurity (Kondo, 1964; Hewson, 1993). In that case, due to spin-flip processes involved in the scattering, a resonant singlet state is formed ("Kondo resonance"), which, being considered perturbatively, leads to logarithmic corrections in the temperature dependences of various physical quantities. It is important that the spin-up and spin-down states of the impurities are degenerate. A magnetic field kills this degeneracy and suppresses the Kondo effect. The scattering potential (11.19) contains off-diagonal matrix elements between electron and hole bands. At $\mu = 0$, these bands are degenerate, and an analog of the Kondo effect arises, making the standard Born approximation insufficient. A finite chemical potential μ plays the same role as the magnetic field in the Kondo effect. The condition (11.28) guarantees that all singularities are suppressed. One can see that this is equivalent to the condition

$$\lambda \ll 1, \tag{11.29}$$

which is the desired criterion of applicability of the standard semiclassical Boltzmann theory. In the vicinity of the neutrality point we are in the "strong-coupling" regime. Note that Eq. (11.21) through (11.23) is probably insufficient in this case. As was emphasized previously, in the situation without any smallness of disorder, other methods have to be applied. They will be briefly discussed later in this chapter (Section 11.7). The main role of the approach considered here is that it justifies the use of the standard Boltzmann equation under the condition (11.29).

Note that these "Kondo" logarithms are related to the divergence of $\partial \Sigma / \partial E$ at the neutrality point Eq. (10.79). From another point of view and in a different context (Dirac fermions in d-wave superconductors), these logarithms were discussed by Lee (1993), Nersesyan, Tsvelik, and Wenger (1994), and Ziegler (1998).

Thus, not too close to the neutrality point, namely at

$$\sigma = \frac{1}{\rho} \gg \frac{e^2}{h}, \tag{11.30}$$

the interband transitions are negligible. If we assume, to be specific, that the Fermi energy lies in the electron band, then only the (1,1) matrix element of the current operator and that of the scattering potential are relevant:

$$\left(j_{\vec{k}x} \right)_{1,1} = ev \cos \varphi_{\vec{k}}, \tag{11.31}$$

$$\left(V'_{\vec{k}\vec{k}'} \right)_{1,1} = V_{\vec{k}\vec{k}'} \frac{1 + \exp\left[i\left(\varphi_{\vec{k}'} - \varphi_{\vec{k}} \right) \right]}{2}. \tag{11.32}$$

Let us consider the most general form of the scattering potential $V_{\vec{k}\vec{k}'}$ in Eq. (11.17):

$$V_{\vec{k}\vec{k}'} = V_{\vec{k}\vec{k}'}^{(0)} + \vec{V}_{\vec{k}\vec{k}'} \cdot \vec{\sigma}. \tag{11.33}$$

Then, the effective scattering potential will be

$$V_{\vec{k}\,\vec{k}'}^{eff} = \left(V'_{\vec{k}\,\vec{k}'}\right)_{1,1} = V_{\vec{k}\,\vec{k}'}^{(0)}\frac{1+\exp\left[i\left(\varphi_{\vec{k}'}-\varphi_{\vec{k}}\right)\right]}{2} + V_{\vec{k}\,\vec{k}'}^{z}\frac{1-\exp\left[i\left(\varphi_{\vec{k}'}-\varphi_{\vec{k}}\right)\right]}{2}$$

$$+\left(V_{\vec{k}\,\vec{k}'}^{x}+iV_{\vec{k}\,\vec{k}'}^{y}\right)\exp\left(-i\varphi_{\vec{k}}\right)+\left(V_{\vec{k}\,\vec{k}'}^{x}-iV_{\vec{k}\,\vec{k}'}^{y}\right)\exp\left(i\varphi_{\vec{k}'}\right).$$

$$(11.34)$$

Thus, under the condition (11.30) we have a single-band problem with the unperturbed Hamiltonian

$$\hat{H}_0 = \sum_{\vec{k}}\hbar vk\hat{\xi}_{\vec{k}}^{+}\hat{\xi}_{\vec{k}},\qquad(11.35)$$

current operator

$$\hat{j}_x = \sum_{\vec{k}}ev\cos\varphi_{\vec{k}}\hat{\xi}_{\vec{k}}^{+}\hat{\xi}_{\vec{k}}\qquad(11.36)$$

and scattering operator

$$\hat{H}' = \sum_{\vec{k}}V_{\vec{k}\,\vec{k}'}^{ef}\hat{\xi}_{\vec{k}}^{+}\hat{\xi}_{\vec{k}'},\qquad(11.37)$$

where we will omit the label "1" for electron operators. In the next section we will present a convenient and general tool that can be used to find the resistivity in this problem.

11.2 The Kubo–Nakano–Mori formula for resistivity

In general, the linearized Boltzmann equation is an integral equation that can only be solved exactly in some special cases (e.g., for contact interaction $u(\vec{r}-\vec{R}_1)$ in Eq. (11.11)). Usually, a variational approach (Ziman, 2001) is used. However, within the Born approximation there is a more straightforward way to calculate transport properties. It is based on the use of the *Kubo–Nakano–Mori* formula (Kubo, 1957; Nakano, 1957; Mori, 1965) for the resistivity. It gives exactly the same result as the solution of Boltzmann equation by the variational approach but in a technically simpler way. Since this method seems not to be well known in graphene community, we will present it here following Mori (1965). It will allow us also to illustrate the idea of coarse graining, which is fundamental for the nonequilibrium statistical mechanics and which was discussed preliminarily in the previous section.

Let us start with the Kubo formula (3.7) for σ_{xx}. It can be rewritten as

$$\sigma_{xx}(\omega) = \beta\int_0^\infty dt\exp\left(i\omega t\right)\left(\hat{j}_x(t),\hat{j}_x\right),\qquad(11.38)$$

where

$$(\hat{A}, \hat{B}) = \frac{1}{\beta} \int\limits_0^\beta d\lambda \left\langle \exp\left(\lambda\hat{H}\right)\hat{A} \exp\left(-\lambda\hat{H}\right)\hat{B}^+ \right\rangle. \tag{11.39}$$

Here, $\hat{H}$ is the Hamiltonian of the system and we put the area of the sample equal to 1. Importantly, if we consider *operators* $\{\hat{A}\}$ as *vectors* in some linear space, Eq. (11.39) determines the *scalar product* in this space and satisfies all of the axioms of the scalar product.

The operator equation of motion is

$$\frac{d\hat{A}(t)}{dt} = i\hat{L}\hat{A}(t), \tag{11.40}$$

where

$$\hat{L}\hat{A} \equiv \left[\hat{H}, \hat{A}\right] \tag{11.41}$$

is the *Liouville (super) operator*. "Super" means that it acts as an operator in the vector space of quantum-mechanical Hermitian operators. Here we put $\hbar = 1$ for simplicity.

Let us assume that $\{\hat{A}\}$ form a set of operators such that their dynamics is closed, that is, $\{\langle\hat{A}(t)\rangle\}$ at any time t is determined by initial conditions $\{\langle\hat{A}(0)\rangle\} \equiv \{\langle\hat{A}\rangle\}$. This implies the coarse-grained dynamics. A technical advantage of Mori's approach is that we use far fewer operators than in the kinetic equation, just current operators, but with almost the same accuracy.

Since Eq. (11.39) defines the scalar product in our vector space, one can introduce a *projection operator* of any set of operators $\{\hat{G}\}$ on the initial set $\{\hat{A}\}$:

$$\hat{P}_0\hat{G} = (\hat{G}, \hat{A}) \cdot (\hat{A}, \hat{A})^{-1} \cdot \hat{A}, \tag{11.42}$$

where the dot denotes the matrix product, e.g.,

$$\left[(\hat{G}, \hat{A}) \cdot \hat{A}\right]_i = \sum_j (G_i, \hat{A}_j)\hat{A}_j \tag{11.43}$$

and i and j label operators within the set $\{\hat{A}\}$. Thus, $\hat{A}(t)$ can be represented as a sum of "projective" and "perpendicular" components with respect to $\{\hat{A}\}$:

$$\hat{A}(t) = \Xi_0(t)\hat{A} + \hat{A}'(t), \tag{11.44}$$

where

$$\Xi_0(t) = (\hat{A}(t), \hat{A}) \cdot (\hat{A}, \hat{A})^{-1} \tag{11.45}$$

and

$$\hat{A}'(t) = (1 - \hat{P}_0)\hat{A}(t).$$
(11.46)

Next, we can derive the equation of motion for $\hat{A}'$. Acting by $(1 - \hat{P}_0)$ on Eq. (11.40) we find

$$\frac{d\hat{A}'(t)}{dt} - i\hat{L}_1\hat{A}'(t) = \Xi_0(t)\hat{f}_1,$$
(11.47)

where

$$\hat{L}_1 = (1 - \hat{P}_0)\hat{L},$$
(11.48)

$$\hat{f}_1 = i\hat{L}_1\hat{A}.$$
(11.49)

It has the formal solution

$$\hat{A}'(t) = \int_0^t ds \Xi_0(s)\hat{f}_1(t - s),$$
(11.50)

where we take into account that $\hat{A}'(0) = 0$ and

$$\hat{f}_1(t) = \exp(i\hat{L}_1 t)\hat{f}_1.$$
(11.51)

Eq. (11.50) represents a convolution. On taking the Laplace transform

$$\hat{A}(z) = \int_0^\infty dt \exp(-zt)\hat{A}(t),$$
(11.52)

we find

$$\hat{A}(z) = \Xi_0(z)\cdot\left[\hat{A} + \hat{f}_1(z)\right].$$
(11.53)

As the next step, we have to repeat the procedure for $\hat{f}_1$. It satisfies the equation of motion

$$\frac{d\hat{f}_1}{dt} = i\hat{L}_1 \hat{f}_1.$$
(11.54)

We can reproduce it as

$$\hat{f}_1(t) = \Xi_1(t)\cdot \hat{f}_1 + \hat{f}'(t),$$
(11.55)

where

$$\Xi_1(t) = \left(\hat{f}_1(t),\hat{f}_1\right)\cdot\left(\hat{f}_1,\hat{f}_1\right)^{-1}$$
(11.56)

and

$$\hat{f}'_1(t) = (1 - \hat{P}_1)\hat{f}_1(t), \tag{11.57}$$

where $\hat{P}_1$ is the projection operator onto $\{\hat{f}_1\}$. Further, we will have for the Laplace transform,

$$\hat{f}_1(z) = \Xi_1(z) \cdot [\hat{f}_1 + \hat{f}_2(z)], \tag{11.58}$$

which is similar to Eq. (11.53), and where

$$\hat{f}_2(t) = \exp\left(i\hat{L}_2 t\right) i\hat{L}_2 \hat{f}_1, \tag{11.59}$$

$$\hat{L}_2 = \left(1 - \hat{P}_1\right)\hat{L}_1. \tag{11.60}$$

Treating $\hat{f}_2$ in a similar way, we introduce a new object $\hat{f}_3$, etc., so that we will have a set of quantities $\{\hat{f}_j(t)\}$ $(\hat{f}_0 = \hat{A})$ defined iteratively as

$$\hat{f}_j(t) = \exp\left(i\hat{L}_j t\right) i\hat{L}_j \hat{f}_{j-1}, \tag{11.61}$$

where

$$\hat{L}_j = \left(1 - \hat{P}_{j-1}\right)\hat{L}_{j-1}, \qquad \hat{L}_0 = \hat{L}, \tag{11.62}$$

and $\hat{P}_j$ is the projection operator onto $\{\hat{f}_j\}$. The Laplace transforms of $\hat{f}_j$ satisfy the chain of equations

$$\hat{f}_j(z) = \Xi_j(z) \cdot [\hat{f}_j + \hat{f}_{j+1}(z)]. \tag{11.63}$$

As a result, we derive a *continued-fraction* representation of the correlators (Mori, 1965):

$$\Xi_0(z) = \cfrac{1}{z - i\omega_0 - \Delta_0^2 \Xi_1(z)}, \tag{11.64}$$

$$\Xi_1(z) = \cfrac{1}{z - i\omega_1 - \Delta_1^2 \Xi_2(z)}, \tag{11.65}$$

etc., where

$$i\omega_j = \left(\dot{f}_j, f_j\right) \cdot \left(f_j, f_j\right)^{-1}, \tag{11.66}$$

$$\Delta_j^2 = \left(f_j, f_j\right) \cdot \left(f_{j-1}, f_{j-1}\right)^{-1}. \tag{11.67}$$

Let us apply this general scheme to the conductivity. We have to choose as the first step $\hat{A} = (\hat{j}_x, \hat{j}_y)$. Next, we have to calculate ω_0. This can easily be done using the identity

$$\left(\hat{A},\hat{B}\right) = \frac{i}{\beta}\int_0^\beta d\lambda \left\langle \exp\left(\lambda\hat{H}\right)\left[\hat{H},\hat{A}\right]\exp\left(-\lambda\hat{H}\right)\hat{B}^+\right\rangle$$

$$= \frac{i}{\beta}\int_0^\beta d\lambda \frac{d}{d\lambda}\left\langle \exp\left(\lambda\hat{H}\right)\hat{A}\exp\left(-\lambda\hat{H}\right)\hat{B}^+\right\rangle \qquad (11.68)$$

$$= \frac{i}{\beta}\left[\left\langle \exp\left(\beta\hat{H}\right)\hat{A}\exp\left(-\beta\hat{H}\right)\hat{B}^+\right\rangle - \left\langle\hat{A}\hat{B}^+\right\rangle\right]$$

$$= \frac{i}{\beta}\left\langle\left[\hat{B}^+,\hat{A}\right]\right\rangle,$$

where we take into account that

$$\left\langle\hat{A}\right\rangle = \mathrm{Tr}\left[\exp\left(-\beta\hat{H}\right)\hat{A}\right]/Z \qquad (11.69)$$

and implement the cyclic permutation under the trace symbol.

In the absence of a magnetic field, the average values of all of the commutators of the current operator are zero (in particular $\langle\hat{\sigma}_z\rangle = 0$), so one can conclude that $\omega_0 = 0$. Also, one can conclude by symmetry arguments that

$$(j_\alpha, j_\beta) = \delta_{\alpha\beta}(j_x, j_x). \qquad (11.70)$$

Let us stop the procedure at the first step, neglecting $\hat{f}_2$ and all higher-order terms. Then, the result for the conductivity (11.38) will be

$$\sigma_{xx}(\omega) = \frac{\beta(j_x, j_x)}{-i\omega + 1/(j_i, j_x)\left(\left[\hat{j}_x\hat{H}\right], \left[\hat{H}, \hat{j}_x\right]\right)_{z=-i\omega}}. \qquad (11.71)$$

Since within the single-band approximation (11.35) through (11.37), the current operator commutes with $\hat{H}_0$, one can replace $\left[\hat{j}_x, \hat{H}\right]$ by $\left[\hat{j}_x, \hat{H}'\right]$ a result, Eq. (11.71) takes the form

$$\sigma_{xx}(\omega) = \frac{\beta(j_x, j_x)}{-i\omega + 1/\tau(\omega)}, \qquad (11.72)$$

where

$$\frac{1}{\tau(\omega)} = \frac{1}{(j_x, j_x)}\int_0^\infty dt\, \exp\left(i\omega t\right)\left(F_x(t-i\lambda), F_x^+\right) \qquad (11.73)$$

and

$$F_x = \left[\hat{j}_x, \hat{H}'\right]. \qquad (11.74)$$

To calculate (j_x, j_x) one can neglect the scattering operator $\hat{H}'$. Then, taking into account that $[\hat{j}_x, \hat{H}] = 0$, we have

$$(j_x, j_x) = \langle \hat{j}_x^2 \rangle. \tag{11.75}$$

By substituting Eq. (11.38) into Eq. (11.75) and using Wick's theorem we find

$$
\begin{aligned}
(j_x, j_x) &= \sum_k e^2 v^2 \cos^2 \varphi_k \left\langle \hat{\xi}_{k1}^+ \hat{\xi}_{k1} \right\rangle \left\langle \hat{\xi}_{k1} \hat{\xi}_{k1}^+ \right\rangle \\
&= \sum_k e^2 v^2 \cos^2 \varphi_k f(\varepsilon_k)[1 - f(\varepsilon_k)] \\
&= \frac{1}{2\beta} \sum_k e^2 v^2 \left(-\frac{\partial f(\varepsilon_k)}{\partial \varepsilon_k} \right),
\end{aligned}
\tag{11.76}
$$

where we average $\cos^2 \varphi_k \to \frac{1}{2}$. At $T \ll |\varepsilon_F|$ the result is

$$\beta(j_x, j_x) = e^2 \frac{N(\varepsilon_F) v^2}{2}. \tag{11.77}$$

On comparing Eq. (11.72) and (11.77) with Eq. (6.23) one can see that Eq. (11.72) is nothing other than the Drude formula, and $\tau(\omega = 0)$ given by Eq. (11.73) is nothing other than the mean-free-path time. At $\omega = 0$ it can be simplified, similarly to the transformation from Eq. (3.7) to Eq. (3.8):

$$\frac{1}{\tau} = \frac{1}{2\langle \hat{j}_x^2 \rangle} \int_{-\infty}^{\infty} dt \langle F_x(t) F_x^+ \rangle. \tag{11.78}$$

This, together with Eq. (6.23), gives us the *Kubo-Nakano-Mori formula* for the resistivity. As has already been mentioned, it is equivalent to the solution of the semiclassical Boltzmann equation by the variational approach (Ziman, 2001).

By substituting Eq. (11.36) and (11.37) into Eq. (11.74) and (11.78) and calculating the average using Wick's theorem, we find, finally (we restore here the Planck constant), the expression for the momentum relaxation rate of Dirac fermions:

$$\frac{1}{\tau} = \frac{2\pi}{\hbar N(\varepsilon_F)} \sum_{\vec{k}\vec{k}'} \delta\left(\varepsilon_{\vec{k}} - \varepsilon_F\right) \delta\left(\varepsilon_{\vec{k}'} - \varepsilon_F\right) \left(\cos \varphi_{\vec{k}} - \cos \varphi_{\vec{k}'} \right)^2 \left| V_{\vec{k}\vec{k}'}^{\text{eff}} \right|^2. \tag{11.79}$$

Together with Eq. (11.34), this allows us to analyze various scattering mechanisms.

11.3 Scattering mechanisms in graphene on a substrate

There are two fundamental experimental facts about the conductivity of graphene on a substrate. First, the dependence of the conductivity on the charge-carrier concentration n typically has a V-shape (Novoselov et al., 2004, 2005a; Zhang et al., 2005). If we introduce the mobility μ via the relation

$$\sigma = ne\mu \qquad (11.80)$$

this means that μ is weakly dependent on the concentration and $\sigma \sim n$ except in the close proximity of the neutrality point. Typical results (Novoselov et al., 2005a) are shown in Fig. 11.1 (note that n is proportional to the gate voltage). This behavior has been confirmed by numerous works by many experimental groups and seems to be universal. It does not depend on the type of substrate, but the value of μ does. Whereas for graphene on SiO_2 one typically has $\mu \approx 10^4$ cm^{2}V^{-1}s^{-1} (Novoselov et al., 2004, 2005a; Zhang et al., 2005), for graphene on hexagonal BN μ can be an order of magnitude higher (Dean et al., 2010).

Second, for graphene on a substrate, the temperature dependence of conductivity is extremely weak. If one tries to separate "extrinsic" (due to defects) and "intrinsic" (e.g., due to electron–phonon interaction) contributions to the mobility using Matthiessen's rule (Ziman, 2001)

$$\frac{1}{\mu(T)} = \frac{1}{\mu_{\text{ext}}} + \frac{1}{\mu_{\text{int}}(T)}, \qquad (11.81)$$

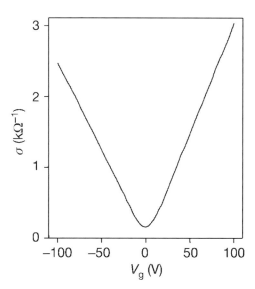

Fig. 11.1 The dependence of the conductivity of graphene on the gate voltage $V_g \sim n$. (Reproduced with permission from Novoselov et al., 2005a.)

assuming that μ_{ext} is temperature-independent and $\mu_{\text{int}}(T) \to 0$ at $T \to 0$, one finds $\mu_{\text{int}} \approx (2 - 4) \times 10^5 \text{ cm}^2\text{V}^{-1}\text{s}^{-1}$ (Morozov et al., 2008), which means that the difference in conductivity between $T \approx 0$ and room temperature is no more than a few percent. We postpone the discussion of this temperature dependence until the next section and focus here on the origin of μ_{ext}.

Importantly, the concentration and temperature dependences of the conductivity for bilayer graphene are more or less the same as for single-layer graphene (Novoselov et al., 2006). To discuss this case, we will use the same semiclassical Boltzmann equation as for the case of single-layer graphene, the only differences being in the dispersion law and the transformation to electrons and holes ($\varphi_k \to 2\varphi_k$ in Eq. (11.19) and (11.34)). In both cases, the inverse relaxation time (11.79) can be estimated as

$$\frac{1}{\tau} \approx \frac{2\pi}{\hbar} N(\varepsilon_F) |\bar{V}(k_F)|^2, \tag{11.82}$$

where $\bar{V}(k_F)$ is a typical value of $V^{\text{eff}}_{\vec{k}\,\vec{k'}}$, at $\left|\vec{k}\right| \approx \left|\vec{k'}\right| \approx k_F$. On substituting Eq. (11.82) into the Drude formula (11.72) and (11.77) we find

$$\sigma(n) \sim \frac{v_F^2}{|\bar{V}(k_F)|^2}, \tag{11.83}$$

where $v_F = v = $ constant for the case of single-layer graphene and

$$v_F = \frac{\hbar k_F}{m} \sim n^{1/2} \tag{11.84}$$

for the case of bilayer graphene. This means that, to explain the experimentally observed behavior $\sigma(n)$, one needs to assume

$$|\bar{V}(k_F)|^2 \approx \text{constant} \tag{11.85}$$

for the case of bilayer graphene and

$$|\bar{V}(k_F)|^2 \sim \frac{1}{k_F^2} \tag{11.86}$$

for the case of single-layer graphene.

For randomly distributed defects, one needs to use Eq. (11.14) (assuming the Born approximation) or the more accurate Eq. (11.15) (assuming only a low concentration of defects). In the latter case, the answer can be expressed in terms of scattering phases; see Eq. (6.23) through (6.26) for the case of single-layer graphene and Eq. (6.54) for the case of bilayer graphene.

Up to now we have not taken into account the screening effects (see Section 7.7). Within the random phase approximation (RPA), the scalar potential $V^{(0)}_{\vec{k}\,\vec{k}'}$ in Eq. (11.34) is replaced by

$$V^{(s)}_{\vec{k}\,\vec{k}'} = \frac{V^{(0)}_{\vec{k}\,\vec{k}'}}{\varepsilon\left(q = \left|\vec{k} - \vec{k}'\right|, \omega = 0\right)}. \tag{11.87}$$

Beyond the RPA, so-called vertex corrections should be taken into account, but we will not discuss them here; this simple theory will suffice just for estimations. At the same time, there is no screening of the vector potential $\vec{V}_{\vec{k}\,\vec{k}'}$ (Gibertini et al., 2010).

Let us restrict ourselves to the case of the scalar potential only and use Eq. (11.14). Thus, Eq. (11.79) will take the form (11.25), with the replacement $u(q) \to u(q)/\varepsilon(q)$. On introducing the new variable $x = \sin(\varphi/2)$ one can rewrite this equation as

$$\frac{1}{\tau(k_F)} = \frac{4k_F}{\pi\hbar v} n_{\text{imp}} \int\limits_0^1 dx x^2 \sqrt{1 - x^2} \left|\frac{u(2k_F x)}{\varepsilon(2k_F x, 0)}\right|^2. \tag{11.88}$$

Note that only $\varepsilon(q, 0)$ with $q < 2k_F$ is involved in Eq. (11.88). In this regime, the RPA coincides with the Thomas–Fermi approximation (see Eq. (7.107)), thus

$$\varepsilon(2k_F x, 0) = \varepsilon_{ext} + \frac{2e^2}{\hbar v} \frac{1}{x} \tag{11.89}$$

and does not depend on k_F.

The behavior (11.86) is provided by Coulomb impurities, where

$$u(q) = \frac{2\pi Z e^2}{q}. \tag{11.90}$$

Moreover, it also takes place with the replacement of the potential u by the *T*-matrix; see Eq. (8.18) and (8.19). Therefore, it is very natural to assume that charge impurities determine the electron mobility in graphene on a substrate (Nomura & MacDonald, 2006; Ando, 2006; Adam et al., 2007; Peres, 2010; Das Sarma et al., 2011). Quantitative estimations for the case of graphene on SiO$_2$ ($\varepsilon_{ext} \approx 2.5$) give (Adam et al., 2007)

$$\sigma(n) \approx \frac{20e^2}{h} \frac{n}{n_{\text{imp}}}, \tag{11.91}$$

where four current channels (two spins and two valleys) are taken into account.

Indeed, an intentional addition of charge impurities (potassium adatoms) to graphene leads to a decrease of the electron mobility, in good agreement with the theory described earlier (Chen et al., 2008). At the same time, there is convincing experimental evidence that this is *not* the main factor restricting electron mobility in standard exfoliated graphene samples on a substrate.

The main argument is that the electron mobility is relatively weakly changed in an environment with a high dielectric constant and, thus, very large ε_{ext}, e.g., after covering graphene with water, ethanol or other polar liquids, or when using substrates with large ε (Ponomarenko et al., 2009). In particular, the mobility in graphene on $SrTiO_3$ (which has a dielectric constant growing from $\varepsilon \approx 300$ at room temperature to $\varepsilon \approx 5,000$ at liquid-helium temperature) is of the same magnitude as that for graphene on SiO_2 and very weakly dependent on temperature (Couto, Sacépé, & Morpurgo, 2011). Of course, the screened Coulomb interaction in such a situation *should be* strongly suppressed and strongly temperature-dependent.

It was suggested by Katsnelson, Guinea, and Geim (2009) that the reason why charged adsorbate adatoms on graphene can be not very important for the electron mobility is their strong tendency to form clusters. Indeed, density-functional calculations (Wehling, Katsnelson, & Lichtenstein, 2009b) show that the more charged the adsorbate species, the weaker its chemical bond with graphene and the lower its migration barriers. This means that strongly bonded and immobile adsorbates have very small charge transfer to graphene and, thus, small effective Z, whereas impurities with $Z \geq 1$ can be kept more or less randomly distributed only at low enough temperatures. This was found to be the case for potassium atoms by Chen et al. (2008). The clusterization suppresses the scattering cross-section per impurity by orders of magnitude (Katsnelson, Guinea, & Geim, 2009).

The effect described here was confirmed experimentally by McCreary et al. (2010). They deposited gold adatoms onto graphene and observed their clusterization, with a simultaneous growth of the electron mobility.

Before discussing other possible scattering mechanisms, we need to say a few words about the case of bilayer graphene. Actually, for any isotopic two-dimensional case, the density of states at the Fermi energy is

$$N(E_F) = \frac{g_v g_s}{2\pi} \int_0^\infty dk\, k\, \delta(\varepsilon_F - \varepsilon(k)) = \frac{g_s g_v}{2\pi} \frac{k_F}{\hbar v_F}, \qquad (11.92)$$

where

$$v_F = \frac{1}{\hbar} \left(\frac{\partial \varepsilon}{\partial k} \right)_{k=k_F} \qquad (11.93)$$

and we have restored the spin and valley degeneracy factors. As a result, the inverse screening radius is, instead of being given by Eq. (7.108) for single-layer graphene,

$$\kappa = g_s g_v \frac{e^2 k_F}{\hbar v_F \varepsilon_{ext}} \tag{11.94}$$

and, thus,

$$\varepsilon(2k_F x, 0) = \varepsilon_{ext} + \frac{g_s g_v e^2}{\hbar v_F} \frac{1}{2x}. \tag{11.95}$$

For the case of bilayer graphene, $\kappa \gg k_F$ since $v_F \to 0$ at $n \to 0$. Actually, this is the case even for single-layer graphene if ε_{ext} is not too large. Therefore, v_F is cancelled out from Eq. (11.83) and we have an estimation

$$\sigma(n) \sim \frac{1}{n_{imp} |u(k_F)|^2}, \tag{11.96}$$

which is valid both for single-layer and for bilayer graphene. This means that for the same type of purely scalar potential scattering, the concentration dependence of the conductivity is the same. Strictly speaking, this is true only within the Born approximation, and for the case of strong scatterers there will be some difference (see later). The numerical coefficients can be different since, in the case of bilayer graphene, one has to make the replacement ($\varphi \to 2\varphi$ in Eq. (11.32) and, as a result, the factor $\cos^2(\varphi/2)$ is replaced by $\cos^2 \varphi$. Thus, for the same electron concentration and the same scatterers, the ratio of the resistivity of single-layer graphene to that of bilayer graphene is

$$\frac{\rho_1}{\rho_2} = \frac{\Phi_1}{\Phi_2}, \tag{11.97}$$

where

$$\Phi_1 = \int_0^1 dx x^4 \sqrt{1 - x^2} |u(2k_F x)|^2$$

$$\Phi_2 = \int_0^1 \frac{dx x^4 (1 - 2x^2)^2}{\sqrt{1 - x^2}} |u(2k_F x)|^2$$

(see Eq. (11.88) and (11.95)).

Another potentially important source of electron scattering is ripples (see Chapter 10). They create both a random vector potential (10.7) and a random scalar potential (10.8). By substituting these expressions into Eq. (11.34) and following the analysis of Sections 10.2 and 10.3 one finds that

$$\left|\bar{V}(k_{\mathrm{F}})\right|^2 \sim F(q \approx k_{\mathrm{F}}), \tag{11.98}$$

where the correlation function F is given by Eq. (10.28). For intrinsic (thermally induced) ripples, one needs to use Eq. (10.29). Thus, for the case of not-too-small doping, when

$$k_F \gg q^*, \tag{11.99}$$

one has (Katsnelson & Geim, 2008)

$$\rho \approx \frac{h}{e^2} \left(\frac{T}{\kappa a}\right)^2 \frac{|\ln(q^* a)|}{n}. \tag{11.100}$$

At room temperature, this has the correct $1/n$ dependence and corresponds to the correct order of magnitude for the mobility $\mu \sim 10^4 \mathrm{cm}^2 \mathrm{V}^{-1}\mathrm{s}^{-1}$. There are two problems, however. First, the mobility is weakly temperature dependent. Therefore, Katsnelson and Geim (2008) suggested that there is a mechanism of freezing (quenching) of the ripples and, hence, that they keep the structure corresponding to some quenching temperature T_{q}. If one makes the replacement $T \to T_{\mathrm{q}}$ of the order of room temperature in Eq. (11.100), it seems to explain μ_{ext} reasonably well. Moreover, if one assumes that the large-scale ripple structure is frozen, but flexural phonons can be excited within the ripples, it can also explain the temperature dependence of μ_{int} (Morozov et al., 2008).

The weak temperature dependence of the ripple structure for graphene on SiO$_2$ has been confirmed by scanning tunneling microscopy (STM) experiments (Geringer et al., 2009). However, the origin of this quenching is still unknown. It was suggested and confirmed by density-functional calculations (Boukhvalov & Katsnelson, 2009b) that ripples can be stabilized by covalently bonded adatoms and admolecules. San-José, González, and Guinea (2011) proposed an intrinsic mechanism of ripple stabilization that is based on the interactions of ripples with conduction electrons. The absence of a detailed theory of the quenching seems to be the weakest point of the idea that the ripples can be the main limiting factor for electron mobility, whereas from the experimental point of view this possibility cannot be excluded. Anyway, as will be discussed in the next section, intrinsic ripples are probably the main limiting factor for the electron mobility in freely suspended graphene samples.

Another important question within this scenario is that of whether the frozen ripples on a substrate have the same structure as intrinsic ripples or not. The results from the first two scanning-probe studies for graphene on SiO$_2$ (Ishigami et al.,

2007; Stolyarova et al., 2007) indicated that these ripples repeat the roughness of the substrate approximately, whereas in the later work by Geringer et al. (2009) for the same system, two types of ripples were found: a first type following the roughness of the substrate and a second type similar to the intrinsic ripples.

It is important to note that the first type seems to be irrelevant for the electron mobility. Indeed, let us consider a general type of correlation function,

$$\left\langle \left[h\left(\vec{r} \right) - h(0) \right]^2 \right\rangle \sim r^{2H}. \tag{11.101}$$

Then,

$$\left\langle \left| h_{\vec{q}} \right|^2 \right\rangle \sim q^{-2(1+H)} \tag{11.102}$$

and for $2H < 1$ the correlation function $F(q)$ in (10.28) has a finite limit at $q = 0$, thus,

$$\left\langle \left| \bar{V}(q = 0) \right|^2 \right\rangle \approx \left(\frac{\hbar v}{a} \right)^2 \frac{z^4}{R^2}, \tag{11.103}$$

where z and R are the characteristic height and radius of ripples, respectively. This leads to a concentration-independent and very small contribution to the resistivity

$$\rho \approx \frac{h}{4e^2} \frac{z^4}{R^2 a^2}. \tag{11.104}$$

For $2H > 1$,

$$\rho \sim n^{1-2H}; \tag{11.105}$$

and for $2H = 1$,

$$\rho \sim \ln^2(k_F a) \tag{11.106}$$

(Katsnelson & Geim, 2008). For the roughness of the substrate, one could expect $2H \approx 1$ (Ishigami et al., 2007). Only frozen ripples with $2H \approx 2$ (such as intrinsic ripples at not too large distances r) are interesting as a scattering mechanism.

Another important potential source of scattering is resonance scattering (see Sections 6.5 and 6.6). They give a concentration dependence of the conductivity that is very close to linear (see Eq. (6.103)), that is, a weakly concentration-dependent mobility

$$\mu \sim \ln^2(k_F a). \tag{11.107}$$

At least in some cases this reproduces the experimental data better than does just constant mobility (Peres, 2010; Wehling et al., 2010a; Couto, Sacepe, & Morpurgo, 2011). This is certainly the case when vacancies are created in graphene by ion bombardment (Chen et al., 2009), but, as discussed in Section 6.5, it is very unlikely that there will be any vacancies in graphene if they are not created intentionally. It was suggested by Wehling et al. (2010a) that the resonant scatterers in real graphene samples could be due to the formation of chemical C–C bonds between graphene and organic pollutants on it. Even a very small concentration of such bonds, $<10^{-4}$, would be sufficient to explain the experimental data.

Zhao et al. (2015) has demonstrated, by straightforward calculation of conductivity via the Kubo formula, that there are some fingerprints of the resonance scatterers in comparison with the two other candidate mechanisms. First, they lead to a much higher degree of electron–hole asymmetry when taking into account the next-nearest-neighbor hopping $t' \approx t/10$ (Kretinin et al., 2013). Second, at high enough concentration of the defects, the mobility in this case demonstrates a shallow minimum as the function of the hole concentration.

For the case of bilayer graphene, within the parabolic-band approximation one could expect $\sigma \sim n$ and $\mu = $ constant for the cases of both resonant and generic impurities (Katsnelson, 2007c); see Section 6.3. Straightforward numerical simulations (Yuan, De Raedt, & Katsnelson, 2010b) show that for the case of resonant scatterers, this is true only if their concentration is very small. When the width of the impurity band exceeds $2|t_\perp|$ there is a cross-over to the behavior typical for single-layer graphene Eq. (11.107).

To conclude this section, we note that one can expect different main scattering mechanisms in different samples. Currently, it seems that for most situations the choice is between resonant scatterers and frozen ripples, but charge impurities can also be relevant if one protects their more or less random distribution and prevents their clusterization.

11.4 Intrinsic mobility and transport properties of suspended graphene flakes

In this section we will consider intrinsic mobility in graphene in relation to electron–phonon interaction (Stauber, Peres, & Guinea, 2007; Mariani & von Oppen, 2008, 2010; Morozov et al., 2008; Castro et al., 2010b; Ochoa et al., 2011). Here, we will follow the last two papers.

The inelastic scattering processes should satisfy the momentum- and energy-conservation laws. For single-phonon processes this means

$$\varepsilon_{\vec{k}} = \varepsilon_{\vec{k}'} \pm \hbar\omega_{\vec{q}}, \tag{11.108}$$

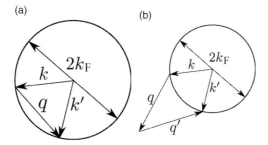

Fig. 11.2 Momentum transfer processes for single-phonon (a) and double-phonon (b) scattering processes.

where

$$\vec{k}' = \vec{k} \pm \vec{q}$$

(see Fig. 11.2(a)). The maximum momentum transfer within a given valley is $q = 2k_F$, and both electron states $\left|\vec{k}\right\rangle$ and $\left|\vec{k}'\right\rangle$ should lie within a layer of the order of T near the Fermi energy. Thus, if

$$T > \hbar\omega_{2k_F},\qquad(11.109)$$

the scattering processes can be considered almost elastic. The scattering probability is proportional to the number of thermally excited phonons (*virtual* phonons do not contribute to the resistivity; see Ziman (2001)) and is negligible at

$$\hbar\omega_{2k_F} \gg T.\qquad(11.110)$$

Up to room temperature, this excludes all optical phonons in graphene from our consideration. It also excludes intervalley scattering processes involving phonons with $\vec{q} \approx \vec{K}$ (see Section 9.8), since for all branches the condition (11.110) is satisfied at $\vec{q} = \vec{K}$ and $T \le 300\,\text{K}$ (see Fig. 9.1). Thus, we are interested only in acoustic phonons at $q \ll a^{-1}$ (in graphene, k_F is always much smaller than a^{-1}). There are three branches of such phonons, longitudinal (L) and transverse (T) in-plane phonons and flexural (F) out-of-plane phonons with the dispersion relations (see Section 9.2)

$$\omega_{\vec{q}}^{\text{L}} = v_L q, \quad v_L = \sqrt{\frac{\lambda + 2\mu}{\rho}},\qquad(11.111)$$

$$\omega_{\vec{q}}^{\text{T}} = v_T q, \quad v_T = \sqrt{\frac{\mu}{\rho}},\qquad(11.112)$$

$$\omega_{\vec{q}}^{F} = \sqrt{\frac{\kappa}{\rho}} q^2, \tag{11.113}$$

where ρ is the mass density. Keeping in mind real parameters for graphene, we can estimate the Bloch–Grüneisen temperature, $T_{BG} = \hbar\omega_{2k_F}$, for the various branches:

$$T_{BG}^{L} = 57\sqrt{n}\,\text{K}, \quad T_{BG}^{T} = 38\sqrt{n}\,\text{K}, \quad T_{BG}^{F} = 0.1n\,\text{K}, \tag{11.114}$$

where n is expressed in units of 10^{12} cm^{-2}. At $T > T_{BG}$ (11.109), phonons can be considered classically. One can see that for *flexural* phonons this is actually the case for any practically interesting temperatures.

The electron–phonon interaction in graphene originates from two sources: the electrostatic potential (10.8), which should be substituted into Eq. (11.17), and the vector potential (10.7), modulating the electron hopping. However, the deformation tensor $\hat{u}_{\alpha\beta}$ should be considered as an operator. It is given by Eq. (9.62), and the operators $\hat{u}_\alpha$ and $\hat{h}$ are expressed in terms of the corresponding phonon operators by Eq. (9.9). The resulting Hamiltonian takes the form (Ochoa et al., 2011)

$$
\begin{aligned}
\hat{H}_{e-ph} = \sum_{\vec{k}\,\vec{k}'} \Bigg(&\hat{a}_{\vec{k}}^{+}\hat{a}_{\vec{k}'} + \hat{c}_{\vec{k}}^{+}\hat{c}_{\vec{k}'} \Bigg) \Bigg\{ \sum_{v\vec{q}} V_{1,\vec{q}}^{v}\left[\hat{b}_{\vec{q}}^{v} + \left(\hat{b}_{-\vec{q}}^{v}\right)^{+} \right] \delta_{\vec{k}',\vec{k}-\vec{q}} + \\
&+\sum_{\vec{q}\,\vec{q}'} V_{1,\vec{q}\vec{q}'}^{F}\left[\hat{b}_{\vec{q}}^{F} + \left(\hat{b}_{-\vec{q}}^{F}\right)^{+} \right]\left[\hat{b}_{\vec{q}'}^{F} + \left(\hat{b}_{-\vec{q}'}^{F}\right)^{+} \right] \delta_{\vec{k}',\vec{k}-\vec{q}-\vec{q}'} \Bigg\} + \\
&+\sum_{\vec{k}\,\vec{k}'} \Bigg\{ \sum_{v\vec{q}} V_{2,\vec{q}}^{v}\hat{a}_{\vec{k}}^{+}\hat{c}_{\vec{k}'}\left[\hat{b}_{\vec{q}}^{v} + \left(\hat{b}_{-\vec{q}}^{v}\right)^{+} \right] \delta_{\vec{k}',\vec{k}-\vec{q}} + \\
&+\sum_{\vec{q}\,\vec{q}'} V_{2,\vec{q}\vec{q}'}^{F}\hat{a}_{\vec{k}}^{+}\hat{c}_{\vec{k}'}\left[\hat{b}_{\vec{q}}^{F} + \left(\hat{b}_{-\vec{q}}^{F}\right)^{+} \right]\left[\hat{b}_{\vec{q}'}^{F} + \left(\hat{b}_{-\vec{q}'}^{F}\right)^{+} \right] \delta_{\vec{k}',\vec{k}-\vec{q}-\vec{q}'} + H.c. \Bigg\}
\end{aligned}
\tag{11.115}
$$

where $v = L; T$, subscripts 1 and 2 label the terms originating from the scalar potential (10.8) and from the vector potential (10.7), respectively; and $\hat{a}_{\vec{k}}$ and $\hat{c}_{\vec{k}}$ are electron annihilation operators for sublattices A and B, respectively. The matrix elements are

$$V_{1,\vec{q}}^{L} = \frac{g}{\varepsilon(q,0)}\,iq\sqrt{\frac{\hbar}{2\rho\Omega\omega_{\vec{q}}^{L}}},$$

$$V_{1,\vec{q}}^{T} = 0,$$

$$V^{\mathrm{F}}_{1,\vec{q}\vec{q}'} = -\frac{g}{\varepsilon\left(|\vec{q}+\vec{q}'|,0\right)} qq' \cos\left(\varphi_{\vec{q}} - \varphi_{\vec{q}'}\right) \frac{\hbar}{4\rho\Omega\sqrt{\omega^{\mathrm{F}}_{\vec{q}}\omega^{\mathrm{F}}_{\vec{q}'}}},$$

$$V^{\mathrm{L}}_{2,\vec{q}} = \frac{\hbar v\beta}{2a} iq \exp\left(2i\varphi_{\vec{q}}\right) \sqrt{\frac{\hbar}{2\rho\Omega\omega^{\mathrm{L}}_{\vec{q}}}},$$

$$V^{\mathrm{T}}_{2,\vec{q}} = -\frac{\hbar v\beta}{2a} q \exp\left(2i\varphi_{\vec{q}}\right) \sqrt{\frac{\hbar}{2\rho\Omega\omega^{\mathrm{T}}_{\vec{q}}}},$$

$$V^{\mathrm{F}}_{2,\vec{q}\vec{q}'} = -\frac{\hbar v\beta}{4a} qq' \exp\left[i\left(\varphi_{\vec{q}} - \phi_{\vec{q}}\right)\right] \frac{\hbar}{2\rho\Omega\sqrt{\omega^{\mathrm{F}}_{\vec{q}}\omega^{\mathrm{F}}_{\vec{q}'}}}, \tag{11.116}$$

where Ω is the sample area and we take into account the screening of scalar potential by the static dielectric function (cf. Eq. (11.87)). Note that all matrix elements tend to zero at $q \to 0$, as usual for the interaction with acoustic phonons (Ziman, 2001).

One can see that the electron–phonon interaction with flexural phonons does not involve single-phonon processes but only two-phonon processes. This follows from the structure of the deformation tensor (9.62). Single-flexural-phonon processes do arise in deformed samples with some external profile $h_0(x, y)$ (Castro et al., 2010b; Ochoa et al., 2011).

The resistivity can be found using the Kubo–Nakano–Mori formula (or, equivalently, by derivation and approximate solution of the Boltzmann equation). First, we have to substitute the operator $\hat{H}_{\mathrm{e-ph}}$ instead of $\hat{H}'$ into Eq. (11.74) and (11.78). The time dependence of the phonon operators is (Vonsovsky & Katsnelson, 1989)

$$\hat{b}_{\vec{q}}(t) = \hat{b}_{\vec{q}} \exp\left(-i\omega_{\vec{q}}t\right),$$

$$\hat{b}^{+}_{\vec{q}}(t) = \hat{b}^{+}_{\vec{q}} \exp\left(i\omega_{\vec{q}}t\right). \tag{11.117}$$

Next, we decouple the electron and phonon operators (this corresponds to the lowest-order approximation in $\hat{H}_{\mathrm{e-ph}}$) and assume that the phonons are in equilibrium:

$$\left\langle \hat{b}^{+}_{\vec{q}}\hat{b}_{\vec{q}} \right\rangle = N_{\vec{q}} = \frac{1}{\exp\left(\hbar\omega_{\vec{q}}/T\right) - 1},$$

$$\left\langle \hat{b}_{\vec{q}}\hat{b}^{+}_{\vec{q}} \right\rangle = 1 + N_{\vec{q}}. \tag{11.118}$$

This means that we neglect the effects of phonon drag, which makes the phonon system a nonequilibrium one in the presence of an electric current. It is known (Ziman, 2001) that this effect is usually not relevant for the resistivity but may be crucially important for the thermoelectric power. We will not consider it here.

At $T > T_{BG}^{L,T}$ the one-phonon scattering can be considered classically, that is, one can put

$$N_{\vec{q}} \approx 1 + N_{\vec{q}} \approx \frac{T}{\hbar\omega_{\vec{q}}} \tag{11.119}$$

and neglect the phonon frequency in the energy-conservation law. The latter can be done, actually, at any temperature, since $\left|\varepsilon_{\vec{k}+\vec{q}} - \varepsilon_{\vec{k}}\right| \gg \hbar\omega_{\vec{q}}$, except in the case $\vec{k} \perp \vec{q}$, which does not contribute to the integral characteristics.

In this case, we have just the same situation as for the scattering by static disorder Eq. (11.79), with

$$\left|V_{\vec{k}\vec{k}'}^{ef}\right|^2 \sim \left\langle\left|\vec{u}_{\vec{k}\vec{k}'}\right|^2\right\rangle = \frac{T}{M\omega_{\vec{k}-\vec{k}'}^2}. \tag{11.120}$$

An accurate calculation gives the result (Castro et al., 2010b)

$$\frac{1}{\tau} \approx \left[\frac{g_{\text{eff}}^2}{v_L^2} + \frac{\beta^2\hbar^2 v^2}{a^2}\left(\frac{1}{v_L^2} + \frac{1}{v_T^2}\right)\right]\frac{k_F T}{2\rho\hbar^2 v}, \tag{11.121}$$

where

$$g_{ef} \approx \frac{g}{\varepsilon(q \approx k_F, 0)} \tag{11.122}$$

is the screened coupling constant. As will be shown later, this contribution is usually much smaller than that due to two-phonon processes (Morozov et al., 2008). This situation is highly unusual; normally, both in a three-dimensional and in a two-dimensional electron gas, single-phonon processes are dominant. It is reminiscent of the case of electron–*magnon* scattering in half-metallic ferromagnets, where single-magnon processes are forbidden and the temperature dependence of the resistivity is determined by two-magnon processes (Irkhin & Katsnelson, 2002).

The energy and momentum conservation for the two-phonon scattering processes can involve phonons with large enough wave vectors (see Fig. 11.2(b)); thus, it is not clear a priori that even at $T > T_{BG}^F$ (which is, actually, always the case) the classical picture is correct. Nevertheless, as we will see later, this is true, and quantum-mechanical treatment of two-phonon scattering gives approximately the same answer (11.100) as the classical consideration of intrinsic ripples (Morozov et al., 2008).

An accurate treatment of the two-phonon processes leads to the expression (Castro et al., 2010b)

$$\frac{1}{\tau} = \frac{1}{32\pi^3\rho^2 v k_F} \int_0^\infty dK \frac{[D(K)]^2 K^2}{\sqrt{k_F^2 - \frac{K^2}{4}}} \int_0^\infty dq \frac{q^3 N_q}{\omega_q}$$

$$\times \int_{|K-q|}^{K|q} dQ \frac{Q^3(N_Q+1)}{\omega_Q \sqrt{K^2 q^2 - \frac{(K^2+q^2-Q^2)^2}{4}}},$$

(11.123)

where we omit the superscript F for ω_q and N_q.

Here

$$[D(K)]^2 = \frac{g^2}{\varepsilon^2(K,0)}\left(1 - \frac{K^2}{4k_F^2}\right) + \left(\frac{\beta\hbar v}{2c}\right)^2.$$

(11.124)

One can see that there is no backscattering $(K = 2k_F)$ for the scalar potential, but there is backscattering for the vector potential, as there should be (see Sections 4.2 and 6.1).

For the case $q^* \ll k_F \ll q_T$, where q^* is the "Ginzburg" vector, as in Eq. (11.99), and q_T is determined by the condition

$$\omega_{q_T}^F = T$$

(11.125)

the result is (Castro et al., 2010b)

$$\frac{1}{\tau} = \frac{\bar{D}^2 T^2}{64\pi\hbar^2\kappa^2 v k_F} \ln\left(\frac{T}{\hbar\omega^*}\right),$$

(11.126)

where $\omega^* = \omega_{q^*}^F$ and $\bar{D}$ is some average value of $D(K)$. The cut-off at $q \approx q^*$ is necessary since, as we know, the harmonic approximation is not applied to the flexural phonons at $q \leq q^*$. Eq. (11.126) agrees with the estimation (11.100). This justifies our statement that at $q_T \gg k_F$, which is equivalent to $T \gg T_{BG}^F$, "two-flexural-phonon" scattering means the same as "scattering by intrinsic ripples." The case of low temperatures where anharmonic coupling of in-plane and out-of-plane modes is crucially important (see Sections 9.3, 9.4) was studied in detail by Mariani and von Oppen (2008, 2010), Castro et al. (2010b), and Gornyi, Kachorovskii, and Mirlin (2012). We will not discuss it since it is not relevant for the current experimental situation.

By comparing Eq. (11.121) and (11.126) one can estimate that the two-phonon processes dominate at

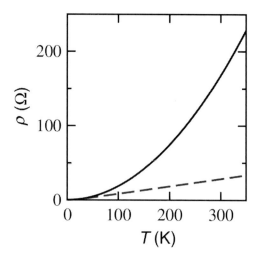

Fig. 11.3 Contributions to the resistivity of single-layer graphene from flexural phonons (solid line) and from in-plane phonons (dashed line). The electronic concentration is $n = 10^{12}$ cm^{-2}.
(Reproduced with permission from Castro et al., 2010b.)

$$T > T_c(\mathrm{K}) \approx 57n(10^{12}\mathrm{cm}^{-2}) \tag{11.127}$$

(Castro et al., 2010b). A quantitative comparison of single-phonon and two-phonon contributions is shown in Fig. 11.3.

The theory for the case of bilayer graphene was developed by Ochoa et al. (2011). Both the temperature dependence and the concentration dependence of the resistivity are the same as for the case of single-layer graphene, accurately to within some numerical coefficients.

As has already been mentioned, for graphene on a substrate, the intrinsic temperature-dependent contribution to the resistivity is negligible in comparison with the extrinsic one. The situation is dramatically different for suspended graphene flakes, for which, after annealing, the defects can be eliminated, and the mobility at liquid helium temperature can be of the order of 10^5–10^6 cm^2 V^{-1} s^{-1} (Bolotin et al., 2008; Du et al., 2008; Castro et al., 2010b; Mayorov et al., 2011a). In this case, the intrinsic mobility dominates completely.

Typical experimental data are shown in Fig. 11.4. Comparison between theory and experiment shows (Castro et al., 2010b) that two-flexural-phonon scattering (or, equivalently, scattering by intrinsic ripples) is probably the main limiting factor for the suspended samples. It restricts the mobility at room temperature to a value of the order of 10^4 cm^2 V^{-1} s^{-1} (see Eq. (11.100)). However, the mobility can be increased by expanding the samples. External deformation suppresses flexural phonons, making them stiffer:

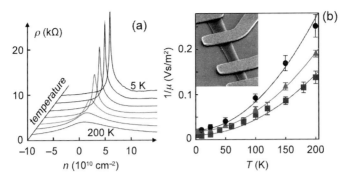

Fig. 11.4 (a) The resistivity of suspended single-layer graphene for $T = 5$, 10, 25, 50, 100, 150, and 200 K. (b) Examples of $\mu(T)$. The inset shows a scanning electron micrograph of one of the suspended devices.
(Reproduced with permission from Castro et al., 2010b.)

$$\rho\omega^2 = \kappa q^4 + 2(\lambda + \mu)q^2 u \qquad (11.128)$$

(cf. Eq. (9.148)). Estimations (Castro et al., 2010b) demonstrate that even small deformations of $u \leq 1\%$ may be sufficient to increase the room-temperature mobility by an order of magnitude.

Interestingly, the situation in other two-dimensional materials seems to be different. In single-layered five-group elements (P, As, Sb), at least, the in-plane phonons are the main limiting factor of intrinsic mobility (Rudenko, Brener, & Katsnelson, 2016; Rudenko et al., 2019). Graphene seems to be unique in this respect (a crucial importance of intrinsic ripples), which opens a way to dramatic-ally improve electron mobility at room temperature, putting it on an atomically flat substrate such as hexagonal boron nitride (hBN); see Chapter 13. This suppresses instability of the flexural phonons (Amorim & Guinea, 2013). For the other two-dimensional materials this probably will not work, since the effect of substrate on in-plane phonons is expected to be much weaker.

11.5 Edge scattering of electrons in graphene

Consider now the case where scattering centers are situated (completely or partially) at the edges of graphene (nano)ribbons. We restrict ourselves only by the case of relatively weak disorder and relatively broad ribbons when

$$Lk_F \gg 1, \qquad (11.129)$$

where L is the width of the ribbon. In this case the semiclassical approach based on the Boltzmann equation is applicable. This approach was broadly used for decades to study surface scattering effects in metals and semiconductors (for review, see Okulov & Ustinov, 1979; Falkovsky, 1983). The situation of graphene is distinct due to the different character of boundary conditions at the edges. In this section we follow the work by Dugaev and Katsnelson (2013).

Let us assume that graphene is situated in the region $0 < x < L$, electric field is parallel to y axis, does not depend on time, and magnetic field is absent. Then, the Boltzmann equation (11.1) – (11.3) takes the form

$$eEv_y \frac{\partial f_0\left(\varepsilon_{\vec{k}}\right)}{\partial \varepsilon_{\vec{k}}} + v_x \frac{\partial \delta f}{\partial x} = -\frac{\delta f}{\tau}, \tag{11.130}$$

where we assume linearization (11.7) and the simplest approximation for the collision integral (11.8):

$$I_{\vec{k}}[f] = -\frac{\delta f}{\tau} \tag{11.131}$$

(for brevity, we skip the argument x and subscript $\vec{k}$ of the distribution function δf). The general solution of Eq. (11.130) for the electrons moving toward right and left edges ($v_x > 0$ and $v_x < 0$, respectively) can be written as

$$\delta f^>(x) = -eEv_y\tau \frac{\partial f_0}{\partial \varepsilon} + C^> e^{-x/l_x}, \tag{11.132}$$

$$\delta f^<(x) = -eEv_y\tau \frac{\partial f_0}{\partial \varepsilon} + C^< e^{(x-L)/l_x}, \tag{11.133}$$

where $l_x = |v_x|\tau$, and $C^>$, $C^<$ are integration constants (dependent on $\vec{k}$), which should be found from boundary conditions. They have the form (Okulov & Ustinov, 1979; Falkovsky, 1983)

$$|v_x|\delta f^>(x=0) = |v_x|\delta f^<(x=0) + \sum_{\vec{k}'} w_L\left(\vec{k}, \vec{k}'\right)[\delta f^<(x=0) - \delta f^>(x=0)], \tag{11.134}$$

$$|v_x|\delta f^<(x=L) = |v_x|\delta f^>(x=L) + \sum_{\vec{k}'} w_R\left(\vec{k}, \vec{k}'\right)[\delta f^>(x=L) - \delta f^<(x=L)], \tag{11.135}$$

where $w_{L,R}(\vec{k}, \vec{k}')$ are scattering probabilities at left (right) edges. It is given by the standard quantum mechanical expression (11.6) with V being the surface scattering potential.

The average current density

$$j = g_s g_v \frac{e}{L} \int\limits_0^L dx \sum_{\vec{k}} v_y [\delta f^<(x) + \delta f^>(x)] \qquad (11.136)$$

can be expressed, by Eq. (11.132) and (11.133) via the constants $C^>$, $C^<$; the former can be found from the boundary conditions (11.134) and (11.135). In general, it requires a solution of integral equations in $\vec{k}$ space, since the constants $C^>$, $C^<$ depend on $\vec{k}$.

Until now, we used just a conventional theory of surface scattering developed for normal metals. Graphene is specific only at the calculations of the scattering probabilities $w(\vec{k}, \vec{k}')$ (for simplicity, we will further assume the same disorder on both edges $w_L(\vec{k}, \vec{k}') = w_R(\vec{k}, \vec{k}') = w(\vec{k}, \vec{k}'))$. It turns out that we have essentially different results for different types of boundary conditions.

Let us assume that the edge scattering is due to defects Eq. (11.14), and the potential of individual defect $u(x,y)$ is atomically sharp in x direction and has a spatial scale a in y direction; therefore, the surface scattering is suppressed for $|k_y - k_y'| > 1/a$. To be specific, we can use the model

$$u_{\vec{k}-\vec{k}'} = V_0 \exp\left[-(k_y - k_y')^2 a^2\right], \qquad (11.137)$$

For the Berry–Mondragon boundary conditions (5.13) and (5.14), the spinor electron wave function near the left boundary reads

$$\left|\vec{k}\right\rangle = A e^{i\vec{k}\cdot\vec{r}} \begin{pmatrix} 1 \\ -i \end{pmatrix}, \qquad (11.138)$$

where A is a renormalization factor. In this case

$$w(\vec{k}, \vec{k}') = \frac{2\pi}{\hbar} n_{imp} V_0^2 \delta(\hbar vk - \hbar vk') \exp\left[-2(k_y - k_y')^2 a^2\right], \qquad (11.139)$$

irrespective to which sublattice, A or B, the scattering centers belong.

On the other hand, for the zigzag boundary conditions similar to Eq. (5.71) (which can be considered as generic ones for the case of terminated honeycomb lattice; see Section 5.3) the A component of the spinor should disappear for $x = 0$ (or, oppositely, for $x = L$ which does not effect the results). The corresponding solution of the Dirac equation has the form:

$$\left|\vec{k}\right\rangle = A e^{ik_y y} \begin{pmatrix} \sin k_x x \\ -i\dfrac{k_x}{k}\cos k_x x + i\dfrac{k_y}{k}\sin k_x x \end{pmatrix}, \qquad (11.140)$$

and the scattering probability is vanishing for the gliding electrons ($k_x \rightarrow 0$). For the case of defects localized in sublattice A

$$V^{(A)}_{\vec{k}\,\vec{k'}} = |A|^2 \int dx dy \sin(k_x x) \sin(k_x' x) e^{i(k_y'-k_y)y} V(x,y) \approx V_A k_x k_x' \exp\left[-(k_y-k_y')^2 a^2\right]$$

(11.141)

where V_A is some constant (proportional to the square radius of action of the potential in x direction). Similarly, for the case of defects localized in sublattice B

$$V^{(B)}_{\vec{k}\,\vec{k'}} = |A|^2 \frac{k_x k_x'}{k^2} \int dx dy \cos(k_x x) \cos(k_x' x) e^{i(k_y'-k_y)y} V(x,y)$$
$$\approx V_B k_x k_x' \exp\left[-(k_y-k_y')^2 a^2\right],$$

(11.142)

where V_B is proportional to $1/k^2$. The total scattering probability in this case is equal to

$$w\left(\vec{k},\vec{k'}\right) = \frac{2\pi}{\hbar} n_{imp} V_1^2 k_x^2 k_x'^2 \delta(\hbar v k - \hbar v k') \exp\left[-2(k_y-k_y')^2 a^2\right], \quad (11.143)$$

where

$$n_{imp} V_1^2 = n_{imp}^{(A)} V_1^{(A)2} + n_{imp}^{(B)} V_1^{(B)2},$$

(11.144)

where $n_{imp(A,B)}$ is the concentration of defects in the corresponding sublattice.

One can see that depending on the type of boundary conditions at the edges, the scattering probability either disappears in the limit of gliding electrons or remains constant; for the case of usual metals the first case is realized (Okulov & Ustinov, 1979; Falkovsky, 1983).

The approach based on separation of the boundary conditions (11.134) and (11.135) from the collision integral (11.131) does not work for the ballistic (Knudsen) regime

$$L \gg l,$$

(11.145)

($l = v\tau$ is the mean free path) when surface scattering is more important than the bulk one. The former should be taken explicitly into account in the collision integral. In particular, in the limit $l \rightarrow \infty$ (the scattering *only* at the edges) the distribution function f does not depend on x, and instead of Eq. (11.130) with the boundary conditions (11.134) and (11.135), we have just one equation

$$ev_y E \frac{\partial f_0}{\partial \varepsilon} = \sum_{\vec{k'}} w\left(\vec{k},\vec{k'}\right)\left(\delta f^{<,>}_{\vec{k'}} - \delta f^{>,<}_{\vec{k}}\right),$$

(11.146)

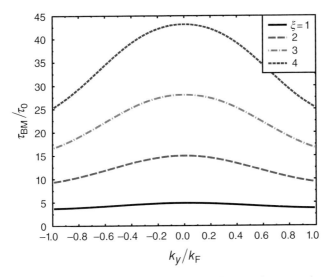

Fig. 11.5 Dependence of effective scattering mean time on the scattering angle for graphene nanoribbon in ballistic regime $l \to \infty$, for the case of Berry–Mondragon (BM) boundary conditions. Here, $\xi = a^2 k_F^2$, $1/\tau_0 = (n_{\text{imp}} V_0^2 k_F)/(2\pi\hbar Lv)$. (Reproduced with permission from Dugaev & Katsnelson, 2013.)

where we assume, again, for simplicity $w_L(\vec{k}, \vec{k}') = w_R(\vec{k}, \vec{k}') = w(\vec{k}, \vec{k}')$. The immediate consequence is that $\delta f \vec{k}^< = \delta f \vec{k}^>$. The integral equation (11.146) can be solved numerically for the models (11.139) and (11.143) (Dugaev & Katsnelson, 2013). The results are shown in Figs. 11.5 and 11.6.

One can see that in the latter case the scattering time is divergent in the limit of gliding electrons $k_x \to 0$ as $\tau_Z \propto k_x^{-2}$, whereas in the former case it is not, and $\tau_{BM} \to const \propto L$. As a result, for the purely ballistic regime the conductivity of graphene nanoribbon scales as L^2 and L for the case of zigzag (terminated graphene lattice) and Berry–Mondragon (strongly chemically functionalized edges), respectively (Dugaev & Katsnelson, 2013).

Similar analysis can be also performed for the case of scattering by curved edges (Dugaev & Katsnelson, 2013), with the same qualitative conclusions.

11.6 Nonlocal transport in magnetic fields

Graphene is unique, in the sense that one can pass continuously from electron conductivity to hole conductivity without crossing an insulator region. This means that by applying some small perturbations one can create two subsystems, an electron one and a hole one, differing by some intrinsic quantum number. The simplest case of such a perturbation is Zeeman splitting

$$\delta = 2\mu_B B, \tag{11.147}$$

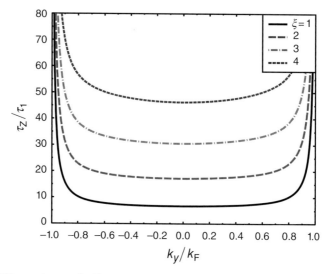

Fig. 11.6 Dependence of effective scattering mean time on the scattering angle for graphene nanoribbon in ballistic regime $l \to \infty$, for the case of zigzag (Z) boundary conditions. Here, $\xi = a^2 k_F^2$, $1/\tau_1 = (n_{\mathrm{imp}} V_1^2 k_F^5)/(2\pi\hbar L v)$.
(Reproduced with permission from Dugaev & Katsnelson, 2013.)

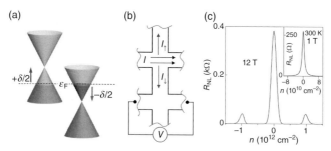

Fig. 11.7 (a) Zeeman splitting at the charge neutrality point. (b) Charge current and spin currents in the presence of the Lorentz force. (c) The nonlocal resistivity predicted by Eq. (11.177) for the quantum Hall regime (main panel) and for weak magnetic fields (inset).
(Reproduced with permission from Abanin et al., 2011.)

which makes the spin-up charge carriers be holes and the spin-down charge carriers be electrons (Fig. 11.7(a)). Similar effects can be brought about by valley polarization, but, for simplicity, we will discuss the effects of spin splitting further. Thus, we have a very strong coupling of spin and charge degrees of freedom: By changing the spin direction one can change the sign of charge! This peculiarity of graphene is probably responsible for one of its salient features, a giant nonlocal spin transport near the neutrality point (Abanin et al., 2011).

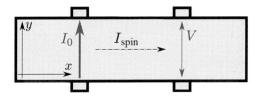

Fig. 11.8 A schematic representation of nonlinear transport (see the text).

The mechanism is the following: Suppose you create a charge current across the sample in the presence of an external magnetic field (it does not necessarily need to be strong enough for the system to be in the quantum Hall regime, since the effect under consideration is actually classical). This charge current consists of spin-up and spin-down components, which are, due to Zeeman splitting, electron and hole ones. In the magnetic field they will be deviated in opposite directions, leading to a *spin current* perpendicular to the original charge current. The spin current can propagate without decay for very large distances, since the time of spin-flip scattering processes τ_s is normally several orders of magnitude larger than the Drude relaxation time τ. Then, due to an inverse mechanism, this spin current creates a voltage. Here we present a phenomenological theory of this effect (Abanin et al., 2011). Previously, similar physics had been discussed for the spin Hall effect in conventional semiconductors (Abanin et al., 2009); however, for the case of graphene the effect is really huge, for the reasons mentioned earlier.

Let us consider the geometry shown in Fig. 11.8. First, let us ignore the spin dependence of the conductivity. The relation between the current density and the electric field $\vec{E} = -\vec{\nabla}\varphi$ is

$$\vec{j} = -\hat{\sigma}\,\vec{\nabla}\,\varphi, \tag{11.148}$$

where

$$\hat{\sigma} = \begin{pmatrix} \sigma_{xx} & \sigma_{xy} \\ -\sigma_{xy} & \sigma_{xx} \end{pmatrix} \tag{11.149}$$

is the conductivity tensor in the presence of a magnetic field, $\sigma_{xy} \sim B$. Eq. (11.149) follows from Onsager's relations

$$\sigma_{\alpha\beta}(B) = \sigma_{\beta\alpha}(-B) \tag{11.150}$$

and the isotropy of macroscopic properties in the *xy*-plane for the honeycomb lattice. Let us assume charge injection into the point $x = 0$, thus the boundary conditions are

$$j_y\left(y = \pm\frac{L}{2}\right) = I_0\delta(x), \qquad (11.151)$$

where L is the sample width.

Owing to the structure of the tensor (11.149), the charge conservation law

$$\vec{\nabla}\vec{j} == 0 \qquad (11.152)$$

is equivalent to the Laplace equation

$$\nabla^2\varphi(x, y) = 0 \qquad (11.153)$$

with a general solution

$$\varphi(x, y) = \int\limits_{-\infty}^{\infty} \frac{dk}{2\pi}[a(k)\cosh(ky) + b(k)\sinh(ky)]\exp(ikx). \qquad (11.154)$$

The coefficients $a(k)$ and $b(k)$ should be found from the boundary condition (11.151), that is,

$$\left(\sigma_{xy}\frac{\partial\varphi}{\partial x} - \sigma_{xx}\frac{\partial\varphi}{\partial y}\right)\Bigg|_{y=\pm\frac{L}{2}} = I_0\delta(x). \qquad (11.155)$$

The solution is straightforward and gives us the voltage distribution

$$V(x) = \varphi\left(x, -\frac{L}{2}\right) - \varphi\left(x, \frac{L}{2}\right) = 2I_0\rho_{xx}\int\limits_{-\infty}^{\infty} \frac{dk}{2\pi}\frac{\exp(ikx)}{k}\tanh\left(\frac{kL}{2}\right), \qquad (11.156)$$

where $\hat{\rho} = \hat{\sigma}^{-1}$ is the resistivity tensor. On calculating the integral explicitly we have the final answer

$$V(x) = \frac{2I_0\rho_{xx}}{\pi}\ln\left(\coth\frac{\pi x}{2L}\right) \approx \frac{4I_0\rho_{xx}}{\pi}\exp\left(-\frac{\pi|x|}{L}\right), \qquad (11.157)$$

where, in the last equality, we assume that $|x| \gg L$. Experimentally, in graphene a rather high nonlocal resistivity

$$R(x) = \frac{V(x)}{I_0} \qquad (11.158)$$

is observed at $|x| \geq 5L$ and even $10L$, which cannot be explained by "just" charge transport ($\exp(-5\pi) \approx 1.5 \times 10^{-7}$). It also cannot be explained by transport via edge states, since it is observed beyond the quantum Hall regime as well.

So, let us come back to our original statement that the transport properties in graphene can be anomalously sensitive to the spin projection. In particular, in the situation shown in Fig. 11.7(a)

$$(\sigma_1)_{xy} = -(\sigma_2)_{xy}, \tag{11.159}$$

where subscripts 1 and 2 will be used for spin up and spin down, respectively.

Let us use, instead of Eq. (11.148), two separate Ohm laws for each spin projection:

$$\vec{j}_i = -\hat{\sigma}_i \vec{\nabla} \varphi_i, \tag{11.160}$$

where $\hat{\sigma}_i$ has the structure (11.149) and

$$\varphi_i = \phi + \frac{n_i}{D_i}, \tag{11.161}$$

where

$$D_i = \frac{dn_i}{d\mu} \tag{11.162}$$

is the thermodynamic density of states (note that here μ is the chemical potential, not the mobility, as in the greatest part of this chapter!), and ϕ is the electrostatic potential. The second term on the right-hand side of Eq. (11.161) describes diffusion processes in the situation in which spin-up and spin-down electron densities n_i are finite. We will assume the electroneutrality condition

$$n_1 = -n_2 = n \tag{11.163}$$

and separate the total current density $\vec{j}_0$ and the spin current density $\vec{j}'$:

$$\vec{j}_{1,2} = \vec{j}_0 \pm \vec{j}'. \tag{11.164}$$

The equation of spin diffusion reads

$$\vec{\nabla} \vec{j}' = -\gamma(n_1 - n_2) = -2\gamma n, \tag{11.165}$$

where $\gamma = \tau_s^{-1}$ is the rate of spin-flip processes. Then we have the following set of equations (together with Eq. (11.165)):

$$\vec{\nabla} \phi + \frac{1}{D_1} \vec{\nabla} n = -\hat{\rho}_1 \left(\vec{j}_0 + \vec{j}' \right), \tag{11.166}$$

$$\vec{\nabla} \phi - \frac{1}{D_2} \vec{\nabla} n = -\hat{\rho}_2 \left(\vec{j}_0 - \vec{j}' \right), \tag{11.167}$$

$$\vec{\nabla} \vec{j}_0 = 0, \tag{11.168}$$

where $\hat{\rho}_i = \hat{\sigma}_i^{-1}$.

One can exclude $\vec{\nabla}\phi$ from these equations and express the spin current as

$$\vec{j}' = -\hat{\sigma}\left[\left(\frac{1}{D_1} + \frac{1}{D_2}\right)\vec{\nabla}n + (\hat{\rho}_1 - \hat{\rho}_2)\vec{j}_0\right], \tag{11.169}$$

where

$$\hat{\sigma} = \hat{\rho}^{-1}, \quad \hat{\rho} = \hat{\rho}_1 + \hat{\rho}_2. \tag{11.170}$$

On substituting Eq. (11.151) into Eq. (11.146) and taking into account Eq. (11.137) we find at last the closed equation for the spin density:

$$\nabla^2 n - \frac{1}{l_s^2}n = -\frac{D_1 D_2}{D_1 + D_2}\vec{\nabla}\left[(\hat{\rho}_1 - \hat{\rho}_2)\vec{j}_0\right], \tag{11.171}$$

where l_s is the spin-diffusion length:

$$\frac{1}{l_s^2} = \frac{2\gamma}{\sigma_{xx}}\frac{D_1 D_2}{D_1 + D_2}. \tag{11.172}$$

It follows from Eq. (11.168) and (11.151) that $j_{0x} = 0$ and j_{0y} does not depend on y:

$$j_{0y} = I_0\delta(x). \tag{11.173}$$

On substituting Eq. (11.173) into Eq. (11.171) we find a rigorous (within our model) equation:

$$\nabla^2 n - \frac{1}{l_s^2}n = -\frac{D_1 D_2}{D_1 + D_2}\left((\hat{\rho}_1)_{xy} - (\hat{\rho}_2)_{xy}\right)I_0\frac{d\delta(x)}{dx}. \tag{11.174}$$

If we assume that $L \ll l_s$ we can neglect the y-dependence of n, and Eq. (11.174) is solved immediately:

$$n(x) = -\frac{D_1 D_2}{2(D_1 + D_2)}\left[\left((\rho_1)_{xy} - (\rho_2)_{xy}\right)\right] \cdot I_0 \operatorname{sgn} x \exp\left(-\frac{|x|}{l_s}\right). \tag{11.175}$$

Finally, taking into account that for the thin strip the current is assumed to be constant in the y-direction, we find

$$V(x) = L\left[(\rho_1)_{xy}j_{1x}(x) + (\rho_2)_{xy}j_{2x}(x)\right] = L\left[(\rho_1)_{xy} - (\rho_2)_{xy}\right]j_x'(x) \tag{11.176}$$

and use Eq. (11.169) for j_x'. The final answer for the nonlocal resistance (11.158) is

$$R(x) = \frac{L}{2l_s} \sigma_{xx} \left[(\rho_1)_{xy} - (\rho_2)_{xy} \right]^2 \exp \left(-\frac{|x|}{l_s} \right). \qquad (11.177)$$

This formula seems to be in good agreement with the experimental data (Abanin et al., 2011). Actually, this derivation is very general. The only peculiarity of graphene is that near the neutrality point the difference $(\rho_1)_{xy} - (\rho_2)_{xy}$ can be huge (see Fig. 11.7(c)).

11.7 Beyond the Boltzmann equation: localization and antilocalization

In general, the semiclassical Boltzmann equation does not suffice to describe the transport properties of a two-dimensional electron gas because of *weak localization* effects (Altshuler et al., 1980). They originate from quantum interference effects between different trajectories passing in opposite directions (Fig. 11.9). The corresponding correction to the conductivity is of the order of

$$\delta\sigma \sim \pm \frac{e^2}{h} \Lambda, \qquad (11.178)$$

where Λ is a "big logarithm": At $T = 0$ it is $\ln(L/a)$. These interference effects are sensitive to the magnetic field (due to the Aharonov–Bohm effect), which results in

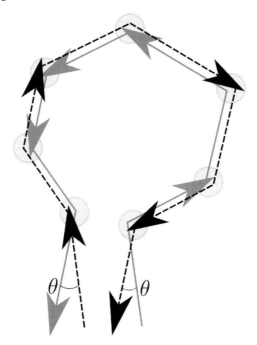

Fig. 11.9 Interference between trajectories with opposite directions of electron motion.

a large magnetoresistivity. Usually, $\delta\sigma < 0$ and suppression of the interference by the magnetic field increases the conductivity (negative magnetoresistance). Inelastic scattering processes also destroy the interference, leading to a cut-off of the logarithm: $\Lambda \rightarrow \ln(\varepsilon_F/T)$. In graphene, the magnetoresistance related to the weak localization is strongly suppressed, in comparison with the case of a conventional electron gas. This was found by Morozov et al. (2006) and explained by them as the effect of random pseudomagnetic fields created by ripples (see Chapter 10). Later, these effects were observed and studied in detail (Tikhonenko et al., 2008, 2009).

Actually, the physics of the weak localization in graphene (McCann et al., 2006) is very complicated. First, the Berry phase π is involved in the interference processes, which changes the sign of localization corrections: Instead of weak localization one can have weak *anti*localization. Second, the effects of trigonal warping break the time-reversal symmetry for a given valley, whereas the inter-valley scattering processes restore it. Since the trajectories in Fig. 11.9 are related by time reversal, this symmetry is very important. As a result, depending on the types of defects in the sample, one can have either weak localization (and negative magnetoresistance) or weak antilocalization (and positive magnetoresistance). This prediction (McCann et al., 2006) has been confirmed experimentally (Tikhonenko et al., 2009). Detailed analysis of the experimental data on weak localization allows us to separate the contributions of three main mechanisms (Section 11.3): static pseudomagnetic fields (e.g., created by ripples), charge impurities, and resonant scattering centers in specific samples (Couto et al., 2014).

Closer to the neutrality point, the localization corrections become of the order of the Boltzmann conductivity, and the semiclassical approach fails completely. This happens in a relatively narrow concentration range that is quite difficult to probe experimentally. Theoretically, the situation also does not look very clear. Earlier works were reviewed by Evers and Mirlin (2008). Here we just mention some important, more recent papers: Bardarson et al. (2007, 2010), Titov et al. (2010), and Ostrovsky et al. (2010). The main results are as follows.

If we do not take into account intervalley scattering (which means that all inhomogeneities are supposed to be smooth), we never have Anderson localization, and the conductivity at the neutrality point remains of the order of minimal metallic conductivity (see Chapter 3) or grows slowly with the sample size (antilocalization). In particular, random pseudomagnetic fields have no effect on the value of the minimal conductivity, since they can be eliminated by a gauge transformation similar to that discussed in Section 3.4 (Ostrovsky, Gornyi, & Mirlin, 2008). The random mass term ($V_z\sigma_z$ in Eq. (11.33)) affects the value of the minimal conductivity very weakly, except when the average mass is not zero ($\langle V_z\rangle \neq 0$); in that case, localization is possible (Bardarson et al., 2010). For a random scalar potential, antilocalization seems to arise (Bardarson et al., 2007).

In the presence of intervalley scattering, Anderson localization takes place in the generic case. However, the most interesting case of resonant scatterers such as vacancies or covalently bonded adatoms (see Chapter 6) requires special consideration, due to the additional "chiral" symmetry (Altland, 2002; Evers & Mirlin, 2008; Ostrovsky et al., 2010; Titov et al., 2010). It seems that in this case the localization radius diverges at the neutrality point, and the conductivity at $n = 0$ remains at the level of the minimal conductivity. All these issues require further study, both theoretically and, especially, experimentally.

11.8 Hydrodynamics of electron liquid in ultra-pure graphene

Huge progress in the quality of graphene samples allows us to reach the regime when electron–electron collisions are a more efficient scattering mechanism than scattering by defects (both in bulk and at the edges)

$$\tau_{ee} \ll \tau_p, \tag{11.179}$$

where τ_{ee}, τ_p are the corresponding mean-free times (τ_p is the time of relaxation of momentum of the electron system as a whole). Since the electron–electron collisions conserve the total momentum of electron systems (in the absence of Umklapp processes; see Ziman, 2001) and, at the same time, provide efficient dissipation of energy, the electron motion under the condition (11.179) can be described as a flow of a viscous liquid. Indeed, at the first step the electron–electron interactions provide effective thermalization and redistribution of the energy and momentum within the system of interacting electrons and, at the second step, the scattering by defects, phonons, edges, etc., decelerate the electron flow as a whole – the situation reminiscent of conventional hydrodynamics (Landau & Lifshitz, 1987; Falkovich, 2011).

Currently, hydrodynamics of electron liquid in graphene is a quickly developing field (Briskot et al., 2015; Narozhny et al., 2015; Torre et al., 2015; Bandurin et al., 2016; Crossno et al., 2016; Levitov & Falkovich, 2016; Lucas et al., 2016; Pellegrino et al., 2016; Bandurin et al., 2018; Guo et al., 2017; Kumar et al., 2017; for review, see Lucas & Fong, 2018). Here, we present a very brief introduction of the basic ideas and results.

The transition from kinetic equation to hydrodynamics is one more example of the coarse-graining approach (Sections 11.1 and 11.2). If we deal with the spatial scales of the system larger than typical microscopic scales such as "thermalization length"

$$l_{ee} = v_F \tau_{ee} \tag{11.180}$$

(in this section we will use the notation v_F for the Fermi velocity of electrons in graphene, to distinguish it from the velocity of flow of electron liquid) and with the processes slow enough in comparison with τ_{ee} then the distribution function (11.13) reaches a local equilibrium characterizing by quasithermodynamic variables such as local temperature, local chemical potential, and local drift velocity weakly dependent on coordinates and time. The behavior of the system under such conditions is determined by the densities of quasiconserving quantities such as number of particles, charge, energy, momentum, etc., and one can hope to have the closed set of equations describing such quantities; for the general scheme, see e.g., Zubarev (1974), Akhiezer and Peletminskii (1981), and Kamenev (2011).

In the case of graphene, if we are interested in the unified description of the system through the neutrality point, the minimal set of macroscopic variables includes density of electrons (e), density of holes (h), and energy density. The corresponding currents are the electric current

$$\vec{j} = e \sum_{\vec{k}} \left(\vec{v}_{\vec{k},e} f_{\vec{k},e} - \vec{v}_{\vec{k},h} f_{\vec{k},h} \right), \tag{11.181}$$

the quasiparticle disbalance current

$$\vec{j}_d = \sum_{\vec{k}} \left(\vec{v}_{\vec{k},e} f_{\vec{k},e} + \vec{v}_{\vec{k},h} f_{\vec{k},h} \right), \tag{11.182}$$

and the energy current

$$\vec{j}_E = \sum_{\vec{k}} \left(\varepsilon_{\vec{k},e} \vec{v}_{\vec{k},e} f_{\vec{k},e} + \varepsilon_{\vec{k},h} \vec{v}_{\vec{k},h} f_{\vec{k},h} \right) \tag{11.183}$$

(Briskot et al., 2015; Narozhny et al., 2015; Lucas et al., 2016). For simplicity, we will consider the case of strong enough doping (to be specific, we will assume the electron doping)

$$\varepsilon_F \gg T. \tag{11.184}$$

In this case, one can consider only one-component electron liquid; in some cases (but not always) one can also separate dynamics of charge from dynamics of energy.

In that case, the former can be described by hydrodynamics of one-component charged uncompressible liquid with the particle density $n = $ const and the velocity field $\vec{v}(\vec{r},t)$, which satisfies the continuity equation

$$\nabla \vec{v} = 0 \tag{11.185}$$

and Navier-Stokes equation (Landau & Lifshitz, 1987; Falkovich, 2011)

$$\rho \left[\frac{\partial \vec{v}}{\partial t} + \left(\vec{v} \nabla \right) \vec{v} + \frac{\vec{v}}{\tau_p} \right] = \vec{F} + \eta \nabla^2 \vec{v} , \qquad (11.186)$$

where $\rho = m^* n$ is the mass density of electron liquid, η is the viscosity, and $\vec{F}$ is the density of external forces (e.g., electromagnetic) acting on electrons. The term $\vec{v} / \tau_p$ describes external "friction" of electrons by defects, phonons, etc.

Importantly, m^* in the definition of the mass density is the *effective* mass of the electron, which determines its acceleration under the effect of external fields. In the case of single-layer graphene (massless Dirac fermions) it reads

$$m^* = \hbar k_F / v_F \qquad (11.187)$$

(compare with Eq. (2.161)). The kinematic viscosity η / ρ can be estimated as the diffusion coefficient of momentum; for the two-dimensional case it is equal to

$$\frac{\eta}{\rho} = \frac{v_F^2}{2\tau_{ee}} . \qquad (11.188)$$

For the case of highly degenerate electron liquid (11.184), the time of electron–electron collisions can be estimated as

$$\frac{\hbar}{\tau_{ee}} \approx \alpha^2 \frac{T^2}{\varepsilon_F} \qquad (11.189)$$

(Abrikosov, 1988; Vonsovsky & Katsnelson, 1989), where α is a dimensionless interaction parameter; for graphene it is of the order of unity. Note that at the neutrality point, in the case opposite to Eq. (11.184), one has

$$\frac{\hbar}{\tau_{ee}} \approx \alpha^2 T \qquad (11.190)$$

(Fritz et al., 2007; Kashuba, 2008). However, in this case the one-component hydrodynamics is not applicable, and one needs to write the coupled set of equations for all three currents (11.181) through (11.183).

For typical experimental situations (doping of the order of 10^{12} cm^{-2}, temperature of the order of room temperature), accurate quantitative estimates of τ_{ee} give the kinematic viscosity $\eta / \rho \approx 0.1$ m^2/s (Principi et al., 2016), which is, roughly, five orders of magnitude smaller than for water, which means that electron liquid in graphene is very viscous. The corresponding Reynolds number (Landau & Lifshitz, 1987; Falkovich, 2011)

$$\mathrm{Re} = \frac{\rho u L}{\eta} , \qquad (11.191)$$

where u is a typical flow velocity and L is a typical spatial scale of the problem, is typically very small for the experiments with graphene (of the order of 10^{-3}, according to the estimate by Torre et al., 2015). In this situation, the nonlinear term $(\vec{v}\,\nabla)\,\vec{v}$ in Eq. (11.186) can be neglected, which dramatically simplifies the situation.

Let us assume that the only external force acting on the system is the electric field, which does not depend on time:

$$\vec{F} = -en\nabla\varphi, \qquad (11.192)$$

where φ is electrostatic potential. We will consider only stationary solutions of Eq. (11.186) and, therefore, skip the term $\partial\vec{v}/\partial t$. As a result we obtain:

$$-\frac{e}{m^*}\nabla\varphi + \frac{\eta}{\rho}\nabla^2\,\vec{v} = \frac{\vec{v}}{\tau_p}. \qquad (11.193)$$

This equation can be rewritten in the form

$$-\frac{\sigma_0}{e}\nabla\varphi + D^2\nabla^2\,\vec{v} = \vec{v}\,, \qquad (11.194)$$

where we introduced Drude conductivity

$$\sigma_0 = \frac{ne^2\tau_p}{m^*} \qquad (11.195)$$

and diffusion length

$$D = \sqrt{\frac{\eta\tau_p}{\rho}}. \qquad (11.196)$$

Taking curl of Eq. (11.194) we find

$$D^2\nabla^2\,\vec{\omega} = \vec{\omega}\,, \qquad (11.197)$$

where

$$\vec{\omega} = \nabla\times\vec{v} \qquad (11.198)$$

is vorticity (in the case of two-dimensional flow, in xy-plane it is directed along z-axis). From Eq. (11.196), D is the diffusion length for vorticity. For the high-quality graphene samples encapsulated in hexagonal boron nitride one can reach the values $\tau_p = 1$ ps and $D \approx 0.3\ \mu$m (Torre et al., 2015).

Calculating divergence of Eq. (11.193) and taking into account Eq. (11.185) one can immediately see that the electrostatic potential satisfies the Laplace equation

$$\nabla^2\varphi = 0. \qquad (11.199)$$

Further, we will consider the same geometry as in Fig. 11.8: graphene stripe of the width L situated along x direction, with possible injection or extraction of the current in some points at $x = \pm x_0$. Our consideration follows the paper by Torre et al. (2015).

Now, we have to complete Eq. (11.194) by the appropriate boundary conditions. Its full derivation is a complicated problem; a related problem of the boundary conditions for Boltzmann equations is discussed in Section 11.5. Torre et al. (2015) used phenomenological Navier boundary conditions, which in general can be written as

$$n_\alpha T_{\alpha\beta}\tau_\beta + \frac{1}{l_b}v_\alpha\tau_\alpha = 0 \qquad (11.200)$$

(Neto et al., 2005; Kelliher, 2006; Bocquet & Barrat, 2007). Here, we introduce the stress tensor (Landau & Lifshitz, 1987) which is equal

$$T_{\alpha\beta} = \frac{\partial v_\alpha}{\partial x_\beta} + \frac{\partial v_\beta}{\partial x_\alpha} \qquad (11.201)$$

(under the condition (11.185), that is, for the case of uncompressible liquid), $\vec{n}$ and $\vec{\tau}$ are normal and tangential vectors to the boundary, l_b is a phenomenological parameter called "boundary slip length," and we assume a summation over repeated indices. It can be related to the electron scattering at the edges, which we discussed in Section 11.5; see Pellegrino, Torre, and Polini (2017) and Kiselev and Schmalian (2019). For the geometry shown in Fig. 11.8 Eq. (11.200) reads

$$\left(\frac{\partial v_x}{\partial y} + \frac{\partial v_y}{\partial x}\right)_{y=\pm L/2} = \mp\frac{v_x(x, y = \pm L/2)}{l_b}. \qquad (11.202)$$

Let us first consider the simplest case where no current is ejected or extracted and electric field E_x is just constant parallel to x axis. Then, Eq. (11.194) reads

$$D^2\frac{d^2v_x(y)}{dy^2} - v_x(y) = \frac{\sigma_0}{en}E_x. \qquad (11.203)$$

This is an ordinary differential equation with constant coefficients, and its general solution is straightforward:

$$v_x(y) = -\frac{\sigma_0}{en}E_x + C_1\cosh\frac{y}{D} + C_2\sinh\frac{y}{D}. \qquad (11.204)$$

It depends on two integration constants C_1 and C_2. We have to choose them from the boundary conditions (11.202). The answer is

$$v_x(y) = \frac{\sigma_0}{ne} E_x \left(1 - \frac{D}{\xi} \cosh \frac{y}{\xi} \right), \tag{11.205}$$

where

$$\xi = l_b \sinh \frac{L}{2D} + D \cosh \frac{L}{2D}. \tag{11.206}$$

The total current can be calculated by integration of Eq. (11.205) over y:

$$I_x = en \int_{-L/2}^{L/2} dy v_x(y) = \sigma_0 E_x \left(1 - \frac{2D^2}{L\xi} \sinh \frac{L}{2D} \right), \tag{11.207}$$

which gives the following expression for the longitudinal conductivity:

$$\sigma_{xx} = \sigma_0 \left(1 - \frac{2D^2}{L\xi} \sinh \frac{L}{2D} \right) \tag{11.208}$$

(Torre et al., 2015).

For the case $l_b \to \infty$ (free-boundary condition), $\sigma_{xx} = \sigma_0$, and hydrodynamic flow has no effect on longitudinal conductivity. In the opposite limit $l_b \to 0$ (no-slip boundary condition, $v_x = 0$ at the edges) the Eq. (11.208) simplifies:

$$\sigma_{xx} = \sigma_0 \left(1 - \frac{2D}{L} \tanh \frac{L}{2D} \right). \tag{11.209}$$

For further simplification, one can consider the case $L \ll D$. Then, we have:

$$\sigma_{xx} = \frac{\sigma_0 L^2}{12 D^2} = \frac{ne^2}{m^*} \tau_{eff}, \tag{11.210}$$

where

$$\tau_{eff} = \frac{L^2}{6 v_F^2 \tau_{ee}} \tag{11.211}$$

(we took into account Eq. (11.196) and (11.188)). In this regime, the effective conductivity obviously *increases* with the temperature increase. However, with the further temperature growth, the vorticity diffusion length D decreases, becomes of the order of L and then smaller than L, and $\tau_{eff} \to \tau_p$ and decreases with the temperature increase. This means that in the hydrodynamic regime (11.179) for thin enough films, the conductivity is a nonmonotonous function of temperature and has a maximum (or, equivalently, resistivity has a minimum). This behavior was predicted by Gurzhi (1968) for the case of usual metals and is sometimes called *Gurzhi effect*. Its experimental observation can be considered as the simplest manifestation of the hydrodynamic regime. Experimentally, this regime is reached not only for graphene (Bandurin et al., 2016; Crossno et al., 2016) but also for the quasi–two-dimensional metal $PdCoO_2$ (Moll et al., 2016).

Much stronger manifestations of the hydrodynamic regime can be observed in the nonlocal transport measurements similar to those described in Section 11.6 (Torre et al., 2015; Bandurin et al., 2016; Levitov & Falkovich, 2016). Sometimes one can observe such a counterintuitive behavior as *negative* local resistance, due to electron counterflows. This can be related to the vortex formation in electron liquid but, dependent on geometry, can be also due to other factors (Pellegrino et al., 2016). The corresponding calculations are too cumbersome to be presented here; they can be found in the cited papers. The effect of negative local resistance was already experimentally observed (Bandurin et al., 2016).

The other bright experimentally observable effect in the hydrodynamic regime for electrons in graphene is a strong violation of the so-called Wiedemann–Franz law related thermal conductivity and conductivity in normal metals (Abrikosov, 1988; Vonsovsky & Katsnelson, 1989; Ziman, 2001). Its detailed discussion would probably be too technical for this book; therefore, we just refer to original theoretical (Lucas et al., 2016) and experimental (Crossno et al., 2016) works.

12

Spin effects and magnetism

12.1 General remarks on itinerant-electron magnetism

Up to now we have not discussed physical phenomena in graphene related to the spin of the electron (here we mean *real* spin and, associated with it, magnetic moment, rather than *pseudo*spin, or the sublattice index, which plays so essential a role throughout the book). The only exception was Zeeman splitting in an external magnetic field but, of course, this is just the simplest (and probably not the most interesting) of the spin effects. In this chapter we will discuss these spin phenomena.

First, due to exchange interactions of purely quantum-mechanical origin, various types of magnetic order can arise (Herring, 1966; Vonsovsky, 1974; Moriya, 1985; Yosida, 1996). The situation with possible magnetic ordering in graphene and other carbon-based materials is highly controversial (see Section 12.2) but, due to the huge interest in the field and its potential practical importance, this issue deserves some discussion. Before doing this, it is worth recalling some general concepts and models of itinerant-electron magnetism.

The simplest model used in the theory of itinerant-electron magnetism is the so-called *Hubbard model* (Gutzwiller, 1963; Hubbard, 1963; Kanamori, 1963). The Hamiltonian reads

$$\hat{H} = \sum_{ij\sigma} t_{ij}\hat{c}_{i\sigma}^{+}\hat{c}_{j\sigma} + U \sum_{i} \hat{n}_{i\uparrow}\hat{n}_{i\downarrow}, \qquad (12.1)$$

where $\hat{c}_{i\sigma}^{+}$ and $\hat{c}_{i\sigma}$ are creation and annihilation operators, respectively, on site i with the spin projection $\sigma = \uparrow, \downarrow$, t_{ij} are the hopping parameters, $\hat{n}_{i\sigma} = \hat{c}_{i\sigma}^{+}\hat{c}_{i\sigma}$ are operators of electron number, and U is the intrasite interaction parameter. The main approximation in the Hubbard model is that we neglect intersite Coulomb interaction. The Hamiltonian (12.1) is a simplification of a more general "polar model" (Schubin & Wonsowski, 1934). One can easily generalize the Hamiltonian (12.1) to the multiband case:

$$\hat{H} = \sum_{ij\sigma\lambda\lambda'} t_{i\lambda, j\lambda'} \hat{c}^+_{i\lambda\sigma} \hat{c}_{j\lambda'\sigma} + \frac{1}{2} \sum_{\substack{i\sigma\sigma' \\ \lambda_1\lambda_2\lambda'_1\lambda'_2}} \langle \lambda_1\lambda_2 | U | \lambda'_1\lambda'_2 \rangle \hat{c}^+_{i\lambda_1\sigma} \hat{c}^+_{i\lambda_2\sigma'} \hat{c}_{j\lambda'_2\sigma'} \hat{c}_{i\lambda'_1\sigma}, \qquad (12.2)$$

where λ is an orbital (band) quantum number.

The simplest theory of itinerant-electron magnetism was proposed by Stoner (1936). It is based just on the mean-field (Hartree–Fock) approximation. Let us make the following replacement in the Hamiltonian (12.1):

$$\hat{n}_{i\uparrow}\hat{n}_{i\downarrow} \rightarrow \hat{n}_{i\uparrow}n_{\downarrow} + \hat{n}_{i\downarrow}n_{\uparrow}, \qquad (12.3)$$

where we assume also that the averages $\langle \hat{n}_{i\sigma} \rangle \equiv n_\sigma$ are not dependent on i (but can be spin dependent). After the standard Fourier transformation, the Hamiltonian (12.1) with the replacement (12.3) takes the form

$$\hat{H} = \sum_{\vec{k}} t_\sigma\left(\vec{k}\right) \hat{c}^+_{\vec{k}\sigma} \hat{c}_{\vec{k}\sigma}, \qquad (12.4)$$

where

$$t_\uparrow\left(\vec{k}\right) = t\left(\vec{k}\right) + Un_\downarrow,$$
$$t_\downarrow\left(\vec{k}\right) = t\left(\vec{k}\right) + Un_\uparrow. \qquad (12.5)$$

This is just a single-electron Hamiltonian, and one can easily find

$$n_\sigma = \sum_{\vec{k}} f_{\vec{k}\sigma}, \qquad (12.6)$$

where

$$f_{\vec{k}\sigma} = f\left(t_\sigma\left(\vec{k}\right)\right) \qquad (12.7)$$

is the Fermi distribution function. One can show straightforwardly that nontrivial solutions with $n_\uparrow \neq n_\downarrow$, corresponding to the ferromagnetic order, exist if

$$\alpha \equiv UN(\varepsilon_F) > 1, \qquad (12.8)$$

where

$$N(\varepsilon) = \sum_{\vec{k}} \delta\left(\varepsilon - t\left(\vec{k}\right)\right) \qquad (12.9)$$

is the density of states (per spin projection). The inequality (12.8) is called the *Stoner criterion*. In the Stoner approximation (12.5), the densities of states for

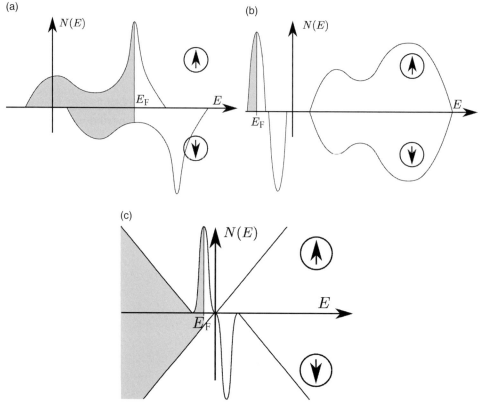

Fig. 12.1 A sketch of the electronic structures for various types of itinerant-electron ferromagnet: (a) the conventional case; (b) and (c) defect-induced half-metallic ferromagnetism in semiconductors and in graphene, respectively.

spin-up and spin-down electrons are just related by a rigid shift (see Fig. 12.1(a)). When $\alpha \to 1$, the saturation magnetization (in units of the Bohr magneton)

$$M(T = 0) = n_\uparrow(T = 0) - n_\downarrow(T = 0) \sim \sqrt{\alpha - 1}. \qquad (12.10)$$

When the temperature increases the magnetization decreases, vanishing at the Curie temperature T_C determined by the condition

$$U \int d\varepsilon \left(-\frac{\partial f}{\partial \varepsilon}\right) N(\varepsilon) = 1. \qquad (12.11)$$

At $\alpha \to 1$,

$$T_C \sim \sqrt{\alpha - 1} \qquad (12.12)$$

in the Stoner approximation.

Using the identity $\hat{n}_{i\sigma}^2 = \hat{n}_{i\sigma}$, one can rewrite the interaction term in the Hubbard Hamiltonian (12.1) as

$$U \sum_i \hat{n}_{i\uparrow}\hat{n}_{i\downarrow} = \frac{U}{2} \sum_i \left(\hat{n}_{i\uparrow} + \hat{n}_{i\downarrow} \right) - \frac{U}{2} \sum_i \left(\hat{n}_{i\uparrow} - \hat{n}_{i\downarrow} \right)^2.$$ (12.13)

The first term is just a renormalization of the chemical potential and can therefore be neglected. The Stoner approximation is exact for some artificial model with infinitely long-range and infinitely weak interaction:

$$\hat{H} = \sum_{ij\sigma} t_{ij}\hat{c}_{i\sigma}^+\hat{c}_{j\sigma} - \frac{U}{4N_0} \left(\hat{N}_\uparrow - \hat{N}_\downarrow \right)^2,$$ (12.14)

where N_0 is the number of sites

$$\hat{N}_\sigma = \sum_i \hat{n}_{i\sigma}.$$ (12.15)

Importantly, two terms on the right-hand side of Eq. (12.14) commute and (using for them the notations $\hat{H}_1$ and $\hat{H}_2$)

$$\exp\left(-\beta\hat{H}\right) = \exp\left(-\beta\hat{H}_1\right)\exp\left(-\beta\hat{H}_2\right).$$ (12.16)

Further, using the Hubbard–Stratonovich transformation

$$\exp\left[\frac{\beta U}{4N_0}\left(\hat{N}_\uparrow - \hat{N}_\downarrow\right)^2\right] = \left(\frac{N_0\beta}{4\pi U}\right)^{1/2} \int_{-\infty}^{\infty} d\Delta \exp\left[-\frac{N_0\beta\Delta^2}{4U} - \frac{\beta\Delta}{2}\left(\hat{N}_\uparrow - \hat{N}_\downarrow\right)\right],$$ (12.17)

one can calculate the partition function by integrating over Δ by the saddle-point method, the latter being exact in the limit $N_0 \to \infty$. This leads *exactly* to Eq. (12.5) through (12.7).

This allows us to understand the physical meaning of the Stoner criterion (12.8). Let us remove

$$\delta N = \frac{\left(\langle\hat{N}_\uparrow\rangle - \langle\hat{N}_\downarrow\rangle\right)}{2} \ll N$$ (12.18)

electrons (N is the total number of electrons) from the states with $\sigma = \downarrow$ below the Fermi energy to the states with $\sigma = \uparrow$ above the Fermi energy (see Fig. 12.2). Each of these electrons increases its band energy by

$$\delta\varepsilon = \delta N \cdot \Delta_1,$$ (12.19)

where

$$\Delta_1 = \frac{1}{N(\varepsilon_F)N_0}$$ (12.20)

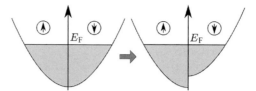

Fig. 12.2 Spontaneous spin polarization in itinerant-electron ferromagnets.

is the average distance between the single-particle energies near the Fermi energy ($N(\varepsilon_F)$ is the density of states *per site*, and the total density of states of the whole system is N_0 times larger). Thus, the increase of the total band energy is

$$\delta E_{\text{band}} = \delta N \, \delta\varepsilon = \frac{\left(\langle\hat{N}_\uparrow\rangle - \langle\hat{N}_\downarrow\rangle\right)^2}{4N(\varepsilon_F)N_0}. \tag{12.21}$$

At the same time, according to Eq. (12.14), the decrease of the interaction energy is

$$\delta E_{\text{int}} = -\frac{U\left(\langle\hat{N}_\uparrow\rangle - \langle\hat{N}_\downarrow\rangle\right)^2}{4N_0}. \tag{12.22}$$

The Stoner criterion (12.8) is nothing but the condition that the spin polarization is energetically favorable

$$\delta E_{\text{band}} + \delta E_{\text{int}} < 0. \tag{12.23}$$

Typically, itinerant-electron ferromagnetism in 3d metals and in their alloys and compounds is related to situations in which, in the paramagnetic case, the Fermi energy ε_F lies close to the peak of the density of states formed by a merging of Van Hove singularities; this is true for the prototype cases like Fe and Ni, as well as for weak itinerant-electron ferromagnets like $ZrZn_2$ (Irkhin, Katsnelson, & Trefilov, 1992, 1993). Actually, this means *some* instability, not necessarily magnetic; it can also be a structural instability (Katsnelson, Naumov, & Trefilov, 1994). This remark will be essential when we discuss the possibility of ferromagnetism in graphene with defects (see the next section).

   In realistic models with a finite radius of interelectron interaction, the Stoner theory of ferromagnetism is not accurate. First, as was shown by Kanamori (1963), the bare Coulomb interaction U in the criterion (12.8) should be replaced by the T-matrix; this statement becomes accurate in the limit of a small concentration of electrons or holes (the gaseous approximation; Galitskii, 1958a, 1958b). For the multiband Hubbard model (12.2), the T-matrix is determined by the equation (Edwards & Katsnelson, 2006)

$$\langle 13|T(E)|24\rangle = \langle 13|U|24\rangle + \sum_{5678} \langle 13|U|57\rangle\langle 57|P(E)|68\rangle\langle 68|T(E)|24\rangle, \tag{12.24}$$

where $|1\rangle = |i_1\lambda_1\rangle$ and

$$\langle 57|P(E)|68\rangle = \int\limits_{-\infty}^{\infty} dx \int\limits_{-\infty}^{\infty} dy \frac{1-f(x)-f(y)}{E-x-y} \rho_{56}(x)\rho_{78}(y), \qquad (12.25)$$

in which $\rho_{12}(x)$ is the corresponding site- and orbital-resolved spectral density and $f(x)$ is the Fermi distribution function. If we have a more or less structureless electron band of width W, $P(E) \sim 1/W$, and, in the limit of strong interaction,

$$U \gg W, \ T(E) \approx W. \qquad (12.26)$$

At the same time, $N(E) \approx 1/W$ and, in general, after the replacement $U \to T(\varepsilon_F)$, $\alpha \approx 1$, in clear contradiction with the original criterion (12.8). Thus, one can conclude that the Stoner theory overestimates the tendency toward ferromagnetism even at temperature $T = 0$.

The situation is essentially different in the cases in which the ferromagnetism is due to some defect-induced (e.g., by an impurity or vacancy) band in a gap, or pseudogap, of the main band (see Fig. 12.1(b) and (c)). This situation is relevant for graphene, as will be discussed in the next section. As was shown by Edwards and Katsnelson (2006), in such cases the T-matrix renormalization is less relevant, and the renormalized interaction $T(\varepsilon_F)$ is close to the bare one, U.

Even more serious problems with the Stoner theory arise at finite temperatures. One can demonstrate that, in general, the main suppression of magnetization is not due to the single-particle excitations but due to collective spin fluctuations (Moriya, 1985). As a result, the Curie temperature is strongly overestimated within the Stoner theory; if iron were a "Stoner ferromagnet" it would have $T_C \approx 4,000$ K instead of the real value of $T_C \approx 1,043$ K (Liechtenstein, Katsnelson, & Gubanov, 1985). For the case of weak itinerant-electron ferromagnets, $\alpha \to 1$, the true behavior is (Moriya, 1985)

$$T_C \sim (\alpha - 1)^{3/4} \qquad (12.27)$$

instead of Eq. (12.12).

At low temperatures, these spin fluctuations are nothing other than spin waves, as in localized (Heisenberg) magnets (Fig. 12.3). Typically, the energy of spin *rotations* is much smaller than that of electron–hole (Stoner) excitations. However, the case of ferromagnetism in a narrow defect-induced band is also special in this sense (Edwards & Katsnelson, 2006).

To explain this important point we first need to describe another basic model of itinerant-electron ferromagnets, the *s–d exchange model* (Vonsovsky, 1946; Zener, 1951a, 1951b, 1951c; Vonsovsky & Turov, 1953). Nowadays, this model is frequently called the *Kondo lattice model*, after the very important work of Kondo (1964) on a magnetic impurity in a metal. I think it is historically more fair to talk about the Kondo *effect* within the s–d exchange (or Vonsovsky–Zener) *model*.

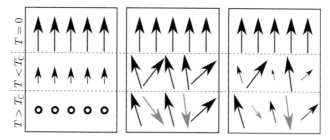

Fig. 12.3 The temperature evolution of ferromagnetic states in the Stoner model (left panel), in the Heisenberg model (middle panel) and in real itinerant-electron ferromagnets (right panel).

Within this model it is postulated that there exist some local magnetic moments described by spin operators $\vec{S}_i$ and that they interact locally with the spins of conduction electrons:

$$\hat{H} = \sum_{ij\sigma} t_{ij} \hat{c}_{i\sigma}^{+} \hat{c}_{j\sigma} - I \sum_{i} \hat{\vec{S}}_i \hat{\vec{s}}_i, \tag{12.28}$$

where

$$\hat{\vec{s}}_i = \frac{1}{2} \sum_{\sigma\sigma'} \hat{c}_{i\sigma}^{+} \vec{\sigma}_{\sigma\sigma'} \hat{c}_{i\sigma'}, \tag{12.29}$$

and I is the s–d exchange interaction constant. Despite the fact that the Hamiltonian (12.28) does not contain the exchange interactions between the localized spins at different sites, it arises as an *indirect* interaction via conduction electrons known as *RKKY* (Ruderman–Kittel–Kasuya–Yosida) *interaction* (Vonsovsky, 1974). Within the lowest order of the perturbation expansion in I, the effective Hamiltonian for localized spins is

$$\hat{H}_{\text{eff}} = -\sum_{i<j} J_{ij} \hat{\vec{S}}_i \hat{\vec{S}}_j, \tag{12.30}$$

where

$$J_{ij} = I^2 \chi_{ij} \tag{12.31}$$

and

$$\chi_{ij} = -\frac{1}{4} T \sum_{\varepsilon_n} G_{ij}^{(0)}(i\varepsilon_n + \mu) G_{ij}^{(0)}(i\varepsilon_n + \mu) \tag{12.32}$$

is the inhomogeneous susceptibility of conduction electrons. Eq. (12.32) is reminiscent of Eq. (6.135) for the interaction between adatoms and can be derived in a similar way. The RKKY interaction (12.31) for the case of graphene has some interesting properties, which will be discussed in the next section.

The criterion of applicability of the expressions (12.30) and (12.31) is the smallness of the spin polarization in the conduction-electron subsystem. For the case of systems with complete spin polarization, such as magnetic semiconductors (Nagaev, 1983, 2001) and half-metallic ferromagnets (Katsnelson et al., 2008), the situation is totally different and, instead, the *double-exchange* mechanism is responsible for the ferromagnetism, with an essentially non-Heisenberg character of exchange interactions (Auslender & Katsnelson, 1982). In this case, typical spin-wave energies are of the order of

$$\hbar\omega_{sw} \approx \frac{n|t|}{M},\tag{12.33}$$

where n is the charge-carrier concentration and M is the magnetization (Edwards, 1967; Irkhin & Katsnelson, 1985a, 1985b). This formula is valid both for s–d exchange and for Hubbard models. In the first case, M is of the order of one and, for small enough n,

$$\hbar\omega_{sw} \ll \varepsilon_F\tag{12.34}$$

since $\varepsilon_F \sim n^{2/3}$ (for the three-dimensional case) and $\varepsilon_F \sim n^{1/2}$ (for the two-dimensional case). If we have a strong polarization in the defect-induced band (see Fig. 12.1(b) and (c)) $M \sim n$ should hold and

$$\hbar\omega_{sw} \sim |t| \gg \varepsilon_F.\tag{12.35}$$

Thus, we have a very unusual situation in which the spin rotations are more energetically expensive than the electron–hole (Stoner) excitations. Also, as was mentioned previously, the *T*-matrix renormalization of the Stoner criterion is not relevant here. As a result, one can conclude that, if it were possible to create ferromagnetism in the defect-induced band of itinerant electrons, this situation would be described by the Stoner model and one could expect much higher Curie temperatures than for conventional magnetic semiconductors (Edwards & Katsnelson, 2006). This is one of the strongest motivations for the search for ferromagnetism in sp-electron systems, including graphene.

12.2 Defect-induced magnetism in graphene

Experimentally, sp-electron magnetism, in particular in carbon-based materials, is one of the most controversial issues in modern materials science (for reviews, see Esquinazi & Höhne, 2005; Makarova & Palacio, 2006; Yazyev, 2010). Typically, the observed experimental magnetic moment (when the existence of ferromagnetism has been claimed) is very small, $\mu \leq 10^{-3} - 10^{-4}\mu_B$ per atom. Keeping in mind that magnetic iron is everywhere on this planet (dust contains a lot of ferrimagnetic magnetite, Fe_3O_4) the question of possible contamination is crucial, and it is very difficult to demonstrate convincingly that the observed magnetism is intrinsic (Nair et al., 2012). To better follow the possible arguments and counterarguments, see, e.g., reviews of the scientific literature on the magnetism of CaB_6 (Edwards & Katsnelson, 2006) and of polymerized fullerenes (Boukhvalov & Katsnelson, 2009c). Importantly, one can prove (Edwards & Katsnelson, 2006) that a Curie temperature of the order of room temperature is thermodynamically incompatible with $\mu < 10^{-2}$ μ_B; thus, if one observes ferromagnetic ordering with $\mu \approx 10^{-3} - 10^{-4}\mu_B$ at room temperature it should be either a mistake or a strongly inhomogeneous situation, with ferromagnetic regions with local $\mu > 10^{-2}$ μ_B in a nonmagnetic surrounding.

The first experimental study of magnetism of *graphene* (actually, graphene paper, a mixture of single-layer and multilayer graphene, was studied) did not reveal any magnetic ordering but, rather, a quite mysterious paramagnetism (Sepioni et al., 2010).

It is natural to ask why we should discuss so controversial an issue at all. Well, first, it is a really hot subject. More importantly, some theoretical results seem to be reliable (actually, there are even some *theorems,* as will be discussed later) and worthy of consideration. They also give us a deeper understanding of the physics of defects (Chapter 6) and edge states (Chapter 5) in graphene.

Let us start with the case of vacancies (Section 6.5) or covalently bonded adsorbates (Section 6.6). As we have seen, their electronic structures are quite similar, so, in the simplest approximation, the vacancy can be considered as a model for the hydrogen adatom or some other "resonant-scattering" center. All these defects create mid-gap states within the graphene pseudogap (see Fig. 6.1). As was discussed in the previous section, a peak in the density of states near the Fermi energy can lead to a magnetic instability. This conclusion is confirmed by straightforward density-functional calculations: The periodic array of vacancies or hydrogen adatoms on graphene has a tendency to undergo spontaneous spin polarization (Yazyev & Helm, 2007). The corresponding electronic structure is shown in Fig. 12.4. For large enough distances between the defects, the magnetic moment per defect is close to the magnitude of the Bohr magneton μ_B. Note that the splitting of mid-gap states induced by hydrogen atom on the top of graphene was observed experimentally (using scanning tunneling microscopy [STM]) by Gonzáles-Herrero et al. (2016). Even though they did not probe spin polarization of the split energy states, the magnetic character of the splitting seems to be the most reasonable interpretation.

Such magnetic moments have been observed experimentally in graphene with vacancies and in fluorinated graphene; however, no magnetic *ordering* has been found (Nair et al., 2012; Nair et al., 2013). In the case of fluorinated graphene or

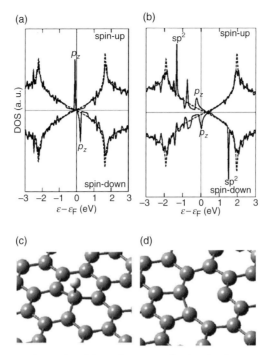

Fig. 12.4 Spin-polarized densities of states of (a) a hydrogen adatom and (b) vacancy in graphene; (c) and (d) shows the corresponding atomic structures (reproduced with permission from Yazyev & Helm, 2007).

graphene with other sp^3 impurities (resonant scattering centers are considered in Sections 6.6 and 11.3) these magnetic moments are supposed to be associated to mid-gap states in p_z band (see later). In the case of vacancies, apart from these magnetic moments, an additional kind of localized magnetic moment is observed, due to dangling bonds on carbon atoms near the vacancies; they can be distinguished by their behavior with the electron or hole doping (Nair et al., 2013).

Note that, from the density functional calculations, bivacancies (Boukhvalov & Katsnelson, 2009c) or couples of neighboring hydrogen atoms (Boukhvalov, Katsnelson, & Lichtenstein, 2008) turn out to be nonmagnetic.

Gonzáles-Herrero et al. (2016) have studied magnetism of hydrogen adatoms on graphene via STM. They did not directly measure spin polarization but detected local magnetic moments by splitting of the hydrogen-induced mid-gap states. They demonstrated that the splitting, which is observed for a single hydrogen adatom, also exists for the couple of A–A hydrogen atoms (that is, both belonging to the same sublattice A) and not for the A–B couples (hydrogen atoms in different sublattices), in agreement with the density functional calculations.

The real meaning of these results is clarified by the *Lieb theorem* (Lieb, 1989), one of the few rigorous results in the theory of itinerant-electron magnetism. The theorem is about the ground state of the single-band Hubbard model (12.1) on a bipartite lattice; the honeycomb lattice is just an example of this generic case. The most general definition of the bipartite lattice is that it consists of two sublattices, A and B, such that all hopping integrals within the same sublattice are zero:

$$\hat{t}_{AA} = \hat{t}_{BB} = 0. \tag{12.36}$$

Therefore, the band part of the Hamiltonian for the bipartite lattice has the structure

$$\hat{H}_0 = \begin{pmatrix} 0 & \hat{t} \\ \hat{t}^+ & 0 \end{pmatrix}, \tag{12.37}$$

with nonzero blocks only between two sublattices. Let us consider the case in which the numbers of sites within the sublattices A and B, N_A and N_B, can be different. This means that we have vacancies, and the numbers of vacancies belonging to A and to B are, in general, not the same. Thus, $\hat{t}$ is an $N_A \times N_B$ matrix.

Before discussing the effects of interactions, let us consider some properties of the single-particle spectrum of the Hamiltonian (12.37) (Inui, Trugman, & Abrahams, 1994; Kogan, 2011). We will assume, to be specific, that $N_B \geq N_A$.

First, there are at least $N_B - N_A$ linearly independent eigenfunctions with the eigenvalue $E = 0$ and all components equal to 0 on the sites of the A sublattice. This is the obvious consequence of the structure (12.37): The system of linear equations

$$t\psi = 0 \tag{12.38}$$

has at least $N_B - N_A$ linearly independent solutions.

Second, for the eigenfunctions $\psi_n = \{\psi_n(i)\}$, corresponding to the nonzero eigenvalues E_n,

$$\begin{pmatrix} -E_n & \hat{t} \\ \hat{t}^+ & -E_n \end{pmatrix} \psi_n = 0, \tag{12.39}$$

there is a symmetry property

$$\psi_{\bar{n}}(i) = \pm \psi_n(i), \tag{12.40}$$

where $\psi_{\bar{n}}$ are the eigenfunctions corresponding to $-E_n$, and the plus and minus signs on the right-hand side of Eq. (12.40) correspond to the cases in which i belongs to A and B, respectively.

Thus, the spectrum of the Hamiltonian (12.37) is symmetric (if E_n is an eigenvalue, $-E_n$ is an eigenvalue, too) and contains at least $N_B - N_A$ solutions with $E = 0$. It turns out that the latter states are unstable with respect to spontaneous spin polarization at arbitrarily small $U > 0$ (Lieb, 1989).

Moreover, the Lieb theorem claims that the ground state of the Hubbard model (12.1) with $U > 0$, the single-particle Hamiltonian (12.37) and the number of electrons equal to the number of sites, $N = N_A + N_B$, is unique (apart from the trivial $(2S + 1)$-fold degeneracy) and has the spin

$$S = \frac{N_B - N_A}{2}.$$
(12.41)

The theorem can be proved in two steps. First, it is shown that the ground state is unique at any U, that is, that the states belonging to different multiplets with spins S and $S' \neq S$ cannot both be eigenstates with the minimal energy. The consequence is that the ground-state spin S cannot be dependent on U. In the opposite case, there will unavoidably be a crossing of the minimal energies with a given spin, $E_0(S; U)$ and $E_0(S'; U)$, at some $U = U_c$. Second, in the limit of large U and $N = N_0$ the Hubbard model (12.1) is equivalent to the Heisenberg model with the Hamiltonian

$$\hat{H}' = \sum_{i<j} \frac{2|t_{ij}|^2}{U} \left(\hat{\vec{s}}_i \hat{\vec{s}}_j - \frac{1}{4} \right)$$
(12.42)

(see, e.g., Yosida, 1996), and for the latter, the result (12.41) can be proved quite straightforwardly and easily (Lieb & Mattis, 1962).

Importantly, the Lieb theorem does not assume the thermodynamic limit $N_0 \to \infty$ and is valid also for small systems. Its applications to the magnetic properties of finite graphene fragments have been discussed by Yazyev (2010).

It follows from the Lieb theorem that if all vacancies sit in the same sublattice their spins are parallel in the ground state. If, oppositely, $N_A = N_B$, the ground state is a singlet, with $S = 0$. This means that the interactions between vacancy-induced magnetic moments are ferromagnetic if the vacancies belong to the same sublattice and antiferromagnetic if they belong to different sublattices. As we see, this result is rigorous within the Hubbard model with half-filling ($N = N_0$). The same conclusion for the covalently bonded adatoms or vacancies follows from the density-functional calculations (Yazyev & Helm, 2007; Boukhvalov & Katsnelson, 2011).

This can be also proved for the RKKY interaction (12.31) within the s–d exchange model (12.28) (Kogan, 2011). By Fourier transformation of Eq. (12.32) it can be represented as

$$\chi_{ij} = -\frac{1}{4} \int_0^\beta d\tau \, G_{ij}^{(0)}(\tau) G_{ji}^{(0)}(-\tau),$$
(12.43)

where $\beta = T^{-1}$ and

$$G_{ji}^{(0)}(\tau) = T \sum_{\varepsilon_n} G_{ij}^{(0)}(i\varepsilon_n + \mu) \exp(-i\varepsilon_n \tau) \qquad (12.44)$$

(see also Cheianov et al., 2009). It can be expressed in terms of the eigenfunctions and eigenenergies of the Hamiltonian $\hat{H}_0$ (Mahan, 1990)

$$G_{ji}^{(0)}(\tau) = \sum_n \psi_n^*(i)\psi_n(j) \exp(-\xi_n \tau)[f(\xi_n) - \theta(\tau)], \qquad (12.45)$$

where $\xi_n = E_n - \mu$ and $\theta(\tau > 0) = 1$, $\theta(\tau < 0) = 0$.

For the case of an undoped bipartite lattice, $\mu = 0$, using Eq. (12.40) one finds

$$G_{ji}^{(0)}(-\tau) = \mp \left[G_{ij}^{(0)}(\tau) \right]^*, \qquad (12.46)$$

where the minus and plus signs correspond to the cases in which i and j belong to the same sublattice and to different sublattices, respectively. As a result,

$$\chi_{ji} = \pm \frac{1}{4} \int_0^\beta d\tau \left| G_{ij}^{(0)}(\tau) \right|^2. \qquad (12.47)$$

On substituting Eq. (12.47) into Eq. (12.30) and (12.31) we come, again, to the conclusion that for the undoped (half-filled) case, the exchange interactions are ferromagnetic within the same sublattice and antiferromagnetic between sites from different sublattices.

To conclude this section, it is worthwhile to warn that the use of the Lieb theorem for graphene derivates with high concentration of sp^3 centers, such as single-side hydrogenated (C_2H) or fluorinated (C_2F) graphene, can lead to incorrect conclusions on their magnetism: In these situations, both real multiband electronic structure and electron–electron interactions beyond the Hubbard model (such as direct-exchange interactions) can be relevant, and instead of ferromagnetic ground state, one can expect complicated noncollinear types of magnetic ordering (Rudenko et al., 2013; Mazurenko et al., 2016).

12.3 Magnetic edges

It is clear from the previous consideration that the possibility of ferromagnetism in graphene-like systems is related to zero-energy modes and other mid-gap states. As was discussed in Chapter 5, the zero-energy modes arise naturally for a generic boundary of a terminated honeycomb lattice (see Eq. (5.70)). One can therefore conclude that the edges should be magnetic (except in the case of armchair edges, for which there are no mid-gap states). This was first suggested by Fujita et al.

(1996) and confirmed by numerous further calculations (e.g., Son, Cohen, & Louie, 2006a; Yazyev & Katsnelson, 2008; for a review, see Yazyev, 2010). If we have nanoribbons with zigzag edges, the atoms at the opposite edges belong to different sublattices. Therefore, one can expect that the exchange interactions between the edges are antiferromagnetic and that the nanoribbon as a whole should have no magnetic moment. Within the framework of the Hubbard model, this just follows from the Lieb theorem. The density-functional calculations by Son, Cohen, and Louie (2006a) show that this interaction can be switched to the ferromagnetic one by applying an external electric field.

One should keep in mind that zigzag edges of graphene are extremely chemically active and can even decompose water (Boukhvalov & Katsnelson, 2008); therefore, all calculations assume partial passivation by hydrogen (one hydrogen atom by carbon atom). In this case, the zero-energy modes are associated with the last row of carbon atoms. Interestingly, even for the case of complete passivation (two hydrogen atoms per carbon atom) zero modes still exist, but associated with the next row; in this case, density-functional calculations also predict ferromagnetism of the edges, although more fragile (Bhandary et al., 2010).

This result seems to be very interesting in the context of *spintronics* based on the coupling between electric and magnetic degrees of freedom of conducting materials (Žutić, Fabian, & Das Sarma, 2004). Possible graphene spintronic devices have been studied theoretically by Kim and Kim (2008) and by Wimmer et al. (2008). As an example, one can mention a simple and elegant way to produce a spin-polarized electric current due to a difference in shapes of the opposite zigzag edges suggested in the latter paper (Fig. 12.5).

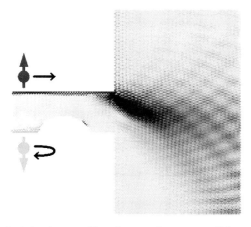

Fig. 12.5 The spin-injection profile of a graphene nanoribbon with a distorted edge for spin injection into a region of n-doped graphene.
(Reproduced with permission from Wimmer et al., 2008.)

However, there are several problems that should be carefully discussed before entertaining any such dreams about applications. First, the Mermin–Wagner theorem (Mermin & Wagner, 1966; Ruelle, 1999) forbids long-range order in low-dimensional systems (such as one-dimensional graphene edges) at finite temperatures. The range of magnetic order is limited by the temperature-dependent spin correlation lengths $\zeta^\alpha (\alpha = x, y, z)$, which define the decay law of the spin correlation

$$\langle \hat{s}_i^\alpha \hat{s}_{i+l}^\alpha \rangle - \langle \hat{s}_i^\alpha \hat{s}_i^\alpha \rangle \exp\left(-\frac{l}{\zeta_\alpha}\right). \tag{12.48}$$

In principle, the spin correlation length ζ imposes limitations on the device sizes. In order to establish this parameter, one has to determine the energetics of spin fluctuations contributing to the breakdown of the ordered ground-state configuration and extract the exchange parameters. This has been done via density-functional calculations by Yazyev and Katsnelson (2008). The total energy of the spin-spiral state (Fig. 12.6) has been calculated and fitted to the classical Heisenberg model. The spin-wave stiffness constant, $D \approx 2,100$ meV·Å², has been found to be several times higher than that in iron or nickel. This confirms the general conclusion (Edwards & Katsnelson, 2006) that defect-induced sp-electron magnetism can be characterized by very high magnon energies (see Section 12.1).

The magnetic correlation length in the presence of spin-wave fluctuations was obtained with the help of the one-dimensional Heisenberg model Hamiltonian

$$\hat{H} = -a \sum_i \hat{\vec{s}}_i \hat{\vec{s}}_{i+1} - d \sum_i \hat{s}_i^z \hat{s}_{i+1}^z, \tag{12.49}$$

where the Heisenberg coupling $a = 105$ meV was found from the fitting of the computational results. The estimated small anisotropy parameter $d/a \approx 10^{-4}$ originates from the weak spin-orbit interaction in carbon (see the next section).

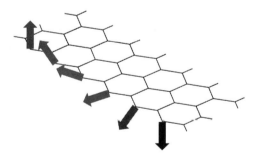

Fig. 12.6 The spin-spiral structure used for the calculation of the exchange coupling constant for a graphene zigzag edge.
(Reproduced with permission from Yazyev & Katsnelson, 2008.)

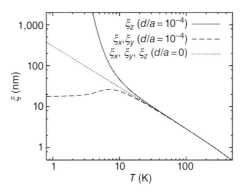

Fig. 12.7 The correlation length of magnetization vector components orthogonal (ξ_z) and parallel ($\xi_{x,y}$) to the graphene plane as a function of temperature for weakly anisotropic ($d/a = 10^{-4}$) and isotropic ($d = 0$) Heisenberg models. (Reproduced with permission from Yazyev & Katsnelson, 2008.)

This simple model Hamiltonian has known analytic solutions (Fisher, 1964). Fig. 12.7 shows the spin correlation lengths calculated for our particular case. Above the cross-over temperature $T_x \approx 10$ K, weak magnetic anisotropy does not play any role and the spin correlation length $\xi \propto T^{-1}$. However, below T_x the spin correlation length grows exponentially with decreasing temperature. At $T = 300$ K the spin correlation length $\xi \approx 1$ nm.

From a practical point of view, this means that the sizes of spintronic devices, based on the magnetic zigzag edges of graphene and operating under normal temperature conditions, are limited to several nanometers. At present, such sizes are very difficult to achieve, which can be regarded as a pessimistic conclusion. Nevertheless, one must keep in mind that the spin stiffness predicted for the magnetic graphene edges is still higher than the typical values for traditional magnetic materials. That is, graphene outperforms d-element-based magnetic materials, and there is room for improvement. Achieving control over the magnetic anisotropy d/a could possible raise the cross-over temperature T_x above 300 K and thus significantly extend ξ. Possible approaches for reaching this goal include chemical functionalization of the edges with heavy-element functional groups or coupling graphene to a substrate.

Another serious problem is the possibility of reconstruction of zigzag edges to some nonmagnetic configuration (see Section 5.6). Theoretically, the result regarding ferromagnetism of zigzag edges at $T = 0$ looks very reliable, but the situation with real edges of real graphene is not so clear. Probably, some chemical protection of the edges can be used to keep the magnetic state stable enough.

Indirect evidence of possible magnetism of graphene edges has been found very recently by scanning tunneling spectroscopy (STS), namely splitting of the edge mid-gap states has been observed (Tao et al., 2011). Spin-polarized STS should be used to prove that this is *spin*-splitting, but this work has not yet been done.

12.4 Spin-orbit coupling

As we discussed earlier, spintronic applications due to an intrinsic magnetism of graphene are still very speculative. At the same time, one can inject spin-polarized current into graphene using ferromagnetic leads, e.g., cobalt, and then manipulate this with current. There is a huge amount of experimental activity in this field (Tombros et al., 2007, 2008; Han et al., 2009a, 2009b; Jo et al., 2011). In this situation, the spin dynamics in graphene is determined by spin-orbit coupling, leading to various spin-relaxation processes, such as Elliott–Yafet, D'yakonov–Perel, Bir–Aronov–Pikus, and other mechanisms (Žutić, Fabian, & Das Sarma, 2004). The main idea is that, in the presence of spin-orbit coupling, *some* of the scattering processes will be accompanied by spin-flips; this is the essential feature of the simplest and most general process, the Elliott–Yafet mechanism. A rough estimation for the spin-flip time τ_s, is given by Elliott's formula (Elliott, 1954)

$$\frac{1}{\tau_s} \approx \frac{(\Delta g)^2}{\tau}, \tag{12.50}$$

where $\Delta g = g - 2$ is the contribution of the orbital moment to the conduction-electron g-factor and τ is the mean-free-path time (that is, the time taken for the relaxation of momentum). The first experiments (Tombros et al., 2007) have already demonstrated that τ_s in graphene is orders of magnitude shorter than one would expect from a naïve estimation of the spin-orbit coupling in graphene. This observation initiated a serious theoretical activity (Castro Neto & Guinea, 2009; Gmitra et al., 2009; Huertas-Hernando, Guinea, & Brataas, 2009; Konschuh, Gmitra, & Fabian, 2010; Jo et al., 2011; Dugaev & Katsnelson, 2014; Kochan, Gmitra, & Fabian, 2014; Kochan, Irmer, & Fabian, 2017). In this section we do not discuss the mechanisms of spin relaxation in graphene; instead, we focus on the quantum-mechanical part of the problem, that is, on the various contributions to spin-orbit coupling and their effects on the electron-energy spectrum.

Spin-orbit coupling is a relativistic effect following from the Dirac equation (here we mean the *real* Dirac equation rather than its analog for graphene) as the second-order perturbation in the fine-structure constant $e^2/(\hbar c)$ (Bjorken & Drell, 1964; Berestetskii, Lifshitz, & Pitaevskii, 1971):

$$\hat{H}_{s-o} = \frac{\hbar}{4m^2c^2} \left(\vec{\nabla} V \times \hat{\vec{p}} \right) \cdot \hat{\vec{\sigma}}, \tag{12.51}$$

where $\hat{\vec{p}} = -i\hbar\vec{\nabla}$, V is the potential energy, and $\hat{\vec{\sigma}}$ are the Pauli matrices acting on the real electron spin (not on the pseudospin, as in the greatest part of the book!). The main contribution originates from regions close to atomic nuclei where $\left|\vec{\nabla}V\right|$ is much larger than it is in interatomic space. As mentioned in Section 1.1, the order of magnitude of the intra-atomic spin-orbit coupling can be estimated from the energy difference of the multiplets $^3P_0$ and $^3P_1$ for the carbon atom (Radzig & Smirnov, 1985),

$$\Delta E_{s-o} \approx 2 \text{ meV}. \tag{12.52a}$$

It is, roughly, 10^{-4} of the π-electron bandwidth, as it would be natural to expect for a quantity proportional to $(e^2/(\hbar c))^2$. The corresponding quantities for carbon analogs in the periodic table, silicon and germanium, are much higher, as is naturally expected for heavier elements,

$$\Delta E_{s-o} \approx 9.6 \text{ meV} \tag{12.52b}$$

and

$$\Delta E_{s-o} \approx 69 \text{ meV}, \tag{12.52c}$$

respectively (Radzig & Smirnov, 1985).

In the representation of valent (2s2p) states of carbon, the Hamiltonian (12.51) can be represented as

$$\hat{H}_{s-o} = \xi \hat{\vec{L}} \cdot \hat{\vec{\sigma}}, \tag{12.53}$$

where

$$\hat{\vec{L}} = \hat{\vec{r}} \times \hat{\vec{p}} \tag{12.54}$$

is the orbital moment operator.

Let us first consider what one can expect for the effective parameters of spin-orbit coupling in the model taking into account only 2s2p states. Our further analysis follows Huertas-Hernando, Guinea, and Brataas (2006) and Yao et al. (2007). Within the basis of sp^3 states of the carbon atom (see Section 1.1) the Hamiltonian (12.53) can be rewritten as

$$\hat{H}_{s-o} = 2\xi \sum_j \left(\hat{c}^+_{jz\uparrow}\hat{c}_{jx\downarrow} - \hat{c}^+_{jz\downarrow}\hat{c}_{jx\uparrow} + i\hat{c}^+_{jz\uparrow}\hat{c}_{jy\downarrow} - i\hat{c}^+_{jz\downarrow}\hat{c}_{jy\uparrow} + i\hat{c}^+_{jx\downarrow}\hat{c}_{jy\downarrow} - i\hat{c}^+_{jx\uparrow}\hat{c}_{jy\uparrow} + \text{H.c.} \right),$$

$$\tag{12.55}$$

where we take into account only intra-atomic matrix elements (j is the site label) and x, y, and z label the $|p_x\rangle$, $|p_y\rangle$, and $|p_z\rangle$ orbitals. Of course, s-orbitals are not involved since $\hat{\vec{L}}|s\rangle = 0$.

Now we have to rewrite the Hamiltonian (12.55) in the representation of σ-orbitals (1.9) and π-orbitals ($|p_z\rangle$). Importantly, in the nearest-neighbor approximation $\left(\hat{H}_{\text{s-o}}\right)_{\pi\pi} = 0$, due to symmetry considerations. First, only the $\hat{L}_z\hat{\sigma}_z$ term survives, due to the mirror symmetry in the graphene plane. Second, there is an additional vertical reflection plane along the nearest-neighbor bonds. Under the reflection in this plane, $\hat{x},\hat{p}_x \to \hat{x},\hat{p}_x$ and $\hat{y},\hat{p}_y \to -\hat{y}, -\hat{p}_y$ and, therefore, $\hat{L}_z \to -\hat{L}_z$ which finishes the proof.

Thus, we have to use second-order perturbation theory, and the effective Hamiltonian of spin-orbit coupling is

$$\left(\hat{H}^{\text{eff}}_{\text{s-o}}\right)_{\pi\pi} = \left(\hat{H}_{\text{s-o}}\right)_{\pi\sigma} \frac{1}{\hat{H}^{(0)}_{\pi} - \hat{H}^{(0)}_{\sigma}} \left(\hat{H}_{\text{s-o}}\right)_{\sigma\pi}, \tag{12.56}$$

where $\hat{H}^{(0)}_{\pi,\sigma}$ are the corresponding band Hamiltonians without spin-orbital coupling. As a result, the effective Hamiltonians for the vicinities of the **K** and **K'** points are (Yao et al., 2007)

$$\hat{H}_{K,K'}\left(\vec{q}\right) = \begin{pmatrix} \mp\xi_1 & \hbar v\left(q_x \mp iq_y\right) \\ \hbar v\left(q_x \pm iq_y\right) & \pm\xi_1 \end{pmatrix} \tag{12.57}$$

instead of Eq. (1.19), where

$$\xi_1 = 2|\xi|^2 \frac{\varepsilon_{2p} - \varepsilon_{2s}}{9V^2_{\text{sp}\sigma}}, \tag{12.58}$$

where ε_{2p} and ε_{2s} are the atomic energy levels for 2p and 2s states, and $V_{\text{sp}\sigma}$ is a matrix element of the hopping Hamiltonian for the σ-block between s and p states. All these energies are of the order of 10 eV; thus the effective spin-orbit coupling constant for the case of flat, defect-free graphene is $\xi_1 \approx 1~\mu eV$ (Huertas-Hernando, Guinea, & Brataas, 2006; Yao et al., 2007).

This evaluation underestimates the effective spin-orbit interaction for graphene in comparison with the first-principle density functional calculations; the latter gives the value $\xi_1 \approx 12~\mu eV$(Gmitra et al., 2009). It turns out that the main contribution originates from the virtual pd-transitions (Konschuh, Gmitra, & Fabian, 2010). It turns out that if we add to the Hamiltonian (12.55) 3d states, then the first-order terms in ξ is no more symmetry forbidden, and instead of Eq. (12.58) we have

$$\xi_1 = 2|\xi_p|^2 \frac{\left(\varepsilon_{2p} - \varepsilon_{2s}\right)}{9V^2_{\text{sp}\sigma}} + \xi_d \frac{9V^2_{\text{pd}\pi}}{2\left(\varepsilon_{3d} - \varepsilon_{2p}\right)^2}, \tag{12.59}$$

where ξ_p and ξ_d are intra-atomic spin-orbit coupling constants for 2p and 3d states, respectively (Konschuh, Gmitra, & Fabian, 2010).

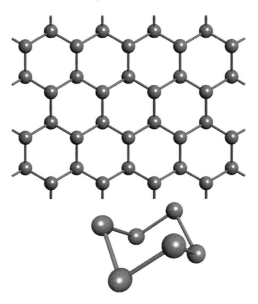

Fig. 12.8 Structure of silicene and germanene: top and side views.

It is interesting to compare graphene with its analogs, silicene and germanene (for review of these graphene-like allotropes of silicon and germanium, see Acun et al., 2015; Le Lay, Salomon, & Angot, 2017). An important difference is that for these two-dimensional materials, the structure is buckled (Fig. 12.8).

This lower symmetry allows the first-order term in ξ_p, even without taking into account d-states (Liu, Jiang, & Yao, 2010). Together with much stronger intra-atomic spin-orbit coupling (see Eq. (12.52)), it leads to the order-of-magnitude larger effective spin-orbit parameters: from 12 µeV for graphene to 12 meV for germanene (Acun et al., 2015). A general symmetry analysis of spin-orbit coupling for various two-dimensional materials can be found in Kochan, Irmer, and Fabian (2017).

Now, we come back to graphene. A special case of spin-orbit coupling is associated with the external electric field perpendicular to the graphene plane (the Rashba effect) (Kane & Mele, 2005a; Huertas-Hernando, Guinea, & Brataas, 2006; Min et al., 2006; Rashba, 2009; Stauber & Schliemann, 2009; Zarea & Sandler, 2009). The potential of the external electric field

$$\hat{H}_E = eEz \tag{12.60}$$

has nonzero matrix elements only between $|s\rangle$ and $|p_z\rangle$ orbitals. In the secondary quantized form, Eq. (12.60) reads

$$\hat{H}_E = z_0 eE \sum_{j\sigma} \left(\hat{c}^+_{jz\sigma} \hat{c}_{js\sigma} + \hat{c}^+_{js\sigma} \hat{c}_{jz\sigma} \right), \tag{12.61}$$

where

$$z_0 = \langle p_z | z | s \rangle \tag{12.62}$$

is of the order of the radius of the carbon atom. The effective Hamiltonian of spin-orbit coupling in the presence of the electric field, apart from Eq. (12.56), contains the cross-term

$$\left(\hat{H}_R\right)_{\pi\pi} = \left(\hat{H}_E\right)_{\pi\sigma} \frac{1}{\hat{H}_{\pi}^{(0)} - \hat{H}_{\sigma}^{(0)}} \left(\hat{H}_{s\text{-o}}\right)_{\pi\pi} + \left(\hat{H}_{s\text{-o}}\right)_{\pi\sigma} \frac{1}{\hat{H}_{\pi}^{(0)} - \hat{H}_{\sigma}^{(0)}} \left(\hat{H}_E\right)_{\sigma\pi} \tag{12.63}$$

(cf. Eq. (10.70)). Taking into account this term, plus Eq. (12.57), we will find the spin-orbit 8 × 8 Hamiltonian

$$\hat{H}_{s\text{-o}} = \xi_1 \hat{\eta}_z \hat{\tau}_z \hat{\sigma}_z + \xi_R \left(\hat{\eta}_x \hat{\tau}_z \hat{\sigma}_y - \hat{\eta}_y \hat{\sigma}_x \right), \tag{12.64}$$

where $\hat{\eta}, \hat{\tau}$ and $\hat{\sigma}$ are Pauli matrices acting on the pseudospin (that is, the sublattice index), valley index, and real-spin projection, respectively. Note that in most of the book the Pauli matrix $\hat{\eta}$ has been written as $\hat{\sigma}$! The *Rashba coupling* ξ_R in Eq. (12.64) is given by (Huertas-Hernando, Guinea, & Brataas, 2006; Min et al., 2006)

$$\xi_R = \frac{2eEz_0}{3V_{sp\sigma}} \xi. \tag{12.65}$$

For the largest values of the electric field which can be created in graphene $E \approx 1$ V nm^{-1}, ξ_R may be of the same order of magnitude as ξ_1 (keeping in mind pd-contribution to the latter Eq. (12.59)).

There are many mechanisms that can dramatically increase the effective spin-orbit coupling in graphene. First, it is very sensitive to the curvature, which can be associated with the ripples (Huertas-Hernando, Guinea, & Brataas, 2006). In curved graphene, there is no longer mirror symmetry in the vertical plane along the nearest-neighbor bonds, and the effective spin-orbit coupling for the π-block does not vanish to first order in ξ; this leads to Rashba-type coupling, with an effective coupling constant of the order of

$$\xi_R \approx \xi a H, \tag{12.66}$$

where H is the mean curvature (9.77) and (9.78). For typical parameters of the ripples this spin-orbit coupling is an order of magnitude larger than the intrinsic one, of the order of $10^{-2} - 10^{-1}$ meV (Huertas Hernando, Guinea, & Brataas, 2006).

Second, the effective spin-orbit coupling can essentially be increased by covalently bonded impurities, such as hydrogen adatoms, which change the state of carbon atoms locally from sp^2 to sp^3 (Castro Neto & Guinea, 2009; Kochan,

Gmitra, & Fabian, 2014). Again, this creates an effective spin-orbit coupling in the π-block already in the first order in ξ, making $\xi_1 \approx \xi$ locally. This makes "resonant impurities" very efficient sources of spin-flip scattering. This conclusion seems to be in agreement with the recent experimental data (Jo et al., 2011).

Finally, let us discuss the effect of the Hamiltonian (12.64) on the electron-energy spectrum of graphene (Kane & Mele, 2005b; Stauber & Schliemann, 2009). This Hamiltonian does not couple the valleys. For the valley K ($\tau_z = +1$), we have a 4×4 matrix (in the basis A$\uparrow$, B$\uparrow$, A$\downarrow$, B$\downarrow$) for the total Hamiltonian:

$$\hat{H} = \begin{pmatrix} -\xi_1 & \hbar v\left(q_x - iq_y\right) & 0 & 0 \\ \hbar v\left(q_x + iq_y\right) & \xi_1 & 2i\xi_R & 0 \\ 0 & -2i\xi_R & \xi_1 & \hbar v\left(q_x + iq_y\right) \\ 0 & 0 & \hbar v\left(q_x + iq_y\right) & -\xi_1 \end{pmatrix},$$

$$(12.67)$$

where we skip the constant energy shift ξ_1 in Eq. (12.57). The equation for the eigenenergies takes the form

$$\det\left(\hat{H} - E\right) = \left(E^2 - \xi_1^2 - \hbar^2 v^2 q^2\right)^2 - 4\xi_R^2(E + \xi_1)^2 = 0. \qquad (12.68)$$

At $\xi_R = 0$, the spectrum is

$$E = \pm\sqrt{\hbar^2 v^2 q^2 + \xi_1^2}, \qquad (12.69)$$

with the gap $\Delta_{s-o} = 2|\xi_1|$. The existence of the gap does not contradict the proof given in Chapter 1 since, in the presence of spin-orbit coupling, the time-reversal operation does not have the form (1.39) but includes the spin reversal.

In the opposite case, $\xi_1 = 0$, the spectrum is

$$E^2 = \hbar^2 v^2 q^2 + 2\xi_R^2 \pm 2\xi_R\sqrt{\hbar^2 v^2 q^2 + \xi_R^2}. \qquad (12.70)$$

This is reminiscent of the spectrum of bilayer graphene in the parabolic-band approximation; see Section 1.4. Two bands have a gap, with the energy $\pm 2|\xi_R|$ at $q = 0$, and two others are gapless, with the parabolic spectrum at $q \to 0$. In general, for finite ξ_1 and ξ_R, the gap exists at $|\xi_1| > |\xi_R|$ and its value is

$$\Delta_{s-o} = 2(|\xi_1| - |\xi_R|). \qquad (12.71)$$

In the regime in which the gap exists, the mass term has opposite signs for the two valleys (see Eq. (12.57)). This results in a very interesting picture of the "quantum spin Hall effect" (Kane & Mele, 2005a, 2005b). This phenomenon is not relevant for real graphene, due to the very small value of the gap. However, this effects can be important for silicene, germanene, or stanene (tin-based analog of graphene; Xu et al., 2013) with their much larger values of the spin-orbit gap. Note that germanene at insulating substrate (MoS_2) is already realized experimentally (Zhang et al., 2016).

These two papers by Kane and Mele were very important in the development of a novel field, namely the physics of topological insulators (Moore, 2009; Hasan & Kane, 2010; Qi & Zhang, 2010; Qi & Zhang, 2011). This is one of the many examples of the huge influence of graphene on our general understanding of physics.

12.5 Spin relaxation due to edge scattering

As we discussed in Section 12.3, theory predicts magnetism at graphene zigzag edges. If we assume that this prediction is correct (experimental situation still looks unclear), then edges should be a very important source of spin relaxation. In this case, the relaxation is determined by not spin-orbit coupling but s–d exchange interactions (12.28), which can be many orders of magnitude stronger. The same happens for the case when some magnetic impurities are situated at the edges. Here we present the main ideas of the corresponding theory (Dugaev & Katsnelson, 2014).

If we assume graphene nanoribbon situated at $0 < x < L$ (the same geometry as in Section 11.5) the corresponding interaction Hamiltonian is

$$\hat{H}_{int} = W(x)\vec{\hat{\sigma}}\vec{m}\,(y),\tag{12.72}$$

where $\vec{m}\,(y)$ is the distribution of magnetic moments along the edges and $W(x)$ is a short-range potential focused near $x = 0$ and $x = L$ (for simplicity, we assume that the edges are equivalent, similar to Section 11.5). For one-dimensional magnets at finite temperatures

$$\langle m_\alpha(y)m_\beta(y')\rangle = \gamma e^{-\lambda|y\,-\,y'|},\tag{12.73}$$

where $\alpha, \beta = x, y$ and $\lambda = 1/\xi$ (cf. Eq. (12.48)). Then, the averaged probability of spin-flip processes due to the interaction (12.72) is

$$
\begin{aligned}
W_{\vec{k}\vec{k}'} &= \frac{2\pi}{\hbar}\left\langle\left|\left\langle\vec{k}'\left|W(x)\left(m_x(y) - im_y(y)\right)\right|\vec{k}\right\rangle\right|^2\right\rangle\delta\left(\varepsilon_{\vec{k}} - \varepsilon_{\vec{k}'}\right)\\
&= \frac{4\pi\gamma}{\hbar}\int dxdy \int dx'dy'\, e^{-\lambda|y-y'|}\left\langle\vec{k}'\left|W(x)\right|\vec{k}\right\rangle\left\langle\vec{k}\left|W(x')\right|\vec{k}'\right\rangle\delta\left(\varepsilon_{\vec{k}} - \varepsilon_{\vec{k}'}\right),
\end{aligned}
$$

$$\tag{12.74}$$

where the wave functions $\left| \vec{k} \right\rangle$ are given by Eq. (11.138) for the case of Berry–Mondragon (BM) boundary conditions and by Eq. (11.140) for the case of zigzag-like (Z) edges.

Explicit calculations give us the answers (Dugaev & Katsnelson, 2014)

$$W^{(BM)}_{\vec{k}\,\vec{k}'} = \frac{2\pi\gamma W_0^2 \lambda \delta\left(\varepsilon_{\vec{k}} - \varepsilon_{\vec{k}'}\right)}{\hbar L_x L^2 \left[\lambda^2 + \left(k_y - k_y'\right)^2\right]}, \tag{12.75}$$

$$W^{(Z)}_{\vec{k}\,\vec{k}'} = \frac{2\pi\gamma W_1^2 \lambda k_x^2 k_x'^2 \delta\left(\varepsilon_{\vec{k}} - \varepsilon_{\vec{k}'}\right)}{\hbar L_x L^2 \left[\lambda^2 + \left(k_y - k_y'\right)^2\right]}, \tag{12.76}$$

where L_x is the length of the ribbon,

$$W_0 = \int dx W(x), \quad W_1 = \int dx W(x) x^2. \tag{12.77}$$

Spin-dependent scattering at the edges can be considered via kinetic equation for the spin-dependent single-electron density matrix (Ustinov, 1980; Katsnelson, 1981). In the simplest case, where this density matrix is diagonal in spin indices, we have just two Boltzmann distribution functions $f_{\vec{k},\uparrow,\downarrow}$, each of them satisfying the same kind of equation as Eq. (11.1). After linearization (11.7), assuming no external electric field and tau-approximation for the bulk collision integral (11.131), we have

$$\left(\frac{\partial}{\partial t} + v_{\vec{k}\alpha}\frac{\partial}{\partial x_\alpha} + \frac{1}{\tau}\right)\delta f_{\vec{k}\sigma} = 0, \tag{12.78}$$

with the boundary conditions of the type of Eq. (11.134) and (11.135). If we assume only spin-flip processes at the boundary, these take the form

$$|v_x|\delta f^>_{\vec{k},\sigma}(x=0) = |v_x|\delta f^<_{\vec{k},\sigma}(x=0) + \sum_{\vec{k}'} W(\vec{k},\vec{k}')[\delta f^<_{\vec{k}',-\sigma}(x=0) - \delta f^>_{\vec{k},-\sigma}(x=0)],$$

$$\tag{12.79}$$

and similar for $x = L$.

Further calculations are quite cumbersome, therefore we present only the main physical result; the details can be found in the paper Dugaev and Katsnelson (2014).

Suppose we create some spin polarization $f_{\vec{k}\uparrow} \neq f_{\vec{k}\downarrow}$ at initial time $t = 0$ and look at the distribution of the polarization along y axis at some instant t. At large enough distances, one can pass from the kinetic equation to the spin diffusion equation. In this case, the typical distance of the polarization propagation is

$$l_d = \sqrt{\langle y^2 \rangle} = \sqrt{D_s t}, \tag{12.80}$$

where D_s is the spin diffusion coefficient. This is the standard approximation, which is always used to interpret the experimental data, assuming the spin-flip scattering processes in the bulk only (Tombros et al., 2007, 2008; Han et al., 2009a, 2009b; Jo et al., 2011). It turns out to also be correct for the case of spin scattering at Berry–Mondragon edges. At the same time, for the most interesting case of zigzag-like edges, where one can expect intrinsic magnetism at the edges, the situation is predicted to be totally different. The probability of scattering processes vanishes for the gliding electrons, according to Eq. (11.143) and (12.76). Therefore, their propagation is ballistic rather than diffusive, which leads to the polarization propagation faster than (12.80). Instead, we have (Dugaev & Katsnelson, 2014):

$$
\begin{aligned}
l_d &\propto t^{5/6}, \quad k_F \xi \gg 1, \\
l_d &\propto t^{3/4}, \quad k_F \xi \ll 1.
\end{aligned}
\tag{12.81}
$$

13

Graphene on hexagonal boron nitride

13.1 Motivation: ripples and puddles

In early days of graphene (2004–2010), amorphous silicon dioxide was a substrate by default. As we discussed in Chapter 11, it already provides pretty high electron mobility. At the same time, the quality of such samples is not extremely good; the most important restriction is charge inhomogeneity in the form of electron and hole puddles which were discovered by Martin et al. (2008). This is a serious obstacle; e.g., to study minimal conductivity (Chapter 3), many-body renormalization of Fermi velocity (Section 8.4), and other effects requires, for their reliable observation, a closeness to the neutrality point. The origin of these puddles was the subject of long discussions (Rossi & Das Sarma, 2008; Polini et al., 2008; Fogler, 2009; Gibertini et al., 2010; Das Sarma et al., 2011; Gibertini et al., 2012). The most obvious factor of the puddle formation is a charge disorder in the substrate; however, this can, in principle, be eliminated. As was shown by Gibertini et al. (2010) and Gibertini et al. (2012), this is, however, not enough to reach the homogeneous state. The reason is the existence of ripples, which seem to be unavoidable for single-layer two-dimensional materials (Chapter 9). They lead to randomness of the deformation tensor and, therefore, to random distributions of pseudomagnetic field (10.7) and pseudoscalar potential (10.8). Whereas the former is not very relevant for the puddles, the latter turns out to be sufficient to explain the observed charge inhomogeneity, even without an assumption on any extrinsic charge disorder.

The calculations in these works were made using a simplified density functional for two-dimensional Dirac electron liquid (Polini et al., 2008). For the case of freely suspended graphene (Gibertini et al., 2010), the distribution of atomic displacements and thus of the deformation tensor, were taken from atomistic simulations of the same kind as discussed in Chapter 9 (Fasolino, Los, & Katsnelson, 2007; Los et al., 2009). The typical picture of corrugations induced by thermal

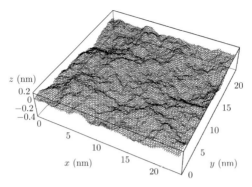

Fig. 13.1 Three-dimensional plot of the corrugated, freely suspended graphene sample from atomistic simulations for the case of room temperature.
(Taken with permission from Gibertini et al., 2010.)

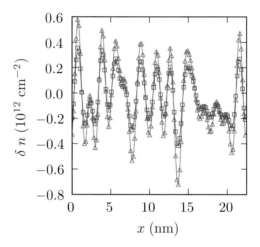

Fig. 13.2 A one-dimensional plot of electron density in nominally undoped, freely suspended graphene sample at room temperature calculated via the density functional (Polini et al., 2008) for the sample shown in Fig. 13.1. The triangles show the results in the Hartree approximation and squares are the results with the exchange-correlation potential taken into account; one can see that the results are qualitatively similar.
(Taken with permission from Gibertini et al., 2010.)

fluctuations is shown in Fig. 13.1 and the corresponding profile of the charge density in Fig. 13.2. One can see that the intrinsic ripples alone provide inhomogeneity of the electron density with the amplitude of the order of 10^{11}–10^{12}cm^{-2}.

To suppress the intrinsic ripples originated from the soft flexural phonons, one can put graphene on a substrate. Unfortunately, SiO_2 is a bad choice because this amorphous layer has unavoidably a strong roughness. The corresponding calculations were made by Gibertini et al. (2012) using the scanning tunneling microscopy (STM) experimental data on a profile of graphene on top of SiO_2 (Geringer et al., 2009) with further numerical solution of the equations (9.65)–(9.67) to restore the whole three-dimensional picture of atomic displacements. It turns out that we have roughly the same electron inhomogeneities as for the freely suspended graphene at room temperature.

Thus, to avoid puddle formation, one needs to suppress ripples by putting graphene on atomically flat crystalline substrate. The best choice turned out to be the hexagonal boron nitride (hBN), which consists of relatively weakly bound layers with the same hexagonal lattice as graphene (Fig. 1.4), but with nonequivalent atoms in sublattices A and B (nitrogen and boron, respectively). Dean et al. (2010) and Meric et al. (2010) have demonstrated that by choosing hBN as a substrate, one can suppress charge inhomogeneities and increase the electron mobility by more than an order of magnitude in comparison with graphene on SiO_2. Now hBN is a substrate by default; graphene samples put on hBN or encapsulated (Mayorov et al., 2011b) in hBN have extraordinary high quality. Some experimental results already referred to in this book were obtained for these samples; see e.g., Yu et al. (2013). Whereas at low temperatures the freely suspended graphene samples can be of a comparable quality, at room temperatures they have a strong restrictions on electron mobility due to scattering by intrinsic ripples (or, equivalently, by flexural phonons); see Eq. (11.126) and Fig. 11.3. Fig. 13.2 also shows an unavoidable charge inhomogeneity in such samples. Graphene on hBN is free of these limitations. Note that graphene can be also epitaxially grown on hBN (Yang et al., 2013; Tang et al., 2013).

At the same time, putting graphene on hBN is not only the way to improve the sample quality and to unveil the intrinsic physics of Dirac fermions near the neutrality point. It creates its own very interesting physics, especially related to a controlled commensurability/incommensurability of a potential acting on electrons. This chapter presents a basic introduction to the related phenomena.

13.2 Geometry and physics of moiré patterns

Importantly, it turns out to be possible to change misorientation angle θ between the graphene layer and hBN substrate in a controllable way; as a result, the so-called moiré patterns arise (Xue et al., 2011; Tang et al., 2013; Yang et al., 2013; Woods et al., 2014; Woods et al., 2016). The consequences, especially for the electronic structure, are very rich. The basic idea is that the electrons in graphene feel both the crystal potential of graphene and of the substrate. In general, these two potentials are not commensurate, they have different periods. For the perfect alignment, when crystallographic axes for graphene and hBN coincide, they are equal to

$$b_1 = b, \quad b_2 = b(1 + \delta), \tag{13.1}$$

where $b = \sqrt{3}a$ is the lattice constant for graphene and $\delta \approx 0.018$ is the misfit of the lattice periods (for hBN it is slightly larger). The real number δ can be approximated as a rational number p/q where p and q are coprime integers (that is, their greatest common divisor is 1). Then, the common period of our system is $qb \gg b$. If we assume that δ is irrational, then we deal with a very challenging problem of electron motion in *quasiperiodic* potential. Probably the best known physical realization of this situation in condensed matter takes place in *quasicrystals* (Shechtman et al., 1984; Guyot, Kramer, & de Boissieu, 1991; Mermin, 1992; Goldman & Kelton, 1993; Lifshitz, 1997; Quilichini, 1997). Quasicrystals are three-dimensional objects that can be formally considered as a projection of six-dimensional *crystals* onto three-dimensional Euclidean space. Graphene on hBN provide us a two-dimensional system with *tunable* quasiperiodic potential, and the tuning can be done just by a rotation of graphene with respect to the substrate. The motion of massless Dirac fermions in such a potential have some important peculiarities, which will be considered in this and the next chapters. Before doing this, we first have to discuss the atomic structure of our systems.

Let us start with just a simple, one-dimensional model of an atomic chain on a slightly incommensurate substrate. This model was suggested by Frank and van der Merwe (1949) as a development of earlier ideas of Frenkel and Kontorowa (1938); for the modern introduction into the field see the book by Braun and Kivshar (2004). The potential energy in Frank–van der Merwe model reads:

$$V = \frac{\mu}{2} \sum_{n=0}^{N-1} (x_{n+1} - x_n + b_1 - b_2)^2 + \frac{W}{2} \sum_{n=0}^{N-1} \left(1 - \cos \frac{2\pi x_n}{b_2}\right), \tag{13.2}$$

where μ is the elastic modulus of the chain, W is the interaction energy with the substrate per atom, N is the total number of atoms in the chain, and x_n are atomic displacements with respect to the equilibrium atomic positions in the substrate, nb_2. The model (13.2) contains competing interactions: Whereas the interaction with substrate (the second term) wants to make the lattice period equal to b_2, the first term reaches the minimum at the interatomic positions in the chain equal to b_1. As a result, generally speaking, the atomic structure will be reconstructed to reach the compromise between these two tendencies.

Introducing the dimensionless quantities

$$\zeta_n = x_n/b_2, \tag{13.3}$$

one can rewrite Eq. (13.2) as

$$V = W l_0^2 \sum_{n=0}^{N-1} \left(\zeta_{n+1} - \zeta_n - \frac{1}{P} \right)^2 + \frac{W}{2} \sum_{n=0}^{N-1} (1 - \cos 2\pi \zeta_n),$$ (13.4)

where

$$P = \frac{b_2}{b_2 - b_1} - \frac{1}{\delta} + 1, \quad l_0 = \sqrt{\frac{\mu b_2^2}{2W}}.$$ (13.5)

Now let us assume that, first, the intrachain interactions are much stronger that the (van der Waals) interactions with the substrate, that is, $l_0 \gg 1$, and, second, that the misfit δ is small, that is, $P \gg 1$. One can show that in this situation one can pass to the continuum limit, replacing the sum in Eq. (13.4) by the integral and finite differences by derivatives:

$$\frac{V}{W} \approx \int_0^N dn \left[l_0^2 \left(\frac{d\zeta}{dn} - \frac{1}{P} \right)^2 + \sin^2 \pi \zeta \right].$$ (13.6)

The Euler equations for the minimization of the functional (13.6) are

$$\frac{d^2 \zeta}{dn^2} = \frac{\pi}{2 l_0^2} \sin 2\pi \zeta,$$ (13.7)

with the solutions expressed in terms of elliptic functions (Abramowitz & Stegun, 1964; Whittaker & Watson, 1927). The general solution reads:

$$\zeta(n) = \frac{1}{2} + \frac{1}{\pi} am \left(\frac{\pi n}{l_0 k} \right), \quad \frac{d\zeta}{dn} = \frac{1}{l_0 k} dn \left(\frac{\pi n}{l_0 k} \right),$$ (13.8)

where $k \leq 1$ is modulus of the elliptic function (note that in Eq. (7.65) $m = k^2$), a parameter dependent on the integration constants. The special case $k = 1$ corresponds to an individual localized defect called misfit dislocation (Frank & van der Merwe, 1949) or crowdion (Frenkel & Kontorowa, 1938). It corresponds to the special boundary condition to Eq. (13.7)

$$\zeta \to 0, \quad n \to -\infty; \quad \zeta \to 1, \quad n \to \infty,$$ (13.9)

which mean that the atoms at $\pm \infty$ lie in two subsequent minima of the substrate potential. The solution

$$\zeta = \frac{2}{\pi} \arctan \left(\exp \frac{\pi n}{l_0} \right)$$ (13.10)

represents a step of atomic displacements localized at the spatial scale of the order of $l_0 b_2 \gg b_2$. At k close to 1, the general solution (13.8) represents a dislocation lattice, with displacement jumps separated by relatively large distances; the latter can be expressed in terms of the elliptic integral (7.65). When k decreases, the period of the lattice decreases, and at $k \to 0$ it transforms to a sine-like modulation.

Frank and van der Merwe (1949) have found a condition of energetic stability of the dislocations. It has the form

$$P < \frac{\pi l_0}{2}. \tag{13.11}$$

Taking into account the definitions (13.5) it can be approximately rewritten as

$$W > \mu(b_2 - b_1)^2, \tag{13.12}$$

which has a very simple physical meaning. The van der Waals interaction with the substrate wants to put each atom of our chain into the minimum of the substrate potential. However, it requires the deformation of the strings. This optimization of the van der Waals interaction is possible if the latter is stronger than the loss of the elastic energy at the deformation. If we have the strong equality ($\gg$ instead of $>$) in Eq. (13.12), the atomic relaxation will result in a state of relatively broad regions where the atoms in the chain fit the substrate, separated by dislocations (13.10). In this situation the incommensurability of the potentials will be all focused in some boundary regions. In the opposite case of relatively weak van der Waals interactions ($<$ instead of $>$ in Eq. (13.12)), the system will not even try to fit the potential of substrate, and one can assume that the effects of atomic relaxation will be relatively small and qualitatively not important.

This is a qualitative explanation of the commensurate-incommensurate transition at the rotation of graphene with respect to hBN substrate discovered by Woods et al. (2014). Unfortunately, for the two-dimensional case we do not have any analytical theory similar to the one-dimensional case considered previously; we have to restrict ourselves to a qualitative analysis and computer simulations (van Wijk et al., 2014; Woods et al., 2016). Before discussing this, first we present a formal geometric theory of two-dimensional moiré structure without taking into account atomic relaxation. We will follow the work by Hermann (2012).

Let us assume that the Bravais lattice for the substrate is built from the lattice vectors $\vec{a}_1$ and $\vec{a}_2$, and for the overlayer, from the lattice vectors $\vec{b}_1$ and $\vec{b}_2$, which are some linear combinations of $\vec{a}_1$ and $\vec{a}_2$:

$$\begin{pmatrix} \vec{b}_1 \\ \vec{b}_2 \end{pmatrix} = \hat{M} \begin{pmatrix} \vec{a}_1 \\ \vec{a}_2 \end{pmatrix}. \tag{13.13}$$

The matrix $\hat{M}$ combines rotations at the misorientation angle θ and possible changes of lattice periods:

$$\hat{M} = \begin{pmatrix} p_1 & 0 \\ 0 & p_2 \end{pmatrix} \cdot \hat{R}(\theta), \quad \hat{R}(\theta) = \begin{pmatrix} \cos\theta & \sin\theta \\ -\sin\theta & \cos\theta \end{pmatrix}; \tag{13.14}$$

for our case of graphene on hBN, the scaling factors are equal: $p_1 = p_2 = 1/(1+\delta)$. Then, the basic reciprocal lattice vectors of the overlayer $\vec{G}_1, \vec{G}_2$ are linear combinations of the basic reciprocal lattice vectors of the substrate $\vec{g}_1, \vec{g}_2$:

$$\begin{pmatrix} \vec{G}_1 \\ \vec{G}_2 \end{pmatrix} = \hat{K} \begin{pmatrix} \vec{g}_1 \\ \vec{g}_2 \end{pmatrix}. \tag{13.15}$$

Keeping in mind that the scalar products $\vec{a}_i\vec{g}_j$ should be equal to the scalar products $\vec{b}_i\vec{G}_j$ (they are both equal to δ_{ij}), one has to choose

$$\hat{K} = \left(\hat{M}^T\right)^{-1}. \tag{13.16}$$

Now consider a crystal potential as a superposition of crystal potentials of substrate and overlayer (at this stage we neglect all the effects of the lattice relaxation). Then, it can be represented as a sum of two Fourier expansions:

$$V\left(\vec{r}\right) = \sum_{mn} \left[u_{mn} \exp\left(im\vec{g}_1\vec{r} + in\vec{g}_2\vec{r} \right) + w_{mn} \exp\left(im\vec{G}_1\vec{r} + in\vec{G}_2\vec{r} \right) \right] \tag{13.17}$$

with the summation over all integer m and n. This sum can be represented as

$$V\left(\vec{r}\right) = \sum_{mn} u_{mn} \exp\left(im\vec{g}_1\vec{r} + in\vec{g}_2\vec{r} \right) \left[1 + \frac{w_{mn}}{u_{mn}} \exp\left(im\vec{\Delta}_1\vec{r} + in\vec{\Delta}_2\vec{r} \right) \right], \tag{13.18}$$

where

$$\begin{pmatrix} \vec{\Delta}_1 \\ \vec{\Delta}_2 \end{pmatrix} = \begin{pmatrix} \vec{G}_1 - \vec{g}_1 \\ \vec{G}_2 - \vec{g}_2 \end{pmatrix} = (\hat{K} - 1) \begin{pmatrix} \vec{g}_1 \\ \vec{g}_2 \end{pmatrix}. \tag{13.19}$$

These vectors can be considered as the reciprocal lattice vectors for the superlattice (moiré lattice) with the elementary vectors

$$\begin{pmatrix} \vec{R}_{M1} \\ \vec{R}_{M2} \end{pmatrix} = \hat{P} \begin{pmatrix} \vec{a}_1 \\ \vec{a}_2 \end{pmatrix}, \tag{13.20}$$

where, taking into account Eq. (13.16),

$$\hat{P} = \left(\left(\hat{K} - 1 \right)^T \right)^{-1} = \left(1 - \hat{M} \right)^{-1} \hat{M}. \tag{13.21}$$

Eq. (13.21) and (13.14) allow us to calculate the lengths of the unit vectors of the moiré lattice and thus to determine its periods. The result is (Herman, 2012):

$$\kappa_1 = \frac{R_{M1}}{a_1} = \frac{p_1 \sqrt{1 + p_2^2 - 2p_2 \cos \theta}}{1 + p_1 p_2 - (p_1 + p_2) \cos \theta + (p_1 - p_2) \cot \omega \sin \theta},$$

$$\kappa_2 = \frac{R_{M2}}{a_2} = \frac{p_2 \sqrt{1 + p_1^2 - 2p_1 \cos \theta}}{1 + p_1 p_2 - (p_1 + p_2) \cos \theta + (p_1 - p_2) \cot \omega \sin \theta}, \tag{13.22}$$

where ω is the angle between the vectors $\vec{a}_1$ and $\vec{a}_2$. For the case $p_1 = p_2 = 1/(1 + \delta)$, we have

$$\kappa = \frac{1}{\sqrt{(1+\delta)^2 + 1 - 2(1+\delta) \cos \theta}} \approx \frac{1}{\sqrt{\delta^2 + \theta^2}}, \tag{13.23}$$

where the last equality is correct for small δ and θ. In this case, the period of the superlattice is much larger than the lattice constant.

Now let us come back to our specific system. The first-principle calculations (Giovannetti et al., 2007; Sachs et al., 2011; Bokdam et al., 2014) show that the interlayer interaction energy is minimal when one carbon atom in the graphene elementary cell is on the top of the boron atom and the other one is in the middle of the boron–nitrogen hexagon. This destroys the equivalence of graphene sublattices and breaks the inversion symmetry; the result is the formation of local gap opening (mass term in the Dirac equation); see Section 1.3. The consequences for the electronic properties will be discussed in the next sections. Note that to give reliable results for the interlayer cohesive energy, the "first-principle calculations" should go beyond the conventional density functional in the form of local density approximation (LDA) or generalized gradient approximations (GGA); it is well known that they cannot correctly describe the van der Waals interactions. Probably the minimal approximation is the *ab initio* random phase approximation (RPA) (Sachs et al., 2011). The corresponding results are shown in Fig. 13.3.

The calculations give an estimate of the constant W in the model (13.2) as 10 meV. Taking into account Eq. (9.157) and $\delta \approx 0.018$, one can also roughly estimate the right-hand side of Eq. (13.12) as tens of meV. This means that for the case of perfect alignment, the competing energies of van der Waals interaction and elastic deformation are comparable. When we rotate the sample, increasing misorientation, we effectively enhance the role of elastic deformations. We do not have any quantitative theory for the two-dimensional case, but both experiment (Woods et al., 2014) and atomistic simulations (van Wijk et al., 2014) demonstrate

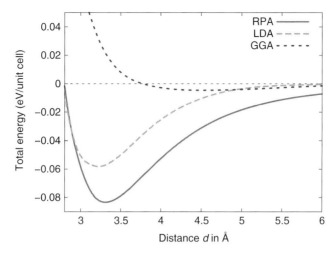

Fig. 13.3 Total energies of graphene-hBN interlayer interaction per unit cell for the optimal stacking.
(Taken with permission from Sachs et al., 2011.)

the commensurate-incommensurate transition. For small enough angles $\theta < \delta$, the atomic relaxation is very strong, and the optimized structure consists of domains of more or less commensurate positions of graphene and hBN atoms separated by "domain walls" with essentially different stacking. For larger angles, one can, in the first approximation, describe the atomic structure as roughly unrelaxed (see Fig. 13.4). This has very important consequences for the electronic properties, which will be discussed in the next section.

13.3 Zero-mass lines and minimal conductivity

The nonequivalence of sublattices A and B for graphene on hBN results in the appearance of mass term (see Eq. (5.1) and (5.2)) and in the local gap opening. Importantly, for the optimized moiré structure, the function Δ can change the sign, which results in the formation of zero-mass lines (Sachs et al., 2011); see Fig. 13.5. According to the density functional calculations, the amplitude of the oscillations of Δ is about tens of meV.

It turns out that the mass term can essentially be enhanced by correlation effects (Song, Shytov, & Levitov, 2013; Bokdam et al., 2014). The *ab initio* GW calculations give for the maximum local gap the value of the order of 300 meV (Bokdam et al., 2014). Nevertheless, qualitatively the picture remains the same as in the density functional (Sachs et al., 2011).

Existence of zero-mass lines (white lines at Fig. 13.5(c)) seems to be the most important detail of this picture (Sachs et al., 2011; Titov & Katsnelson, 2014).

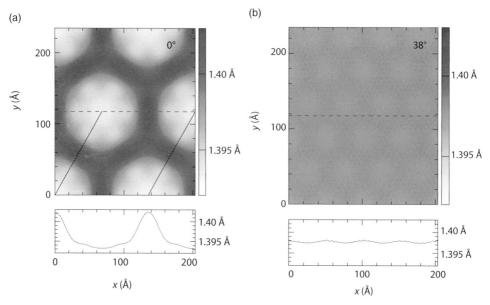

Fig. 13.4 Distribution of carbon–carbon bond lengths in a graphene layer on a rigid hBN substrate, according to atomistic simulations (van Wijk et al., 2014). (a) Misorientation angle $\theta = 0$, one can see a formation of domains of local commensurability. (b) Misorientation angle $\theta = 38°$, atomic relaxation is almost negligible. (Taken with permission from van Wijk et al., 2014.)

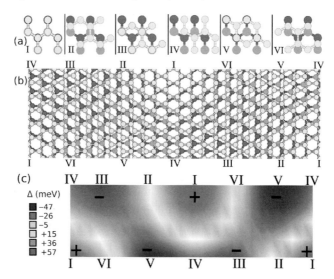

Fig. 13.5 (a) Top view of various stacking configurations for graphene at hBN; carbon, boron and nitrogen atoms are shown in light gray, dark gray, and gray, respectively. Configuration V is the most energetically favorable. (b) The fragment of calculated moiré pattern for the perfect alignment ($\theta = 0$). (c) The corresponding distribution of the calculated "mass" term Δ; see Eq. (5.1). (Reproduced with permission from Sachs et al., 2011.)

These lines produce the basis of unidirectional linear-dispersion modes (Volkov & Pankratov, 1986; Ludwig et al., 1994; Tudorovskiy & Katsnelson, 2012), similar to edge modes in quantum Hall regime (Section 5.8).

Consider the Schrödinger equation with the Hamiltonian (5.1):

$$\left(-i\hbar v \vec{\sigma} \nabla + \sigma_z \Delta(x,y)\right)\Psi(x,y) = E\Psi(x,y). \tag{13.24}$$

First, consider the case of one-dimensional geometry, $\Delta = \Delta(y)$. Then, we can try the solution of Eq. (13.24) as

$$\Psi(x,y) = \exp\left(\frac{ip_x x}{\hbar}\right)\begin{pmatrix} \psi_1(y) \\ \psi_2(y) \end{pmatrix}. \tag{13.25}$$

For further simplifications, it is convenient to make a rotation in pseudospin space (Tudorovskiy & Katsnelson, 2012)

$$\eta_1 = \frac{\psi_1 + \psi_2}{\sqrt{2}}, \quad \eta_2 = \frac{\psi_1 - \psi_2}{\sqrt{2}}. \tag{13.26}$$

In this notation, Eq. (13.24) takes the form

$$\begin{pmatrix} vp_x - E & \hbar v\dfrac{\partial}{\partial y} + \Delta \\ \hbar v\dfrac{\partial}{\partial y} - \Delta & vp_x + E \end{pmatrix}\begin{pmatrix} \eta_1 \\ \eta_2 \end{pmatrix} = 0. \tag{13.27}$$

Similar to the transition from Eq. (2.39) to (2.42), one can replace the matrix first-order differential equation (13.27) by the scalar second-order differential equation

$$\left[-\frac{d^2}{dy^2} + \frac{\Delta^2(y)}{\hbar^2 v^2} \pm \frac{1}{\hbar v}\frac{d\Delta(y)}{dy}\right]\eta_{1,2} = \lambda\eta_{1,2}, \tag{13.28}$$

in which sign $+$ corresponds to the subscript 1 and sign $-$ to the subscript 2,

$$\lambda = \frac{E^2 - v^2 p_x^2}{\hbar^2 v^2}. \tag{13.29}$$

The connection formulas between η_1 and η_2 are given by the equations

$$(E + vp_x)\eta_2 = \left(\Delta - \hbar v\frac{d}{dy}\right)\eta_1,$$

$$(E - vp_x)\eta_1 = \left(\Delta + \hbar v\frac{d}{dy}\right)\eta_2. \tag{13.30}$$

Eq. (13.28) has a very special property: For any function $\Delta(y)$ changing from negative to positive values, it always has a solution with $\lambda = 0$. Actually, this equation is one of the prototype examples of supersymmetry in quantum mechanics, and the existence of this zero mode is a consequence of the supersymmetry (Gendenshtein & Krive, 1985). The solution can be found directly from the first-order differential equation (13.30). If we assume that $\Delta > 0$ at $y \to +\infty$ and $\Delta < 0$ at $y \to -\infty$, the solution has the energy $E = vp_x$, $\eta_1 = 0$ and

$$\eta_2 \propto \exp\left(-\int_0^y dy' \frac{\Delta(y')}{\hbar v}\right). \tag{13.31}$$

In the opposite case $\Delta < 0$ at $y \to +\infty$ and $\Delta > 0$ at $y \to -\infty$ the solution has the energy $E = -vp_x$, $\eta_2 = 0$ and

$$\eta_1 \propto \exp\left(\int_0^y dy' \frac{\Delta(y')}{\hbar v}\right). \tag{13.32}$$

In both cases, the corresponding modes have linear dispersion, unidirectional (their group velocity is either parallel or antiparallel to x axis) and fully polarized in pseudospin. For the second valley, the propagation direction and sublattice polarization are opposite. The analogy with topologically protected zero modes considered in Section 2.3 is quite straightforward.

These modes exist only if there is a straight line where the function $\Delta(y)$ changes sign (zero-mass line). One can also build the corresponding solution if the zero-mass line is curved (Tudorovskiy & Katsnelson, 2012). Let us assume that this line is given by the equation

$$(x, y) = \vec{R}(\tau), \tag{13.33}$$

where τ is the path along the line, that is, $\left| d\vec{R}/d\tau \right| = 1$. In the vicinity of the line (13.33) one can assume

$$(x, y) = \vec{R}(\tau) + \zeta \vec{n}(\tau), \tag{13.34}$$

where $\vec{n}(\tau)$ is the unit normal vector to the line (13.33). In curvilinear coordinates (τ, ζ) the equation (13.23) takes the form

$$\left(\frac{-i\hbar v \vec{\sigma} \vec{R}'(\tau)}{1 - \zeta k(\tau)} \frac{\partial}{\partial \tau} - i\hbar v \vec{\sigma} \vec{n}(\tau) \frac{\partial}{\partial \zeta} + \sigma_z \Delta\right) \Psi = E\Psi, \tag{13.35}$$

where $\vec{R}'(\tau) = d\vec{R}/d\tau$ and $k(\tau) = -\vec{R}'(\tau) \cdot \vec{n}'(\tau)$ is the curvature at the point τ (DoCarmo, 1976). Since the Jacobian of the transformation (13.34) is not unity:

$$J = \frac{\partial(x, y)}{\partial(\tau, \zeta)} = 1 - k(\tau)\zeta, \tag{13.36}$$

one has to introduce a new wave function

$$\Phi = \sqrt{J}\Psi, \tag{13.37}$$

which satisfies a "conventional" normalization condition

$$\int d\tau d\zeta \langle \Phi | \Phi \rangle = 1. \tag{13.38}$$

In these new variables the modified Dirac equation (13.35) reads

$$\left(\frac{-i\hbar v \vec{\sigma} \, \vec{R}'(\tau)}{1 - \zeta k(\tau)} \frac{\partial}{\partial \tau} - i\hbar v \vec{\sigma} \, \vec{n} \,(\tau) \frac{\partial}{\partial \zeta} + \sigma_z \Delta - \frac{i\hbar v k(\tau) \, \vec{\sigma} \, \vec{n} \,(\tau)}{2[1 - \zeta k(\tau)]} - \frac{i\hbar v \vec{\sigma} \, \vec{R}'(\tau) \zeta k'(\tau)}{2[1 - \zeta k(\tau)]^2} \right) \Phi = E\Phi. \tag{13.39}$$

Further analysis of this equation confirms a robustness of the unidirectional "zero" modes for the curved lines, at least, assuming that the curvature radius is large enough in comparison with the "Compton wavelength" $\Lambda = \hbar v / \Delta$ (Tudorovskiy & Katsnelson, 2012).

In the case when we have two parallel zero-mass lines situated at $y = \pm b$, one can calculate the tunneling amplitude between the lines. One can expect from Eq. (13.31) and (13.32):

$$T \approx \exp\left[-\frac{1}{\hbar v} \int\limits_{-b}^{b} dy |\Delta(y)| \right]; \tag{13.40}$$

a more accurate expression with a preexponential factor can be found in Tudorovskiy and Katsnelson (2012).

If we assume for estimate $\Delta \approx 0.1$ eV the Compton wavelength $\Lambda \approx 10$ nm, that is, comparable with a typical moiré period $\xi = \kappa a$; see Eq. (13.23). This means that quantum effects such as tunneling are essential at the consideration of low-energy spectrum of graphene at hBN. One can, nevertheless, consider a model under the assumption

$$\lambda \gg \Lambda, \tag{13.41}$$

which allows an accurate formal treatment. In this limit, the electronic transport in graphene on hBN at the neutrality point can be described as a classical percolation along zero-mass lines (Titov & Katsnelson, 2014). For the physical introduction to the percolation theory, we refer to the books of de Gennes (1979) and Shklovskii & Efros (1984); a more advanced review can be found in Isichenko (1992).

For the commensurate case, that is, for small misorientation angle, the super-structure of graphene is optimized to minimize the van der Waals interaction

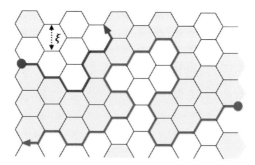

Fig. 13.6 Percolation model. Gray and white hexagons represent the regions with positive and negative masses, respectively, with counterpropagating modes at the boundaries between them (shown by arrows).
(Reproduced with permission from Titov & Katsnelson, 2014.)

energy with the hBN substrate. In this case, the regions of positive and negative mass do not have equal areas, and there is an average macroscopic gap; according to GW calculations (Bokdam et al., 2014) it is about 30 meV. Therefore, perfectly aligned graphene is supposed to be insulating at low temperatures. Such behavior was indeed observed for some samples but not for all of them (Woods et al., 2014); probably in other samples the effect is hidden by some disorder in the substrate. In the incommensurate phase, the average macroscopic gap is expected to be 0, as we will discuss in detail in Section 13.5. Therefore, one can assume that the regions of negative and positive masses have equal areas, with zero-mass lines in between (Fig. 13.6). In this situation, the zero-mass lines form a *critical percolation cluster* (de Gennes, 1979; Shklovskii & Efros, 1984; Isichenko, 1992).

In two dimensions, very powerful tools can be used to study the critical percolation, such as conformal field theory (Di Francesco, Mathieu, & Sénéchal, 1997) and Schramm–Loewner evolution (Kemppainen, 2017). As a result, a number of rigorous results have been obtained (Saleur & Duplantier, 1987; Isichenko, 1992; Cardy, 2000; Smirnov, 2001; Beffara, 2004; Kager & Nienhuis, 2004; Hongler & Smirnov, 2011).

Let us calculate the minimal conductivity of graphene on hBN in the model of classical percolation. Landauer–Büttiker formula for the conductance (3.16) takes the form

$$G = \frac{2e^2}{h} \langle N_{line} \rangle, \tag{13.42}$$

where $\langle N_{line} \rangle$ is the average number of the percolation paths connecting the sample edges at $x = 0$ and $x = L_x$, and we take into account that the transmission probability for each classical path is 1, and that there are two channels per path, due to spin degeneracy. Indeed, as follows from our analysis after Eq. (13.30) only one valley has zero mode propagating in the needed direction; therefore, the valley degeneracy factor 2 should not be taken into account, contrary to the consideration in Chapter 3.

For rectangular geometry in the limit $L_x \gg L_y$, where L_y is the sample width (cf. Eq. (3.18)), $\langle N_{line} \rangle$ is given by the Cardy formula derived by the tools of conformal field theory (Cardy, 2000)

$$\langle N_{line} \rangle = \frac{\sqrt{3}}{2} \frac{L_x}{L_y};$$ (13.43)

mathematically rigorous proof of this formula was given by Hongler and Smirnov (2011). Substituting Eq. (13.43) into Eq. (13.42) one can find for the conductivity

$$\sigma = \sqrt{3} \frac{e^2}{h}.$$ (13.44)

Importantly, the critical percolation path in two dimensions cannot be considered as a conventional line directly connecting, more or less, the sample edges. It is a fractal object, that is, a very thick and meandrous line. In particular, the hull Hausdorff dimensionality of the critical percolation cluster (roughly speaking, the fractal dimensionality of the percolation path) is equal to 7/4 (Saleur & Duplantier, 1987).

13.4 Berry curvature effects

Appearance of the mass term in the Dirac Hamiltonian modifies our consideration of the Berry phase in comparison with the massless case (see Section 2.4). Now we have the effective Hamiltonian (2.87) with the vector

$$\vec{R}\left(\vec{k}\right) = 2(\hbar v k_x, \hbar v k_x, \Delta).$$ (13.45)

The corresponding expression for the Berry curvature vector (2.86) is:

$$\vec{V}\left(\vec{k}\right) = \frac{(\hbar v)^2}{2\left((\hbar v k)^2 + \Delta^2\right)^{3/2}} (\hbar v k_x, \hbar v k_y, \Delta)$$ (13.46)

(cf. Eq.(2.89)). Here we consider, to be specific, only electron band; for the hole band, the sign will be opposite.

The appearance of z-component of the Berry curvature essentially changes the electron dynamics in the presence of external electric and magnetic fields (Xiao, Chang, & Niu, 2010; Gorbachev et al., 2014). For simplicity, we will restrict ourselves here to the case of electric field $\vec{E}(t)$ only. We will use the gauge (7.2). Then, the effect of the electric field on the state $|n\,\vec{k}\rangle$ will be just a replacement $|n\,\vec{k}\rangle \rightarrow |n,\vec{k}(t)\rangle$, with

$$\vec{k}(t) = \vec{k} - \frac{e}{\hbar c}\vec{A}(t), \qquad (13.47)$$

which leads immediately to the equation of motion

$$\frac{d\vec{k}(t)}{dt} = \frac{e}{\hbar}\vec{E}(t). \qquad (13.48)$$

Now, we have to write the expression for the average value of the group velocity in the state $|u(t)\rangle$, which solves the Cauchy problem for the time-dependent Schrödinger equation (2.76) and (2.78):

$$\vec{v}_n = \frac{1}{\hbar}\langle u(t)|\frac{\partial \hat{H}_{eff}}{\partial \vec{k}}|u(t)\rangle. \qquad (13.49)$$

The further derivation (Xiao, Chang, & Niu, 2010) is a modification of the general consideration of the Berry phase in Section 2.4. We start with a general, formally exact expansion of the solution in the basis $|n,\vec{k}(t)\rangle$:

$$|u(t)\rangle = \sum_m a_m(t)\exp\left[-\frac{i}{\hbar}\int_0^t dt' E_m\left(\vec{k}(t)\right)\right]|m,\vec{k}(t)\rangle. \qquad (13.50)$$

Substituting Eq. (13.50) into Eq. (2.76), and taking into account Eq. (2.77), we find

$$\frac{da_n(t)}{dt} = -\sum_m a_m(t)\exp\left\{\frac{i}{\hbar}\int_0^t dt'\left[E_n\left(\vec{k}(t')\right) - E_m\left(\vec{k}(t')\right)\right]\right\}$$

$$\times \left\langle n,\vec{k}(t)|\nabla_{\vec{k}}|m,\vec{k}(t)\right\rangle\frac{d\vec{k}(t)}{dt}. \qquad (13.51)$$

The right-hand side of Eq. (13.51) is proportional to the electric field, due to Eq. (13.48). Assuming that the electric field is weak, one can solve Eq. (13.51) by iterations. Note that the term with $m = n$ can be excluded by some phase shift (actually, this is the Berry phase (2.81)), which is obviously irrelevant at the calculation of (13.49). In the linear approximation, taking into account Eq. (13.50) and integrating on time by part, one obtains:

$$|u(t)\rangle = \exp\left[-\frac{i}{\hbar}\int_0^t dt' E_m\left(\vec{k}(t')\right)\right]$$

$$\times \left\{|n, \vec{k}(t)\rangle - i\hbar \frac{d\vec{k}(t)}{dt} \sum_{m \neq n} |m, \vec{k}(t)\rangle \frac{\langle m, \vec{k}(t)|\nabla_{\vec{k}}|n, \vec{k}(t)\rangle}{E_n\left(\vec{k}(t)\right) - E_m\left(\vec{k}(t)\right)}\right\}.$$

$$(13.52)$$

At last, we use the identity (2.85) and substitute Eq. (13.52) into Eq. (13.49). The result is expressed in terms of the Berry curvature vector (Xiao, Chang, & Niu, 2010):

$$\vec{v}_n\left(\vec{k}\right) = \frac{1}{\hbar}\left[\nabla_{\vec{k}} E_n\left(\vec{k}\right) + e\,\vec{E} \times \vec{V}_n\left(\vec{k}\right)\right]. \qquad (13.53)$$

The term with the vector product represents *anomalous group velocity*. It is relevant for the two-dimensional transport only if the vector $\vec{V}_n\left(\vec{k}\right)$ has a non-vanishing component perpendicular to the plane, which is possible only at $\Delta \neq 0$ (see Eq. (13.46)).

In this case, one can expect the current that is normal to the direction of the electric field, like in the Hall effect. Assuming that $\vec{E} \parallel 0x$, one can calculate from Eq. (13.53):

$$j_y = 2e \sum_{\alpha} \int \frac{d^2k}{(2\pi)^2} v_y\left(\vec{k}\right) f_\alpha\left(\vec{k}\right) = \sigma_{xy} E, \qquad (13.54)$$

where $f_\alpha\left(\vec{k}\right)$ is the Fermi function for electron and hole bands ($\alpha = e, h$)

$$\sigma_{xy} = 2\frac{e^2}{h} \sum_{\alpha} \int \frac{d^2k}{2\pi} V_z^\alpha\left(\vec{k}\right) f_\alpha\left(\vec{k}\right), \qquad (13.55)$$

and the factor 2 is the spin degeneracy; note that $V_z^e\left(\vec{k}\right) = -V_z^h\left(\vec{k}\right) = V_z\left(\vec{k}\right)$. Appearance of the "Hall-like" off-diagonal conductivity assumes broken time-reversal symmetry; we already know that in the Dirac model with nonzero mass this symmetry is broken, indeed (see Eq. (5.8)). It is restored if we take into account the contribution of two valleys; one can prove that they are different by the signs of $V_z\left(\vec{k}\right)$. As a result, the total charge current in y-direction vanishes, but *valley* current and therefore valley Hall effect arises, with the corresponding valley conductivity

$$\sigma_{xy}^v = \sigma_{xy}^K - \sigma_{xy}^{K'} = 4\frac{e^2}{h} \sum_{\alpha} \int \frac{d^2k}{2\pi} V_z^\alpha\left(\vec{k}\right) f_\alpha\left(\vec{k}\right). \qquad (13.56)$$

Substituting Eq. (13.46) into Eq. (13.55) and assuming zero temperature, we obtain (Gorbachev et al., 2014):

$$\sigma_{xy}^{v} = \frac{2e^2}{h} \times \begin{cases} 1, & |\mu| < \Delta, \\ \Delta/|\mu|, & |\mu| \geq \Delta, \end{cases} \tag{13.57}$$

where μ is the chemical potential. Valley Hall current can be transformed to *nonlocal* charge current and thus detected, similar to our discussion in Section 11.6. This effect was observed by Gorbachev et al., (2014), which can be considered as a bright, experimental manifestation of Berry curvature for massive two-dimensional Dirac fermions.

Another interesting effect related to the broken inversion symmetry and appearance of the mass term in graphene at hBN is the optical second-harmonic generation (Vandelli, Katsnelson, & Stepanov, 2019). Note that in this case the Dirac approximation is not sufficient, and one needs to work with the lattice model or, at least, take into account the trigonal warping term (1.34).

13.5 Electronic structure of moiré patterns

Now, let us build the effective Hamiltonian describing the electronic structure of graphene at hBN. We start with the consideration of a purely geometric moiré structure without atomic relaxation. As explained in Section 13.2, this is a reasonable approximation for the incommensurate moiré pattern, with large enough misorientation angle. The effects of the atomic relaxation will be included later.

If we completely neglect electron–electron interactions and consider purely single-electron tight-binding Hamiltonian it can be represented in the form

$$\hat{H} = \hat{H}_{BN} + \hat{H}_g + \hat{H}_\perp, \tag{13.58}$$

where $\hat{H}_{BN}$ is the tight-binding Hamiltonian for hBN substrate, $\hat{H}_g$ is the Hamiltonian for graphene overlayer, and $\hat{H}_\perp$ is the interlayer-hopping Hamiltonian connecting the substrate and the overlayer. The basis vectors of our two subsystems are connected by Eq. (13.13) and (13.14), and the reciprocal lattice vectors, Eq. (13.15) and (13.16). For the case of triangular Bravais lattice, we are interested in the minimal reciprocal lattice vectors, which are given by the expression (note that we rotate our coordinate frame by $60°$ in comparison with that used in Eq. (1.12))

$$\vec{G}_{m=0,\dots,5} = \hat{R}\left(\frac{2\pi m}{6}\right)\vec{G}_0, \quad \vec{G}_0 = \left[1 - \frac{1}{1+\delta}\hat{R}(\theta)\right]\left(0, \frac{4\pi}{3a}\right), \tag{13.59}$$

see Eq. (13.14).

In the lowest (that is, second) order in a small interlayer hopping, the effective Hamiltonian for graphene overlayer can be written as

$$\hat{H}_{eff} = \hat{H}_g + \hat{H}_\perp \frac{1}{E - \hat{H}_{BN}} \hat{H}_\perp, \tag{13.60}$$

cf. Eq. (12.63). We will postpone the derivation of the Hamiltonian $\hat{H}_\perp$ between two misoriented layers (Lopes dos Santos, Peres, & Castron Neto, 2007; Bistritzer & MacDonald, 2011; Kindermann & First, 2011; Mele, 2011) until the next chapter, where we will consider twisted bilayer graphene. Here, we present just the result, the effective Hamiltonian (13.60) for the closed vicinity of Dirac points K and K' (Kindermann, Uchoa, & Miller, 2012; Wallbank et al., 2013; Diez et al., 2014). In the basis (1.27), the answer reads (Wallbank et al., 2013):

$$\hat{H}_{eff} = -i\hbar v \tau_0 \otimes \vec{\sigma} \nabla + U_0 \tau_0 \otimes \sigma_0 f_1\left(\vec{r}\right) + U_3 \tau_z \otimes \sigma_z f_2\left(\vec{r}\right)$$
$$+ U_1 \tau_z \otimes \vec{\sigma} \left(\vec{e}_z \times \nabla f_2\left(\vec{r}\right)\right) + U_2 \tau_z \otimes \vec{\sigma} \nabla f_2\left(\vec{r}\right), \tag{13.61}$$

where $\vec{e}_z = (0, 0, 1)$ is the unit vector along z-direction, and τ_0, σ_0 are unit matrices in the valley and sublattice spaces, respectively,

$$f_1\left(\vec{r}\right) = \sum_{m=0}^{5} \exp\left(i\vec{G}_m\vec{r}\right), \quad f_2\left(\vec{r}\right) = \sum_{m=0}^{5}(-1)^m \exp\left(i\vec{G}_m\vec{r}\right), \tag{13.62}$$

and we have taken into account only the largest terms, conserving inversion symmetry. All contributions U_i in our model are of the order of $t_\perp^2/\Delta_{BN} \approx 1$ meV where $\Delta_{BN} \approx 6$ eV is the energy gap in hBN (Watanabe, Taniguchi, & Kanda, 2004). Note however that the term U_0, describing modulation of scalar potential, has much larger contribution (of the order of 60 meV) from direct Coulomb interaction between electrons in substrate and graphene overlayer (Wallbank et al., 2013).

The term proportional to U_3 represents the local gap opening. This term is proportional to the function $f_2(\vec{r})$, with average value equal to 0; this justifies our assumption in Section 13.3 that the average mass vanishes in the incommensurate phase of the moiré pattern. Note that our too-simplified tight-binding model probably essentially underestimated this term. At least the first-principle calculations cited earlier give an order-of-magnitude larger amplitude of the local gap fluctuations.

The terms proportional to U_1 and U_2 originate from the modulation of hopping parameters and describe oscillating pseudomagnetic field (10.2). Both these terms and the mass term are proportional to τ_z, that is, have opposite signs for the valleys K and K'.

In the model (13.61) we deal with Dirac fermions under the action of periodic scalar potential, vector potential, and mass term. Probably the most interesting effect is the formation of the secondary families of the Dirac points related to the boundaries of the new "moiré" Brillouin zone, with the reciprocal lattice vectors determined by Eq. (13.59) (note that at small δ and θ their length is very large: $|G| \gg \pi/a$). This effect was predicted by Park et al. (2008). Following this work, let us first consider the model with purely scalar periodic potential in one dimension:

$$\hat{H} = \hbar v \left[-i\sigma_x \frac{\partial}{\partial x} - -i\sigma_y \frac{\partial}{\partial y} + \frac{V(x)}{\hbar v} \right], \tag{13.63}$$

$$V(x) = V(x + \xi). \tag{13.64}$$

To be specific, we assume that the average value of the periodic potential is 0 and that it is even: $V(x) = V(-x)$. The unitary transformation $\hat{H}' = \hat{U}_1^+ \hat{H} \hat{U}_1$, with the matrix

$$\hat{U}_1 = \frac{1}{\sqrt{2}} \begin{pmatrix} e^{-i\alpha(x)/2} & -e^{i\alpha(x)/2} \\ e^{-i\alpha(x)/2} & e^{i\alpha(x)/2} \end{pmatrix} \tag{13.65}$$

and

$$\alpha(x) = \frac{2}{\hbar v} \int_0^x dx' V(x') \tag{13.66}$$

gives us the new Hamiltonian

$$\hat{H}' = \hbar v \begin{pmatrix} -i\dfrac{\partial}{\partial x} & -e^{i\alpha(x)} \dfrac{\partial}{\partial y} \\ e^{-i\alpha(x)} \dfrac{\partial}{\partial y} & i\dfrac{\partial}{\partial x} \end{pmatrix}. \tag{13.67}$$

Its diagonal part has two eigenstates propagating along x-direction and opposite to it:

$$|\Phi_1\rangle = e^{ikx} \begin{pmatrix} 1 \\ 0 \end{pmatrix}, \quad |\Phi_2\rangle = e^{-ikx} \begin{pmatrix} 0 \\ 1 \end{pmatrix}, \tag{13.68}$$

with the same eigenenergy $E = \hbar v k$ (note that the direction of the pseudospin vector and, therefore, of the x axis is now different from the original one, due to the unitary transformation). The off-diagonal part depending on the periodic potential induces back-scattering connecting these two waves.

Let us assume for simplicity that α is small. Then, we know from a general theory of electrons in a weak periodic potential (Vonsovsky & Katsnelson, 1989) that the latter can produce a strong effect only for the wave vectors close to

$$G_n = \frac{2\pi}{\varsigma} n, \quad n = \pm 1, \pm 2, \ldots \tag{13.69}$$

One can expand the phase factor in the Taylor series

$$e^{i\alpha(x)} = \sum_{n=-\infty}^{+\infty} f_n e^{iG_n x}, \tag{13.70}$$

the coefficients f_n are all real for the even function $V(x)$.

For small enough f_n and for the wave vectors $k \approx G_n$ one can project the Hamiltonian (13.67) onto a two-dimensional basis

$$|\tilde{\Phi}_1\rangle = e^{i(q_x + G_n/2)x + iq_y y} \begin{pmatrix} 1 \\ 0 \end{pmatrix}, \quad |\tilde{\Phi}_2\rangle = e^{i(q_x - G_n/2)x + iq_y y} \begin{pmatrix} 0 \\ 1 \end{pmatrix}, \tag{13.71}$$

the vector $\vec{q}$ is supposed to be small. The result is

$$\hat{H}\prime = \hbar v \left(q_x \sigma_z + f_n q_y \sigma_y \right) + \frac{\hbar v \pi n}{\varsigma}. \tag{13.72}$$

After the second unitary transformation

$$\hat{U}_2 = \frac{1}{\sqrt{2}} \begin{pmatrix} 1 & 1 \\ -1 & 1 \end{pmatrix} \tag{13.73}$$

the Hamiltonian (13.72) takes the form

$$\hat{H}\prime\prime = \hbar v \left(q_x \sigma_x + f_n q_y \sigma_y \right) + \frac{\hbar v \pi n}{\varsigma}. \tag{13.74}$$

It has the eigenenergies

$$E_\pm^{(n)}(\vec{q}) = \pm \hbar v \sqrt{q_x^2 + f_n^2 q_y^2} + \frac{\hbar v \pi n}{\varsigma}, \tag{13.75}$$

corresponding to the anisotropic conical points, with essentially different group velocities along x and y directions (we have to recall that these directions originate from initial ones by two rotations, (13.65) and (13.73)).

Numerical solution of the two-dimensional analog of the problem (13.63) and (13.64) shows the appearance of the additional conical points near the moiré Brillouin zone face centers $\vec{G}_m/2$; see Eq. (13.59) (Park et al., 2008). The same conclusion also remains correct in the full model (13.61), with modulated scalar potential, vector potential, and mass term (Wallbank et al., 2013).

When we go beyond the weak-coupling approximation and connect more than two waves, the higher-order sequences of conical points arise. In the presence of quantized magnetic field, each of these sequences results in a sequence of the

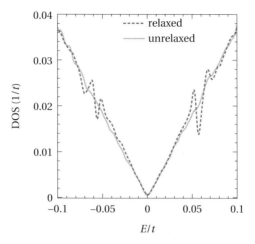

Fig. 13.7 Calculated density of states (DOS) for graphene on hBN for perfect alignment ($\theta = 0$); energy is in the units of the nearest-neighbor hopping parameter $t = 2.7$ eV. Minima of DOS correspond to additional conical points. One can see that the atomic relaxation dramatically increases the effect of moiré on the electronic structure.
(Reproduced with permission from Slotman et al., 2015.)

corresponding Landau levels, leading to a very complicated "fractal" structure of the energy spectrum, reminiscent of the so-called Hofstadter butterfly (Hofstadter, 1976). The cloning of conical point and Hofstadter butterfly effects were experimentally observed in graphene at hBN (Dean et al., 2013; Hunt et al., 2013; Ponomarenko et al., 2013).

Slotman et al. (2015) simulated the electronic structure of graphene at hBN, taking into account the atomic relaxation effects; the latter turn out to be very important for small enough misorientation angles. The corresponding results are shown in Fig. 13.7.

The other important conclusion from these simulations is that the results are very sensitive to the amplitude of the mass term; to have a reasonable agreement with the available experimental data, many-body enhancement of the mass term (Song, Shytov, & Levitov, 2013) should be probably taken into account. To illustrate this sensitivity we show in Fig. 13.8 the computational results for the optical conductivity for two different values of the amplitude of the mass term.

Apart from the cloning of the conical points, periodic fields acting on massless Dirac fermions renormalize the value of the electron velocity in the Dirac point (Tan, Park, & Louie, 2010; Dugaev & Katsnelson, 2012). This effect will be important for our consideration of twisted bilayer graphene in the next chapter. It can already be seen from Eq. (13.75). Here, we consider a more general case, with periodic modulations of both scalar and (pseudo)vector potentials, but neglecting

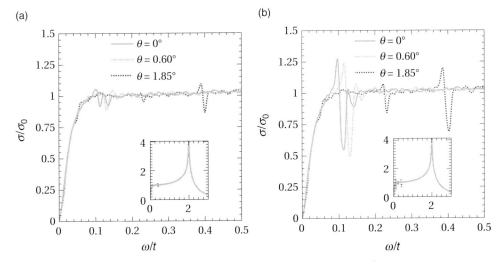

Fig. 13.8 Calculated optical conductivity (in the units of $\sigma_0 = \pi e^2/2h$, Eq. (7.36)) for the values of the mass term taken from the calculations (Bokdam et al., 2014) (a) and for that term enhanced by a factor of 2 (b). (Reproduced with permission from Slotman et al., 2015.)

the mass term (Dugaev & Katsnelson, 2012). The case of modulation of only the vector potential was considered by Tan, Park, and Louie (2010).

Instead of Eq. (13.63), let us consider a more general Hamiltonian (cf. Eq. (10.1))

$$\hat{H} = \vec{\sigma}\left(-i\hbar v\nabla - \vec{A}(x)\right) + V(x), \qquad (13.76)$$

both functions $\vec{A}(x)$ and $V(x)$ are supposed to be periodic, with the period ξ. The corresponding Schrödinger equation for the spinor wave function $\Psi = (\phi, \chi)^T$ reads

$$\begin{pmatrix} E - V & i\hbar v\partial_- + A_- \\ i\hbar v\partial_+ + A_+ & E - V \end{pmatrix} \begin{pmatrix} \varphi \\ \chi \end{pmatrix} = 0, \qquad (13.77)$$

where $\partial_\pm = \partial/\partial x \pm i\partial/\partial y$, $A_\pm = A_x \pm iA_y$.

We want to build Bloch functions with the wave vector $k \ll \pi/\xi$. To this aim, from the conventional k·p perturbation theory (Tsidilkovskii, 1982; Vonsovsky & Katsnelson, 1989), we first need to find the solutions of Eq. (13.77) for $\vec{k} = 0$:

$$(E - V)\varphi + i\hbar v\frac{d\chi}{dx} + A_-\chi = 0,$$

$$(E - V)\chi + i\hbar v\frac{d\varphi}{dx} + A_+\varphi = 0. \qquad (13.78)$$

We are looking for the solutions with small energy. If we put in Eq. (13.78) $E = 0$, one can derive the second-order differential equation for the function φ

$$\frac{d^2\varphi}{dx^2} - \left(\frac{d\ln V}{dx} + \frac{2iA_x}{\hbar v}\right)\frac{d\varphi}{dx} + \left(\frac{V^2}{\hbar^2 v^2} - \frac{i}{\hbar v}\frac{dA_+}{dx} + \frac{i}{\hbar v}\frac{d\ln V}{dx}A_+ - \frac{A_+A_-}{\hbar^2 v^2}\right)\varphi = 0$$

(13.79)

and the expression for the second component of the spinor

$$\chi = \frac{i\hbar v}{V}\frac{d\varphi}{dx} + \frac{A_+}{V}\varphi$$

(13.80)

(Dugaev & Katsnelson, 2012).

Now consider the case $\vec{A}(x) = 0$; then Eq. (13.79) and (13.80) are dramatically simplified. One can straightforwardly check that they have the solutions

$$\varphi_{1,2}(x) = \exp\left[\pm\frac{i}{\hbar v}\int_0^x dx' V(x')\right],$$

(13.81)

$$\chi_{1,2}(x) = \mp\exp\left[\pm\frac{i}{\hbar v}\int_0^x dx' V(x')\right].$$

In k·p approximation, the normalized basic functions can be chosen as

$$\Psi_1 = \frac{e^{\vec{k}\cdot\vec{r}}}{\sqrt{2S}}\binom{\varphi_1}{\chi_1}, \quad \Psi_2 = \frac{e^{\vec{k}\cdot\vec{r}}}{\sqrt{2S}}\binom{\varphi_2}{\chi_2},$$

(13.82)

where S is the sample area. The Hamiltonian (13.63) in this basis has the form

$$\hat{H} = \hbar v\begin{pmatrix} -k_x & \gamma_1 k_x - \gamma_2 k_y \\ \gamma_1 k_x - \gamma_2 k_y & k_x \end{pmatrix},$$

(13.83)

where

$$\gamma_1 = \int_0^\xi \frac{dx}{\xi}\cos\left[\frac{2}{\hbar v}\int_0^x dx' V(x')\right],$$

(13.84)

$$\gamma_2 = \int_0^\xi \frac{dx}{\xi}\sin\left[\frac{2}{\hbar v}\int_0^x dx' V(x')\right].$$

This Hamiltonian represents the anisotropic conical point with the velocities

$$v_x = v\sqrt{1 + \gamma_1^2}, \quad v_y = v\gamma_2. \tag{13.85}$$

For the case of a purely vector potential ($V = 0$), Eq. (13.79) and (13.80) have two solutions:

$$\varphi_1(x) = \exp\left[\frac{i}{\hbar v}\int_0^x dx' A_+(x')\right], \quad \chi_1(x) = 0 \tag{13.86}$$

and

$$\varphi_2(x) = 0, \quad \chi_2(x) = \exp\left[\frac{i}{\hbar v}\int_0^x dx' A_-(x')\right]. \tag{13.87}$$

The effective Hamiltonian in this basis is

$$\hat{H} = \frac{\hbar v}{\sqrt{\Lambda}}\begin{pmatrix} 0 & k_x - ik_y \\ k_x + ik_y & 0 \end{pmatrix}, \tag{13.88}$$

where

$$\Lambda = \int_0^\xi \frac{dx_1}{\xi}\exp\left[-\frac{2}{\hbar v}\int_0^{x_1} dx' A_y(x')\right]\int_0^\xi \frac{dx_2}{\xi}\exp\left[\frac{2}{\hbar v}\int_0^{x_2} dx' A_y(x')\right] \tag{13.89}$$

(Tan, Park, & Louie, 2010; Dugaev & Katsnelson, 2012). This Hamiltonian describes isotropic conical (Dirac) point with the effective velocity

$$\tilde{v} = v/\sqrt{\Lambda}. \tag{13.90}$$

The expression for Λ can be rewritten as

$$\Lambda = \int_0^\xi \frac{dx_1}{\xi}\int_0^\xi \frac{dx_2}{\xi}\cosh\left[\frac{2}{\hbar v}\int_{x_1}^{x_2} dx' A_y(x')\right], \tag{13.91}$$

which makes obvious $\Lambda > 1$ and, therefore, periodic (pseudo)magnetic field always diminishes the Fermi velocity.

Similarly, one can derive a general expression for the effective (anisotropic) Fermi velocities in the presence of both scalar and vector potentials. Also, one can consider the "Fock" renormalization of the *anisotropic* Fermi velocity, similar to Eq. (8.95). All of these results can be found in Dugaev and Katsnelson (2012).

13.6 Magnetic bands in graphene superlattices

Moiré superstructures in graphene on hBN provide a unique opportunity to better understand the physics of Bloch electrons in magnetic field. For "normal" crystal lattices, we are always in the regime (2.2), which is equivalent to the assumption

$$\Phi_{el} \ll \Phi_0, \tag{13.92}$$

where $\Phi_{el} = BS_{el}$ is the magnetic flux per elementary cell, S_{el} is the area of the elementary cell, and Φ_0 is the flux quantum (2.52). For the moiré superlattices with the period $\xi = \sqrt{3}a\kappa \gg a$ (see Eq. (13.23)), Φ_{el} is enhanced by the factor κ^2 and can be comparable with Φ_0 in relatively easily achievable magnetic fields of the order of 10 T.

Let us come back to our general consideration of Section 2.1 to study what happens with electronic states when the condition (13.92) is violated. As a first step, we need to study in a bit more detail the generalized translation operators (2.15), which form the *magnetic translation group* (Brown, 1964; Zak, 1964). Similar to Eq. (2.18), we derive:

$$\exp\left(\frac{i}{\hbar}\vec{R_i}\hat{\vec{\Pi}}\right)\exp\left(\frac{i}{\hbar}\vec{R_j}\hat{\vec{\Pi}}\right) = \exp\left[\frac{i}{\hbar}\left(\vec{R_i}+\vec{R_j}\right)\hat{\vec{\Pi}}\right]\exp\left[\frac{ie}{\hbar c}\left(\vec{R_i}\times\vec{R_j}\right)\vec{B}\right]. \tag{13.93}$$

Note that if magnetic field is perpendicular to the crystal plane, then

$$\left(\vec{R_i}\times\vec{R_j}\right)\vec{B} = BS_{ij}, \tag{13.94}$$

where S_{ij} is the area of a parallelogram build on the vectors $\vec{R_i}$ and $\vec{R_j}$ and

$$S_{ij} = n_{ij}S_{el}, \tag{13.95}$$

where n_{ij} is some integer number.

Let us assume first that

$$\Phi_{el} = p\Phi_0, \tag{13.96}$$

where p is an integer. Then, for any translation, vectors $\vec{R_i}$ and $\vec{R_j}$

$$\exp\left[\frac{ie}{\hbar c}\left(\vec{R_i}\times\vec{R_j}\right)\vec{B}\right] = \exp\left(\frac{ie\Phi_{el}}{\hbar c}n_{ij}\right) = \exp\left(2\pi ipn_{ij}\right) = 1, \tag{13.97}$$

and the magnetic translation operators

$$\hat{T}\left[\vec{R_i}\right] = \exp\left(\frac{i}{\hbar}\vec{R_i}\hat{\vec{\Pi}}\right) \tag{13.98}$$

commute:

$$\hat{T}\left(\vec{R}_i\right)\hat{T}\left(\vec{R}_j\right) = \hat{T}\left(\vec{R}_i + \vec{R}_j\right) = \hat{T}\left(\vec{R}_j\right)\hat{T}\left(\vec{R}_i\right). \tag{13.99}$$

At the same time, each of them commutes with the Hamiltonian, as was discussed in Section (2.1); see Eq. (2.13) and (2.16). This means that they have a common system of the eigenfunctions; that is, we can choose the solution of the stationary Schrödinger equation $\psi(\vec{r})$ such that

$$\hat{T}\left(\vec{R}_i\right)\psi\left(\vec{r}\right) = \tau\left(\vec{R}_i\right)\psi\left(\vec{r}\right), \tag{13.100}$$

where $\tau\left(\vec{R}_i\right)$ is the eigenvalue of the operator $\hat{T}\left(\vec{R}_i\right)$. Since the latter is unitary (which follows from its definition (13.98)) $\left|\tau\left(\vec{R}_i\right)\right| = 1$, and the only expression consistent with Eq. (13.99) is $\tau\left(\vec{R}_i\right) = e^{\vec{k}\vec{R}_i}$, $\vec{k}$ is a real vector. Therefore Eq. (13.100) is an analog of the Bloch theorem (2.75).

Taking into account Eq. (2.15), one can rewrite Eq. (13.100) as

$$\psi\left(\vec{r} + \vec{R}_i\right) = \exp\left(\frac{i\vec{R}_i}{\hbar}\hat{\vec{p}}\right)\psi\left(\vec{r}\right) = \exp\left[i\vec{k}\vec{r} - \frac{ie}{2\hbar c}\left(\vec{R}_i \times \vec{B}\right)\vec{r}\right]\psi\left(\vec{r}\right), \tag{13.101}$$

and

$$\left|\psi\left(\vec{r} + \vec{R}_i\right)\right|^2 = \left|\psi\left(\vec{r}\right)\right|^2. \tag{13.102}$$

This means that under the condition (13.96), the stationary electronic states can be represented as waves propagating through the whole crystal without scattering or localization, exactly like the Bloch states in an ideal crystal in the absence of magnetic field. Obviously, the same statement is correct if, instead of Eq. (13.96), we have the condition

$$\Phi_{el} = \frac{p}{q}\Phi_0, \tag{13.103}$$

with integer p and q. Indeed, in this situation we can just consider the elementary cell with translation vectors multiplied by q. This statement was already used in Section (2.9).

For the irrational elementary flux (in the units of flux quantum) the translation operators (13.98) do not commute, which makes the use of noncommutative geometry natural (Bellissard, van Elst, & Schulz-Baldes, 1994). In this case, electronic states and their energies form a complicated fractal structure known as Hofstadter butterfly, as was mentioned in the previous section.

For two-dimensional free-electron gas, classical electron motion in the magnetic field is a Larmor rotation and, therefore, is restricted by some finite region; in quantum case this leads to Landau quantization of the electron energy spectrum. In general, this should also be the case in the presence of periodic crystalline potential. At the same time, under the condition (13.103), the electron motion should be infinite, similar to the motion of Bloch electrons without magnetic field. This leads to a very interesting effect, which was experimentally observed by Krishna Kumar et al. (2018) and Krishna Kumar et al. (2017). It turns out that the longitudinal conductivity of graphene superstructures on hBN oscillates with the magnetic field, reaching local maxima at the condition (13.103); the pronounced maxima are observed even for relatively large values as $p = 4$ and $q = 11$. These oscillations are completely different from the conventional magneto-oscillation effects considered in Section 2.8. They are observed at relatively high temperatures, such as 100–200 K.

14

Twisted bilayer graphene

14.1 Geometry and atomic structure

In this chapter we continue our consideration of large-scale periodic superstructures originated in misoriented Van der Waals heterostructures and discuss a particular but very important case of twisted bilayer graphene (or similar problem of graphene on graphite). In these situations, the lattice constants of both layers are identical, and the period of moiré pattern is determined by Eq. (13.22) with $p_1 = p_2 = 1$:

$$\kappa = \frac{1}{2|\sin(\theta/2)|}. \tag{14.1}$$

Generally speaking, the resulting structure is incommensurate except some special values of the misorientation angle θ. For theory and simulations, these special cases are important because they allow us to use well-developed solid-state theory for crystals; the case of *quasicrystals* is much more difficult and will be briefly touched on in the next section. Even for this case practical calculations are possible only with the use of long-periodic crystalline approximants.

To build the supercell for the two layers, which exists in the commensurate case, we will use the following procedure (Shallcross et al., 2010).

Suppose we succeed with the building of a common crystal lattice for two layers rotated one with respect to the other by angle θ. Then, each vector of this common lattice can be represented as a linear combination of the unit cell vectors in each layer with integer coefficients:

$$\vec{r} = m_1\vec{a}_1 + m_2\vec{a}_2 = n_1\hat{R}(\theta)\vec{a}_1 + n_2\hat{R}(\theta)\vec{a}_2, \tag{14.2}$$

where n_1, n_2, m_1, m_2 are integer, $\hat{R}(\theta)$ is the rotation matrix (13.14), and the lattice vectors are given by Eq. (1.10). Multiplying Eq. (14.2) by $\vec{a}_1$ and by $\vec{a}_2$ and keeping in mind that $\vec{a}_1^2 = \vec{a}_2^2 = 3a^2$, $\vec{a}_1\vec{a}_2 = 3a^2/2$ one obtains

$$m_1 + \frac{m_2}{2} = n_1 R_{11} + n_2 R_{12},$$

$$\frac{m_1}{2} + m_2 = n_1 R_{21} + n_2 R_{22},$$

(14.3)

where

$$R_{ij} = \frac{1}{3a^2} \vec{a}_i \cdot \hat{R}(\theta) \vec{a}_j.$$

(14.4)

Solving linear equations (14.4) and substituting Eq. (1.10) and (13.14) into Eq. (14.4) we find

$$\begin{pmatrix} m_1 \\ m_2 \end{pmatrix} = \begin{pmatrix} \cos\theta - \frac{1}{\sqrt{3}}\sin\theta & -\frac{2}{\sqrt{3}}\sin\theta \\ \frac{2}{\sqrt{3}}\sin\theta & \cos\theta + \frac{1}{\sqrt{3}}\sin\theta \end{pmatrix} \begin{pmatrix} n_1 \\ n_2 \end{pmatrix}.$$

(14.5)

All matrix elements in Eq. (14.5) should be rational which requires

$$\frac{1}{\sqrt{3}}\sin\theta = \frac{k_1}{k_3},$$

(14.6)

$$\cos\theta = \frac{k_2}{k_3},$$

(14.7)

where all k_i are integer. They should satisfy Diophantine equation

$$3k_1^2 + k_2^2 = k_3^2.$$

(14.8)

Its solution is equivalent to finding all rational points lying at the ellipse $3x^2 + y^2 = 1$. The general solution is

$$k_1 = 2pq, \quad k_2 = 3q^2 - p^2, \quad k_3 = 3q^2 + p^2,$$

(14.9)

with arbitrary integer p and q (for the proof see Shallcross et al., 2010). Substituting Eq. (14.9) into Eq. (14.7) we find

$$\cos\theta = \frac{3q^2 - p^2}{3q^2 + p^2}.$$

(14.10)

To find the supercell lattice vectors we need to substitute Eq. (14.6) and (14.7) into Eq. (14.5):

$$\begin{pmatrix} m_1 \\ m_2 \end{pmatrix} = \frac{1}{k_3} \begin{pmatrix} k_2 - k_1 & -2k_1 \\ 2k_1 & k_2 + k_1 \end{pmatrix} \begin{pmatrix} n_1 \\ n_2 \end{pmatrix}.$$

(14.11)

The solution of these Diophantine equations is quite cumbersome, and we refer the reader to Shallcross et al. (2010). The answer depends on the parameter

$$\delta = \frac{3}{\gcd(p, 3)},$$

(14.12)

where $\gcd(m, n)$ is the greatest common divisor of the natural numbers m and n, that is, $\delta = 1$ if 3 divides p, and $\delta = 3$ otherwise. For $\delta = 1$ the unit vectors of the supercell can be chosen as

$$\vec{t}_1 = \frac{1}{\gamma}\begin{pmatrix} p+3q \\ -2p \end{pmatrix}, \quad \vec{t}_2 = \frac{1}{\gamma}\begin{pmatrix} 2p \\ -p+3q \end{pmatrix},$$

(14.13)

where

$$\gamma = \gcd(3q+p, 3q-p).$$

(14.14)

Depending on p and q, γ can take one of the following values: 1, 2, 3, or 6 (Shallcross et al., 2010). For $\delta = 3$ we have instead of Eq. (14.13)

$$\vec{t}_1 = \frac{1}{\gamma}\begin{pmatrix} -p-q \\ 2q \end{pmatrix}, \quad \vec{t}_2 = \frac{1}{\gamma}\begin{pmatrix} 2q \\ -p+q \end{pmatrix}.$$

(14.15)

The number of atoms in the elementary supercell is given by the expression

$$N = 4\frac{\left|\vec{t}_1 \times \vec{t}_2\right|_z}{\left|\vec{a}_1 \times \vec{a}_2\right|_z} = \frac{12}{\delta\gamma^2}(3q^2 + p^2),$$

(14.16)

where the factor 4 originates from two layers and two atoms per elementary cell in each layer. Keeping in mind Eq. (14.10) one can rewrite Eq. (14.16) as

$$N = 4\frac{\left|\vec{t}_1 \times \vec{t}_2\right|_z}{\left|\vec{a}_1 \times \vec{a}_2\right|_z} = \frac{12}{\delta\gamma^2}\frac{p^2}{\sin^2\frac{\theta}{2}}.$$

(14.17)

At the same time, the number of atoms per moiré period (14.1) is

$$N_m = 4\kappa^2 = \frac{1}{\sin^2\frac{\theta}{2}},$$

(14.18)

with the same origin of factor 4. For some cases (e.g., $p = 1$ and odd q) $N = N_m$, but, in general, the supercell contains several moiré periods and $N > N_m$ (Shallcross et al., 2010).

Some illustrations of the superlattice are shown in Fig. 14.1.

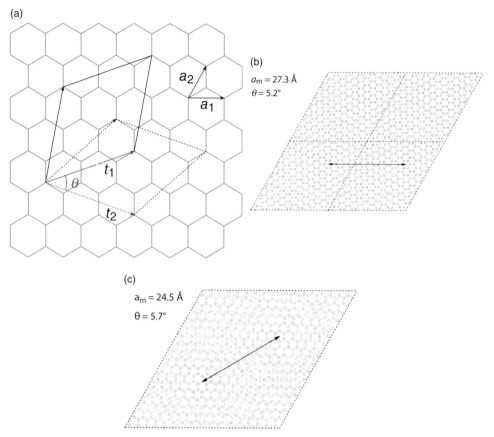

Fig. 14.1 (a) Schematic construction of supercell for twisted bilayer graphene or graphene on graphite. (b) Four supercells with one moiré pattern in each of them. (c) One supercell with three moiré patterns in it. The distance between moiré patterns a_m is indicated by the arrow.
(Reproduced with permission from van Wijk et al., 2015.)

Similar to the case of graphene on hBN (Section 13.2) one can expect that, generally speaking, atomic relaxation should be essential. Unfortunately, for the two-dimensional case, we do not yet have an analytical theory similar to Frank & van der Merwe (1949) for the one-dimensional situation. The issue was investigated by atomistic computer simulations (van Wijk et al., 2015). Some of the results are shown in Fig. 14.2. The obtained picture reminds vortex lattice with atomic displacements circulated around each center of the moiré pattern.

Atomic relaxation also results in the modulation of interlayer distances (Fig. 14.3). At the same time, the modulation of the in-layer bond lengths is negligibly small (van Wijk et al., 2015). Importantly, relaxation effects in the interlayer distances decrease with the increase in misorientation angle.

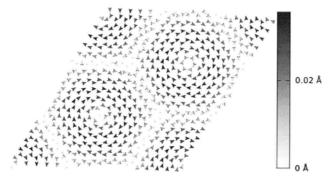

Fig. 14.2 Atomic displacements at the relaxation for graphene on graphite for the case corresponding to Fig. 14.1(c) ($a_m = 24.5$ Å).
(Reproduced with permission from van Wijk et al., 2015.)

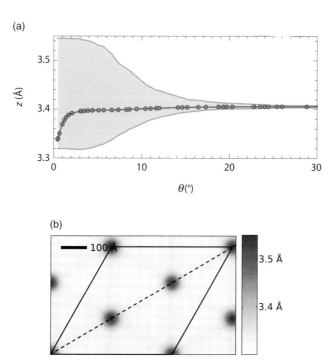

Fig. 14.3 (a) Average interlayer distance after relaxation for different samples of graphene on graphite as a function of misorientation angle; the shaded area indicates the spreading between minimum and maximum of the interlayer distance. (b) Distribution of interlayer distances for the case $n = 216$, $m = 1$, $\theta = 0.46°$, $a_m = 302.6$ Å.
(Reproduced with permission from van Wijk et al., 2015.)

14.2 Dodecagonal graphene quasicrystal

If cosθ is irrational, the equation (14.10) for integer n,m does not have any solutions, and we deal with incommensurate, or quasiperiodic case. The simplest example is $\theta = 30°$, $\cos\theta = \sqrt{3}/2$. In this case, twisted bilayer graphene forms a two-dimensional quasicrystal, which was experimentally realized by Yao et al. (2018) and Ahn et al. (2018).

As follows from Fig. 14.3(b), for large misorientation angles, the atomic relaxation effects can be neglected, and just bare (unrelaxed) structure should provide a very good approximation for the real bilayer. It is shown in Fig. 14.4. One can clearly see dodecagonal symmetry with the local rotation axis of 12th order ($p = 12$). Only rotational symmetries of the order of $p = 2, 3, 4$, and 6 are allowed by translational symmetry and may occur in crystals (Vonsovsky & Katsnelson, 1989).

Three-dimensional quasicrystals are well known and relatively well studied (Shechtman et al., 1984; Guyot, Kramer, & de Boissieu, 1991; Mermin, 1992; Goldman & Kelton, 1993; Lifshitz, 1997; Quilichini, 1997). In the case of twisted bilayer graphene we deal with two-dimensional quasicrystals. Contrary to the three-dimensional case, in this new system one can study ultrarelativistic particles in incommensurate external potential. Indeed, direct measurements of angle-resolved photoelectron spectra (ARPES) show 12 Dirac (conical) points connected by intervalley scattering (Umklapp processes). This field is very young, and now we only have first attempts to theoretically consider electronic structure of such

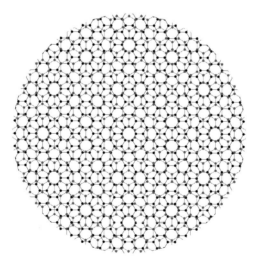

Fig. 14.4 Atomic structure for the bilayer graphene twisted by $\theta = 30°$; the atoms in two different layers are shown in black and gray, respectively.
(Courtesy Guodong Yu and Shengjun Yuan.)

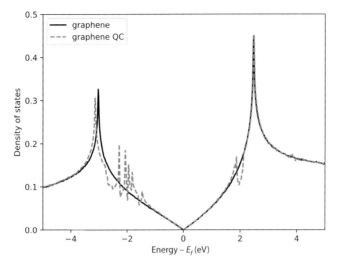

Fig. 14.5 Electronic density of states for the bilayer graphene twisted by $\theta = 30°$ (gray dashed line) in comparison with single-layer graphene (black solid line); the calculations are done within a tight-binding model.
(Courtesy Guodong Yu and Shengjun Yuan.)

systems (Moon, Koshino, & Son, 2019; Yu et al., 2019). In Fig. 14.5 the electronic density of states of graphene quasicrystal is shown, one can see numerous additional Van Hove singularities associated to the formation of Dirac points of next generations; compare with Section 13.5.

14.3 Electronic structure of twisted bilayer graphene

The electronic structure of twisted bilayer graphene can be considered in a close analogy to the case of graphene on hBN (Section 13.5), with obvious modifications. In particular, in Eq. (13.6), instead of second-order contributions of the interlayer hopping we have just the first-order term: $\hat{H}_\perp \frac{1}{E - \hat{H}_{BN}} \hat{H}_\perp \rightarrow \hat{H}_\perp$.

Essential simplifications happen for the case of a small misorientation angle θ in the continuum approximation, that is, for the electronic states in the vicinity of conical points K and K' Eq. (1.13) (Lopes dos Santos, Peres, & Castro Neto, 2007; Shallcross et al., 2010; Bistritzer & MacDonald, 2011). For small θ, intervalley scattering processes are negligible, and we can restrict ourselves to considering the vicinity of the K point only. The position of the K point in the second (rotated) layer is given by the vector

$$\vec{K}^\theta = \hat{R}(\theta)\,\vec{K} = \vec{K} + \Delta\vec{K}, \qquad (14.19)$$

where

$$\Delta \vec{K} \approx \theta \hat{z} \times \vec{K},$$ (14.20)

and $\hat{z}$ is the unit vector in z direction; Eq. (14.20) is valid in the first order in θ.

Then, without taking into account interlayer hopping, the Hamiltonian of two nonconnected graphene layers can be derived from Eq. (3.1) by the corresponding changes of the coordinate system:

$$\hat{H}_0 = \hbar v \sum_{\vec{k}} \left[\hat{\Phi}_{1\vec{k}}^+ \, \vec{\sigma} \left(\frac{\vec{k} + \Delta \vec{K}}{2} \right) \hat{\Phi}_{1\vec{k}} + \hat{\Phi}_{2\vec{k}}^+ \, \vec{\sigma}^\theta \left(\frac{\vec{k} - \Delta \vec{K}}{2} \right) \hat{\Phi}_{2\vec{k}} \right],$$ (14.21)

where

$$\hat{\Phi}_{i\vec{k}} = \hat{\Psi}_{i,\,\vec{k} \pm \Delta \vec{K}/2}$$ (14.22)

are spinor electron annihilation operators in the corresponding layer $i = 1,2$ and we shift the k-space $\vec{k} \to \vec{k} \pm \Delta \vec{K}/2$, with plus sign for the first layer and minus sign for the second layer,

$$\vec{\sigma}^\theta = \exp\left(i\theta\sigma^z/2\right) \vec{\sigma} \exp\left(-i\theta\sigma^z/2\right)$$ (14.23)

are the rotated Pauli matrices.

To model the interlayer-hopping Hamiltonian, we will take into account, following Lopes dos Santos, Peres, and Castro Neto (2007), only the hoppings from each site in layer 1 to the neighboring sites in layer 2 for each sublattice. The vectors $\vec{\rho}^{\alpha\beta}$ between the atom situated in the sublattice $\alpha = $ A, B of the first layer near the point $\vec{r}$ and the neighboring atom belonging to the sublattice $\beta = $ A, B, and therefore the corresponding matrix elements $t_\perp^{\alpha\beta}\left(\vec{r}\right) = t_\perp\left(\left|\vec{\rho}^{\alpha\beta}\right|\right)$, are dependent on $\vec{r}$, which is clearly seen in Fig. 14.1. The total single-electron Hamiltonian in this model takes the form (Lopes dos Santos, Peres, & Castro Neto, 2007)

$$\hat{H} = \hat{H}_0 + \sum_{\vec{k}\vec{G}} \left[\hat{\Phi}_{1\vec{k}}^+ \hat{t}_\perp\left(\vec{G}\right) \hat{\Phi}_{2\vec{k}} + h.c. \right],$$ (14.24)

where $\vec{G}$ are reciprocal vectors of the supercell, and $\hat{t}_\perp\left(\vec{G}\right)$ is the matrix with the matrix elements

$$t_\perp^{\alpha\beta}\left(\vec{G}\right) = \int d^2 r \, t_\perp^{\alpha\beta}\left(\vec{r}\right) \exp\left[i\vec{K}^\theta \vec{\rho}^{\alpha\beta}\left(\vec{r}\right) - i\vec{G}\vec{r}\right].$$ (14.25)

For small θ, further simplifications are possible (Lopes dos Santos, Peres, & Castro Neto, 2007; Bistritzer & MacDonald, 2011).

Now we can use the Hamiltonian (14.24) to find the effective energy spectrum for the layer 1 in the presence of interlayer hopping. It turns out that in the continuum model considered here, the only effect is the renormalization of the Fermi velocity, the spectrum remains isotropic and no gap is open. The perturbative result in $t_\perp$ for this renormalization reads (Lopes dos Santos, Peres, & Castro Neto, 2007):

$$\frac{\tilde{v}}{v} = 1 - 9\alpha^2, \ \alpha = \frac{t_\perp}{\hbar v \Delta K}. \tag{14.26}$$

A more accurate consideration within the consequent k·p perturbation theory gives the result also valid for $\alpha \leq 1$ (Bistritzer & MacDonald, 2011):

$$\frac{\tilde{v}}{v} = \frac{1 - 3\alpha^2}{1 + 6\alpha^2}. \tag{14.27}$$

For $\alpha = 1/\sqrt{3}$, the Fermi velocity tends to 0. Estimating $t_\perp$ from experimental data for graphene, one can find that it happens at the "magic angle" $\theta \approx 1.05°$; numerical analysis beyond Eq. (14.27) demonstrates existence of other magic angles at large α: $\theta \approx 0.5°$; 0.35°; 0.24°; 0.2°, etc. (Bistritzer & MacDonald, 2011). Other calculations give slightly different values of the largest magic angle, $\theta \approx 1.5°$ (Suárez Morell et al., 2010) and $\theta \approx 1.08°$ (Fang & Kaxiras, 2016).

Numerical calculations cited previously demonstrate that at the magic angles the electron velocity (almost) vanishes, not only at the conical points but along the whole lines, that is, a flat band is formed. Importantly, for undoped case, it lies exactly at the Fermi energy. Experimentally, a formation of the flat bands in magic-angle twisted bilayer graphene was studied by Kim et al. (2017).

This situation is very special. In principle, flat bands at the Fermi energy can originate from electron–electron interactions. This was postulated and studied phenomenologically, within Landau Fermi-liquid theory, in terms of a so-called "Fermi condensation" (Khodel & Shaginyan, 1990; Volovik, 1991; Nozieres, 1992; Zverev & Baldo, 1999). A specific microscopic scenario of the flat band formation due to many-body effects (closeness of the Fermi energy to Van Hove singularity in two-dimensional strongly correlated systems) was considered by Irkhin, Katanin, and Katsnelson (2002) and by Yudin et al. (2014); it was suggested in the first of these papers that this phenomenon can be important for physics of high-temperature superconductivity in cuprates. In the case of twisted bilayer graphene, we deal with the formation of flat electron bands already in single-electron approximation. This case is not unique, and sometimes flat bands are induced and protected by topological considerations (Heikkilä, Kopnin, & Volovik, 2011).

Irrespective to the origin of the flat band at the Fermi energy, its existence unavoidably assumes the crucial role of the many-body effects. The prototype example is the physics of quantum Hall systems (Prange & Girvin, 1987). Indeed, a formation of Landau levels instead of continuum single-electron spectrum projects kinetic energy to 0 and opens a way for numerous many-body instabilities, including formation of incompressible electron liquid, responsible for fractional quantum Hall effect, quantum Hall ferromagnetism, different types of charge ordering, etc. If we recall the discussion of the Stoner criterion of itinerant-electron magnetism in Section 12.1, the key point there was Eq. (12.21), meaning that for a very high density of states at the Fermi energy $N(\varepsilon_F)$ one can reoccupy electron states near the Fermi energy with a very small energy cost. The flat band means $N(\varepsilon_F) \to \infty$. Apart from ferromagnetism, a high density of states at the Fermi energy is favorable for superconductivity (Volovik, 2018) and for different types of lattice instability (Katsnelson, Naumov, & Trefilov, 1994). These different types of instabilities (and many others, such as spin density wave, charge density wave, etc.) influence one another and suppress competing instability channels, which results in a very complicated phase diagram (Irkhin, Katanin, & Katsnelson, 2001; Katanin & Kampf, 2003). The problem of interacting electrons with flat bands is extremely complicated, but we can be sure that *something* interesting happens.

And it does! Cao et al. (2018a) discovered an insulating behavior in twisted bilayer graphene at the magic angle $\theta \approx 1.1°$, which is supposedly caused by many-body effects (Mott insulator, Wigner crystal ...). In the doped case, the system becomes superconducting (Cao et al., 2018b). Currently this is a subject of very intensive experimental and especially theoretical studies.

15

Many-body effects in graphene

15.1 Screening and effective interactions

Most of the consideration in this book assumes the picture of noninteracting electrons (except some sections in Chapters 7, 8, and 12). The reason for this is that, as we already know after 15 years of graphene research, this picture describes, at least qualitatively and in many cases even quantitatively, basic electronic phenomena in graphene. At the same time, this fact itself requires a justification, since electron–electron interaction in graphene is by no means weak (see Eq. (7.90)); moreover, even at the Hartree–Fock level it can result in an essential reconstruction of the electron energy spectrum (Section 8.4). Here, we systematically discuss the role of electron–electron interactions in graphene. In this chapter we will work with the full electronic structure of honeycomb lattice rather than with the Dirac approximation. The reason is that many-body effects involve *virtual* electron–hole excitations, which are distributed over the whole band, even if the *real* electrons or holes live in the vicinity of the conical points. In particular, these virtual excitations determine screening of the effective interactions.

The simplest way to describe these effects is the use of random phase approximation (RPA), similarly to how we did it in Sections 7.6 and 7.7 for the case of Dirac electrons. Let us first introduce this approximation for the general case. Here we will follow Vonsovsky & Katsnelson (1989).

Let the single-electron Hamiltonian have eigenstates $\psi_\nu(\vec{r})$, the corresponding eigenenergies E_ν, and the equilibrium occupation numbers $f_\nu = f(E_\nu)$, where $f(E)$ is the Fermi function. Then, under the action of perturbation described by the potential energy

$$V\left(\vec{r},t\right) = V\left(\vec{r}\right)\exp\left(-i\omega t + \delta t\right)\big|_{\delta\to+0},\qquad(15.1)$$

the single-particle density matrix (2.170) will have, in the linear approximation in V, the nonequilibrium contribution $\hat{\rho}'\exp\left(-i\omega t + \delta t\right)\big|_{\delta\to+0}$ with

$$\rho'_{vv'} = \frac{f_v - f_{v'}}{E_{v'} - E_v + \hbar(\omega + i\delta)} V_{vv'} \tag{15.2}$$

(cf. Eq. (2.175)). The electron-density operator

$$\hat{N}\left(\vec{r}\right) = \delta\left(\vec{r} - \vec{r}'\right) \tag{15.3}$$

has matrix elements

$$N_{vv'} = \int d\vec{r}' \, \psi_v^*\left(\vec{r}'\right) \delta\left(\vec{r} - \vec{r}'\right) \psi_{v'}\left(\vec{r}'\right) = \psi_v^*\left(\vec{r}\right) \psi_{v'}\left(\vec{r}\right). \tag{15.4}$$

Then, using Eq. (2.176), one finds for the perturbation of electron density, the expression $\delta n(\vec{r}) \exp\left(-i\omega t + \delta t\right)_{\delta \to +0}$ where

$$\delta n\left(\vec{r}\right) = \int d\vec{r}' \, \Pi\left(\vec{r}, \vec{r}'\right) V\left(\vec{r}'\right), \tag{15.5}$$

$$\Pi\left(\vec{r}, \vec{r}'\right) = \sum_{vv'} \frac{f_v - f_{v'}}{E_{v'} - E_v + \hbar(\omega + i\delta)} \psi_{v'}^*\left(\vec{r}\right) \psi_v\left(\vec{r}\right) \psi_v^*\left(\vec{r}'\right) \psi_{v'}\left(\vec{r}'\right) \tag{15.6}$$

is the polarization operator (cf. Eq. (7.75) and (7.76)).

At last, we assume a purely electrostatic (Hartree-like) relation between the external potential $U(\vec{r}\ t)$ and the total potential $V(\vec{r}\ t)$, similar to Eq. (7.104):

$$V\left(\vec{r}\right) = U\left(\vec{r}\right) + e^2 \int d\vec{r}' \frac{\delta n\left(\vec{r}'\right)}{\left|\vec{r} - \vec{r}'\right|}. \tag{15.7}$$

Substituting Eq. (15.5) into Eq. (15.7) we obtain

$$V\left(\vec{r}\right) = \int d\vec{r}' \, \varepsilon^{-1}\left(\vec{r}, \vec{r}'\right) U\left(\vec{r}'\right), \tag{15.8}$$

where $\hat{\varepsilon}^{-1}$ is the inverse to the operator of dielectric permittivity

$$\varepsilon\left(\vec{r}, \vec{r}'\right) = \delta\left(\vec{r} - \vec{r}'\right) + e^2 \int d\vec{r}'' \frac{\Pi\left(\vec{r}'', \vec{r}'\right)}{\left|\vec{r} - \vec{r}''\right|}. \tag{15.9}$$

In principle, this function determines the screening of *external* potential rather than the potential of interelectron Coulomb interaction

$$v_C\left(\vec{r}, \vec{r}'\right) = \frac{e^2}{\left|\vec{r} - \vec{r}'\right|}; \tag{15.10}$$

however, within the RPA we neglect this difference and determine the effective potential of interelectron interaction as

$$\hat{v}_{eff} = \hat{\varepsilon}^{-1}\hat{v}_C \qquad (15.11)$$

(for details, see Giuliani & Vignale, 2005).

For Dirac electrons and for the static case ($\omega = 0$), the Coulomb potential is weakened by the dielectric constant (7.89). The first-principle calculation for the case of finite wave vectors q (van Schilfgaarde & Katsnelson, 2011) shows that the screening drops quite quickly with the wave vector (Fig. 15.1).

In the rest of this chapter we will discuss many-body effects in graphene for the model, including only p_z electronic states. This model contains only one electron state per site and can be described, in the simplest approximation, by the Hamiltonian

$$\hat{H} = \sum_{ij\sigma}{}' t_{ij}\hat{c}_{i\sigma}^{+}\hat{c}_{j\sigma} + U\sum_{i}\hat{n}_{i\uparrow}\hat{n}_{i\downarrow} + \frac{1}{2}\sum_{ij}{}' V_{ij}\hat{n}_i\hat{n}_j, \qquad (15.12)$$

where $\hat{c}_{i\sigma}^{+}, \hat{c}_{i\sigma}$ are electron creation and annihilation operators on site i with spin projection σ

$$\hat{n}_{i\sigma} = \hat{c}_{i\sigma}^{+}\hat{c}_{i\sigma}, \quad \hat{n}_i = \sum_{\sigma}\hat{n}_{i\sigma} \qquad (15.13)$$

are electron occupation number operators, and sum with prime means the summation over $i \neq j$. Without the last term, this model coincides with the Hubbard model (12.1) with Hubbard on-site interaction U; V_{ij} are intersite interaction parameters. Note that even in the single-band approximation, the full many-body Hamiltonian

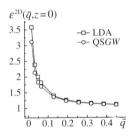

Fig. 15.1 Fourier component of two-dimensional static dielectric permittivity describing a screening in graphene plane ($z = 0$) as a function of dimensionless wave vector $\bar{q} = qa_{lat}/2\pi$ ($a_{lat} = \sqrt{3}a$ is the lattice constant for graphene) along (10) direction. Squares and circles show different methods of electronic structure calculations (local density approximation versus quasiparticle self-consistent GW); one can see that the results are insensitive to this difference. (Reproduced with permission from van Schilfgaarde & Katsnelson, 2011.)

contains many more terms, such as exchange interactions, many-body contribution to hopping, etc. (Schubin & Wonsowski, 1934; Vonsovsky & Katsnelson, 1979).

The point is that when we eliminate from the Hamiltonian all other states except p_z, we have to take into account their effect on the interactions within the p_z band via screening. Currently, the only practical way to do this from the first principles is to use RPA, that is, Eq. (15.6), (15.9), and (15.11). However, we should exclude from the double sum in Eq. (15.6) the transitions from p_z states to p_z states because these kinds of processes will be taken into account when we treat the Hamiltonian (15.12) by other methods (such as quantum Monte Carlo, see Section 15.4), and we want to avoid double counting. This corresponds to the constrained RPA (cRPA) method (Aryasetiawan et al., 2004). The corresponding calculations for the case of graphene were performed by Wehling et al. (2011).

First of all, we cannot expect any change of asymptotics of the effective potential $V_{ij} = V(r_{ij})$ in the limit of large distances $r_{ij} \to \infty$. Indeed, after elimination of the virtual transitions around conical point we have $\Pi(q, \omega = 0) \propto q^2$ at $q \to 0$ (see Eq. (7.94)), and $v_c(q) \propto 1/q$ (see Eq. (7.79)); therefore, the dielectric permittivity $\varepsilon(q, \omega = 0) \to 1$ at $q \to 0$, which means inefficient screening at large distances. This can also be proven in a purely phenomenological way, within the electrodynamics of continuous media (Emelyanenko & Boinovich, 2008; Wehling et al., 2011).

At the same time, for small distances the screening effect is quite essential. According to Wehling et al. (2011), it decreases the parameter U from its bare (that is, Hartree–Fock) value 17.0–9.3 eV, the nearest-neighbor interaction parameter V_{01} from 8.5 eV to 5.5 eV, and the next-nearest-neighbor interaction parameter V_{02} from 5.4 eV to 4.1 eV. Starting from the third neighbors ($r \geq 2a$) the static screened potential (Wehling et al., 2011) can be approximated by a simple formula (Astrakhantsev et al., 2018)

$$V(r) = \frac{A}{(r/a) + C},\qquad (15.14)$$

with $A = e^2/a = 10.14$ eV, $C = 0.82$.

The effective interaction (15.11) is actually frequency dependent. However, this dependence is not very essential for the frequency range of the order of 10 eV, that is, roughly half the width of the p_z band (Wehling et al., 2011). Further, we will discuss only statically screened effective interaction.

The screening of the effective p_z–p_z interaction by other graphene bands does not look very strong (it changes from the factor of the order of 2 for the on-site interaction parameter U and decreases to 1 with the interatomic distance increase). Nevertheless, as we will see in Section 15.4, it results in important physical consequences.

15.2 Mapping onto the Hubbard model

The simpler the Hamiltonian, the more accurately one can study its properties. In particular, the Hubbard model (12.1) is definitely simpler than the more complete model (15.12), and for many applications it would be nice to have the Hubbard model for graphene derived from the first principles. At the same time, long-range interactions in graphene are important and by no means small, and we cannot just neglect the terms with V_{ij}. The effective Hubbard model for graphene was built by Schüler et al. (2013) by the use of Peierls–Feynman–Bogolyubov variational principle (Peierls, 1938; Feynman, 1955; Bogolyubov, 1958; Feynman, 1972). This is a very useful tool, providing a general method how to map in an optimal way, a more complicated Hamiltonian to a simpler one. Since it is not very well known in the condensed-matter community, we present here a proof of this variational principle following a recent book by Czycholl (2008).

We want to map the Hamiltonian $\hat{H}$ on the Hamiltonian $\hat{H}^*$, dependent on some trial parameters. We are going to find these parameters from a minimization of a trial free energy of the system. The equilibrium (Gibbs) density matrices for these systems are

$$\hat{\rho} = e^{\beta(F - \hat{H})}, \quad \hat{\rho}^* = e^{\beta(F^* - \hat{H}^*)}, \tag{15.15}$$

where β is the inverse temperature, and F and F^* are the corresponding free energies. Let us assume that the density matrices (15.15) are diagonal in the basis $|n\rangle$ and $|a\rangle$, respectively, with the eigenvalues ρ_n and ρ_a^*. Of course, the set $|n\rangle$ diagonalizes the Hamiltonian $\hat{H}$, and the set $|a\rangle$ diagonalizes the Hamiltonian $\hat{H}^*$.

One can prove an important inequality:

$$Tr\hat{\rho}^* \ln \hat{\rho} \leq Tr\hat{\rho}^* \ln \hat{\rho}^*. \tag{15.16}$$

Indeed,

$$Tr\hat{\rho}^*(\ln \hat{\rho} - \ln \hat{\rho}^*) = \sum_a \rho_a^*(\ln \langle a|\hat{\rho}|a\rangle - \ln \rho_a^*). \tag{15.17}$$

Taking into account that the set $|n\rangle$ is complete, one can write

$$\ln \langle a|\hat{\rho}|a\rangle = \sum_n |\langle a|n\rangle|^2 \ln \rho_n. \tag{15.18}$$

Substituting Eq. (15.18) into Eq. (15.17) we have

$$Tr\hat{\rho}^*(\ln \hat{\rho} - \ln \hat{\rho}^*) = \sum_{an} \rho_a^* |\langle a|n\rangle|^2 \ln \frac{\rho_n}{\rho_a^*} \leq \sum_{an} \rho_a^* |\langle a|n\rangle|^2 \left(\frac{\rho_n}{\rho_a^*} - 1\right)$$

$$= \sum_n \rho_n - \sum_a \rho_a^* = 1 - 1 = 0, \tag{15.19}$$

where we take into account that for any positive x, one has $\ln x \leq x - 1$. This proves Eq. (15.16).

Substituting Eq. (15.15) into Eq. (15.16), we have a very important inequality

$$F \leq F^* + \left\langle \hat{H} - \hat{H}^* \right\rangle^*, \tag{15.20}$$

where $\langle \ldots \rangle^*$ means the average with the operator $\hat{\rho}^*$. One can choose the parameters entering $\hat{H}^*$ to make the right-hand side of Eq. (15.20) as small as possible. Thus, we will have the best estimation of the free energy F from above. This is the variational principle that we wanted to prove. Note that if we choose $\hat{H}^* = \hat{H}_0$, we come to the conclusion that the first-order perturbation correction always give a rigorous estimation for the free energy from above.

Now let us proceed with the Hamiltonian (15.12). As a trial Hamiltonian, we will use the Hubbard model (12.1) but with some effective on-site interaction constant U^*:

$$\hat{H}^* = \sideset{}{'}\sum_{ij\sigma} t_{ij} \hat{c}_{i\sigma}^+ \hat{c}_{j\sigma} + U^* \sum_i \hat{n}_{i\uparrow} \hat{n}_{i\downarrow}. \tag{15.21}$$

In principle, we can also assume a renormalization of the hopping parameters $t_{ij} \to t_{ij}^*$, but let us try to keep the scheme as simple as possible (for a more general consideration, see in 't Veld, 2019). Substituting Eq. (15.12) and (15.21) into Eq. (15.20) we have

$$F \leq F_t = F^* + (U^* - U) \sum_i \left\langle \hat{n}_{i\uparrow} \hat{n}_{i\downarrow} \right\rangle^* + \frac{1}{2} \sideset{}{'}\sum_{ij} V_{ij} \left\langle \hat{n}_i \hat{n}_j \right\rangle.^* \tag{15.22}$$

To find the best possible value U^*, we have to minimize the right-hand side of Eq. (15.22). The necessary condition is

$$\frac{dF_t}{dU^*} = \frac{dF^*}{dU^*} - \sum_i \left\langle \hat{n}_{i\uparrow} \hat{n}_{i\downarrow} \right\rangle^* + (U^* - U) \frac{d}{dU^*} \sum_i \left\langle \hat{n}_{i\uparrow} \hat{n}_{i\downarrow} \right\rangle^*$$
$$+ \frac{1}{2} \sideset{}{'}\sum_{ij} V_{ij} \frac{d}{dU^*} \left\langle \hat{n}_i \hat{n}_j \right\rangle^* = 0. \tag{15.23}$$

According to the Hellmann–Feynman theorem, for any Hamiltonian dependent on a parameter

$$\frac{dF(\lambda)}{d\lambda} = \left\langle \frac{d\hat{H}(\lambda)}{d\lambda} \right\rangle, \tag{15.24}$$

therefore we have:

$$\frac{dF^*}{dU^*} = \sum_i \langle \hat{n}_{i\uparrow} \hat{n}_{i\downarrow} \rangle^* \tag{15.25}$$

and

$$U^* = U + \frac{1}{2} \sum_{ij}' V_{ij} \frac{\partial_{U^*} \langle \hat{n}_i \hat{n}_j \rangle^*}{\sum_l \partial_{U^*} \langle \hat{n}_{l\uparrow} \hat{n}_{l\downarrow} \rangle^*}, \tag{15.26}$$

where $\partial_{U^*} = d/dU^*$. According to the particle number conservation

$$\sum_i \hat{n}_{i\uparrow} \hat{n}_{i\downarrow} = \sum_i \hat{n}_{i\uparrow} \left(N_e - \hat{n}_{i\downarrow} - \sum_{j \neq i} \hat{n}_j \right) = N_e N_{e\uparrow} - N_{e\uparrow} - \sum_{ij}' \hat{n}_{i\uparrow} \hat{n}_j, \tag{15.27}$$

where we take into account that $\hat{n}_{i\uparrow}^2 = \hat{n}_{i\uparrow}$, $N_{e\uparrow} = \sum_i \hat{n}_{i\uparrow}$ is the total number of spin-up electrons. Equivalently, we have

$$\sum_i \hat{n}_{i\uparrow} \hat{n}_{i\downarrow} = N_e N_{e\downarrow} - N_{e\downarrow} - \sum_{ij}' \hat{n}_{i\downarrow} \hat{n}_j. \tag{15.28}$$

Summing up Eq. (15.27) and (15.28) we obtain

$$\sum_i \hat{n}_{i\uparrow} \hat{n}_{i\downarrow} = \frac{N_e(N_e - 1)}{2} - \frac{1}{2} \sum_{ij}' \hat{n}_i \hat{n}_j. \tag{15.29}$$

At last, substituting Eq. (15.29) into Eq. (15.26) we find

$$U^* = U - \frac{\sum_{ij}' V_{ij} \partial_{U^*} \langle \hat{n}_i \hat{n}_j \rangle^*}{\sum_{ij}' \partial_{U^*} \langle \hat{n}_i \hat{n}_j \rangle^*}, \tag{15.30}$$

which means that the effective Hubbard-U parameter is smaller than the initial U-value by an averaged intersite interaction. Interestingly, if we assume that the correlation function $\langle \hat{n}_i \hat{n}_j \rangle^*$ is nonvanishing only at the neighboring sites, we have a very simple result:

$$U^* = U - V_{01}. \tag{15.31}$$

Calculations for graphene (as well as for some other sp materials) were performed by Schüler et al. (2013). The density-density correlation function in the Hubbard model $\langle \hat{n}_i \hat{n}_j \rangle^*$ was calculated using the quantum Monte Carlo method (see later, Section 15.4). As a result, the effective Hubbard parameter for graphene is $U^* \approx 1.6t$, instead of initial value $U \approx 3.6t$. Surprisingly, this value is quite close to the naïve estimate (15.31). It should be compared with the total bandwidth $W = 6t$ (Fig. 7.2): $U^* \approx W/4$. In this sense, graphene should be considered as a moderately correlated system.

15.3 Renormalization of the electron spectrum beyond Dirac approximation

The long-range character of interelectron interactions already leads to an essential renormalization of the Fermi velocity at the Hartree–Fock level (Section 8.4). In Dirac approximation we can calculate this renormalization with only logarithmic accuracy; see Eq. (8.93). Astrakhantsev, Braguta, and Katsnelson (2015) have performed the calculations of the Hartree–Fock effective Hamiltonian for the hexagonal lattice. Instead of Eq. (8.91), the additional contribution to the single-electron Hamiltonian is

$$\hat{h}\left(\vec{k}\right) = \frac{\hat{\sigma}_x}{2} \sum_{\vec{k}'} \frac{S\left(\vec{k}'\right)}{\left|S\left(\vec{k}'\right)\right|} \tanh \frac{t\left|S\left(\vec{k}'\right)\right|}{2T} V\left(\vec{k} - \vec{k}'\right) \equiv \delta h\left(\vec{k}\right)\hat{\sigma}_x, \qquad (15.32)$$

where $S(\vec{k})$ is given by Eq. (1.15), the bare energy spectrum by Eq. (1.16) (we take into account only the nearest-neighbor hopping t), $V(\vec{k})$ is the Fourier component of static electron-electron interaction, and we put chemical potential to the neutrality point: $\mu = 0$.

The renormalized electron spectrum is determined by the expression

$$E\left(\vec{k}\right) = \pm\left|tS\left(\vec{k}\right) + h\left(\vec{k}\right)\right|. \qquad (15.33)$$

The computational results are shown in Fig. 15.2.

In the vicinity of Dirac points, Eq. (15.33) results in a renormalization of the Fermi velocity

$$v^R(T)/v = 1 + \frac{e^2}{4\hbar\varepsilon} \ln\left(\frac{\Lambda}{T}\right), \qquad (15.34)$$

where ε is the dielectric constant (7.89), and for the case of graphene on substrate with dielectric constant ε_s, we have $\varepsilon_{ext} = \frac{1+\varepsilon_s}{2}$ (see Eq. (7.84)). For the case of graphene at hBN, the parameter Λ is equal to 3.2 eV for the case of bare Coulomb interaction and 2.4 eV for the case of screened Coulomb interaction (Astrakhantsev, Braguta, & Katsnelson, 2015).

The results can easily be generalized for the case of finite doping and compared with available experimental data for graphene on hBN (Elias et al., 2011; Yu et al., 2013). The agreement is quite good. This is not surprising since, due to a relatively

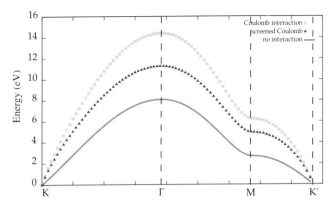

Fig. 15.2 Energy spectrum of electrons with Coulomb (squares), screened Coulomb (triangles) interactions, and no interaction (solid line) for $T = 0.1$ eV. (Reproduced with permission from Astrakhantsev, Braguta, & Katsnelson, 2015.)

large ε_{ext} in this situation, interelectron interaction can be considered perturbatively. For the case of freely suspended graphene, with a large interaction parameter (7.90), much more sophisticated methods should be used. We will consider them in the next section.

15.4 Quantum Monte Carlo results

What shall we do to solve quantum many-body (or quantum field) problem without explicit small parameters? One of the most powerful modern techniques is the so called quantum Monte Carlo (QMC) approach. A systematic derivation and even explanation of this method goes far beyond the mathematical level adopted in this book, therefore we just refer to monographs by Creutz (1983) and Gubernatis, Kawashima, and Werner (2016) and to reviews by De Raedt and Lagendijk (1985) and Foulkes et al. (2001). In particular, this method is the main technical tool applied in quantum chromodynamics, a theory of strongly interacting quarks and gluons (Creutz, 1983). The same approach was applied to graphene (Hands & Strouthos, 2008; Drut & Lähde, 2009a; 2009b; Ulybyshev et al., 2013; Ulybyshev & Katsnelson, 2015; Boyda et al., 2016; Astrakhantsev et al., 2018). Importantly, for the case of zero doping ($\mu = 0$) and when taking into account only the nearest-neighbor hopping, due to electron–hole symmetry, QMC calculations for graphene are free of fermionic sign problem, which restricts an accuracy of QMC calculations in a general case (De Raedt & Lagendijk, 1985).

Already the first calculations (Drut & Lähde, 2009a) led to a dramatic conclusion that freely suspended graphene ($\varepsilon_{ext} = 1$) is not semimetal: Many-body effects transform it to an antiferromagnetic insulator with spontaneously broken chiral symmetry, that is, symmetry between sublattices. The calculations were performed assuming a bare Coulomb interelectron interaction. Ulybyshev et al. (2013) has demonstrated that this conclusion is wrong, with the screening of the Coulomb interaction considered in Section 15.1 playing a crucial role. It turns out that freely suspended graphene is situated at the conducting side of the semimetal–insulator transition, in a full agreement with available experimental evidences. It becomes an insulator if one increases the electron–electron interactions by a factor approximately 1.4; this is more or less the screening effect for intermediate distances. Ulybyshev and Katsnelson (2015) have shown that graphene can be made insulating by creating about 1% of empty sites which physically means either vacancies or sp^3 impurities like hydrogen or fluorine, as was discussed in Sections 6.5 and 6.6.

QMC method, being formally exact, allows us to clarify a controversial issue on many-body renormalization of optical conductivity discussed in Section 7.3. Boyda et al. (2016) have demonstrated that this renormalization is either absent or numerically small; at least, theories predicted strong renormalization are in a clear contradiction with the QMC computational results (Fig. 15.3).

At last, QMC calculations were used by Astrakhantsev et al. (2018) to study the screening of static potential. The results are surprising: Despite that the interaction constant (7.90) for freely suspended graphene is large, the RPA expression (7.89) agrees very well with the QMC results. This is strange indeed, since the straight-forward calculation of the higher-order corrections to the dielectric constant (Sodemann & Fogler, 2012) predict quite strong deviations:

$$\varepsilon = 1 + \frac{\pi}{2}\alpha + 0.778\alpha^2 + \dots \qquad (15.35)$$

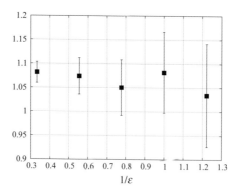

Fig. 15.3 Renormalization of optical conductivity in comparison with its non-interacting value (7.36). The factor $1/\varepsilon$ determines the enhancement of interelec-tron interaction potential: $V(r) \rightarrow V(r)/\varepsilon$. One can see that the dependence on the interaction strength is rather weak.
(Adapted from Boyda et al., 2016.)

Anyway, it means that RPA works for graphene much better than one could naïvely expect; however, the reasons for this good luck are currently unclear.

15.5 Many-body renormalization of minimal conductivity

We see that the interelectron–electron interaction cannot destroy semimetallic state in graphene via spontaneously broken chiral symmetry. However, this is not the final answer on the question whether an ideal, totally isolated, undoped graphene will be conducting at zero temperature or not. As discussed in Chapter 3, its finite conductivity arises from a very special mechanism, that is, electron tunneling via zero modes of Dirac operator represented by evanescent waves (Fig. 3.1). In many-body systems, any tunneling is, generally speaking, fragile. If we have a two-level quantum mechanical system interacting with an environment (such as phonon thermal bath), its tunneling probability between two states is suppressed by an interaction with the environment (Caldeira & Leggett, 1983). If we have a system of *interacting* electrons, the electron–hole continuum plays the role of dissipative environment (Guinea, 1984). Guinea and Katsnelson (2014) applied this consideration to the problem of minimal conductivity in graphene. Again, the mathematics required to follow this theory are beyond the scope this book (everything is based on Feynman path integral formulation of quantum mechanics); therefore, we will present here only the initial formula and the final answer.

As a result of virtual electron–hole excitations, the probability amplitude of electron motion along the trajectory $x(\tau)$ is multiplied by the suppression factor e^{-S}, where

$$S = \frac{1}{2} \int\limits_{-\infty}^{\infty} d\tau \int\limits_{0}^{\beta} d\tau' \int\limits_{-\infty}^{\infty} \frac{dq}{2\pi} \exp\left[iq(x(\tau) - x(\tau'))\right] V^2(q) \hat{T}\left[\hat{n}_q(\tau)\hat{n}_{-q}(\tau')\right], \quad (15.36)$$

$V(q)$ is the one-dimensional Fourier component of electron–electron interaction, β is inverse temperature, $\hat{n}_q(\tau)$ is the density operator, and $\hat{T}$ is the symbol of time ordering. Using the fluctuation-dissipation theorem (Zubarev, 1974; Vonsovsky & Katsnelson, 1989; Giuliani & Vignale, 2005) the time-ordered correlator $\hat{T}\left[\hat{n}_q(\tau)\hat{n}_{-q}(\tau')\right]$ can be expressed via imaginary part of dielectric function; for the case of undoped graphene, the latter is given by Eq. (7.83) and (7.86).

Due to the long-range character of Coulomb interaction, the integral over q in Eq. (15.36) is logarithmically divergent at $q \to 0$ and $\beta \to \infty$. For the geometry considered in Section 3.2 and for zero temperature, the result reads (Guinea & Katsnelson, 2014)

$$S = \frac{L_x}{8\pi L_y} \frac{\alpha^2}{4\sqrt{2} + \alpha} \ln \frac{L_x}{a} \equiv \varsigma \ln \frac{L_x}{a}, \tag{15.37}$$

where $\alpha = e^2/\hbar v \varepsilon$. This means that the tunneling probability (3.15) is multiplied by the factor $e^{-2S} = (a/L_x)^{2\varsigma}$.

If temperature is not too low,

$$T > \frac{\hbar v}{L_x}, \tag{15.38}$$

the divergence is cut by the finite β rather than by the finite width of the sample, and $\ln \frac{L_x}{a} \rightarrow \ln \frac{t}{T}$ in Eq. (15.37). As a result, the conductance of the sample (3.16) and therefore the conductivity is estimated as

$$\sigma \approx \frac{e^2}{h} \left(\frac{T}{t} \right)^{2\varsigma}. \tag{15.39}$$

The power-law temperature dependence of the conductivity at the neutrality point was observed in an ultraclean graphene sample by Amet et al. (2013). The issue definitely requires more experimental and theoretical studies but it can be that the many-body effects do transform semimetallic graphene to insulating but without broken symmetry and gap formation, due to suppression of the transport via evanescent waves.

References

Abanin, D. A., Lee, P. A., & Levitov, L. S. (2006). *Phys. Rev. Lett.* **96**, 176803

Abanin, D. A., Morozov, S. V., Ponomarenko, L. A., et al. (2011). *Science* **332**, 328

Abanin, D. A., Shytov, A. V., Levitov, L. S., & Halperin, B. I. (2009). *Phys. Rev. B* **79**, 035304

Abergel, D. S. L., Apalkov, V., Berashevich, J., Ziegler, K., & Chakraborty, T. (2010). *Adv. Phys.* **59**, 261

Abergel, D. S. L., Russel, A., & Fal'ko, V. I. (2007). *Appl. Phys. Lett.* **91**, 063125

Abraham, F. F., & Nelson, D. R. (1990). *Science* **249**, 393

Abramowitz, M., & Stegun, L. A. (1964). *Handbook of Mathematical Functions.* New York: Dover

Abrikosov, A. A. (1988). *Fundamentals of the Theory of Metals.* Amsterdam: North Holland

(1998). *Phys. Rev. B* **58**, 2788

Acun, A., Zhang, L., Bampoulis, P., et al. (2015). *J. Phys.: Condens. Matter* **27**, 443002

Adam, S., Hwang, E., Galitski, V. M., & Das Sarma, S. (2007). *Proc. Natl. Acad. Sci. USA* **104**, 18392

Adhikari, S. K. (1986). *Am. J. Phys.* **54**, 362

Aharonov, Y., & Bohm, D. (1959). *Phys. Rev.* **115**, 485

Aharonov, Y., & Casher, A. (1979). *Phys. Rev. A* **19**, 2461

Ahn, S. J., Moon, P., Kim, T.-H., et al. (2018). *Science* **361**, 782

Akhiezer, A. I., & Peletminskii, S. V. (1981). *Methods of Statistical Physics.* Oxford: Pergamon

Akhmerov, A. R., & Beenakker, C. W. J. (2008). *Phys. Rev. B* **11**, 085423

Aleiner, I. L., & Efetov, K. B. (2006). *Phys. Rev. Lett.* **97**, 236801

Alonso-Gonzáles, P., Nikitin, A. Y., Gao, Y., et al. (2017). *Nature Nanotech.* **12**, 31

Altland, A. (2002). *Phys. Rev. B* **65**, 104525

Altshuler, B. L., Khmelnitskii, D., Larkin, A. I., & Lee, P. A. (1980). *Phys. Rev. B* **22**, 5142

Amet, F., Williams, J. R., Watanabe, K., Taniguchi, T., & Goldhaber-Gordon, D. (2013). *Phys. Rev. Lett.* **110**, 216601

Amorim, B., & Guinea, F. (2013). *Phys. Rev. B* **88**, 115418

Anderson, P. W. (1958). *Phys. Rev.* **112**, 900

(1970). *J. Phys. C* **3**, 2346

Ando, T. (2006). *J. Phys. Soc. Japan* **75**, 074716

Ando, T., Fowler, A. B., & Stern, F. (1982). *Rev. Mod. Phys.* **54**, 437

Ando, T., Nakanishi, T., & Saito, R. (1998). *J. Phys. Soc. Japan* **67**, 2857

Ando, T., Zheng, Y., & Suzuura, H. (2002). *J. Phys. Soc. Japan* **71**, 1318

Aronovitz, J. A., & Lubensky, T. C. (1988). *Phys. Rev. Lett.* **60**, 2634

Aryasetiawan, F., Imada, M., Georges, A., et al. (2004). *Phys. Rev. B* **70**, 195104

Astrakhantsev, N. Yu., Braguta, V. V., & Katsnelson, M. I. (2015). *Phys. Rev. B* **92**, 245105

Astrakhantsev, N. Yu., Braguta, V. V., Katsnelson, M. I., Nikolaev, A. A., & Ulybyshev, M. V. (2018). *Phys. Rev. B* **97**, 035102

Atiyah, M. F., Patodi, V. K., & Singer, I. M. (1976). *Math. Proc. Cambridge Philos. Soc.* **79**, 71

Atiyah, M. F., & Singer, I. M. (1968). *Ann. Math.* **87**, 484
    (1984). *Proc. Natl. Acad. Sci. USA* **81**, 2597

Auslender, M. I., & Katsnelson, M. I. (1982). *Teor. Mat. Fiz.* **51**, 436
    (2007). *Phys. Rev. B* **76**, 235425

Balandin, A. A. (2011). *Nature Mater.* **10**, 569

Balandin, A. A., Chosh, S., Bao, W., et al. (2008). *Nano Lett.* **8**, 902

Balatsky, A. V., Vekhter, I., & Zhu, J.-X. (2006). *Rev. Mod. Phys.* **78**, 373

Balescu, R. (1975). *Equilibrium and Nonequilibrium Statistical Mechanics*. New York: Wiley

Bandurin, D. A., Shytov, A. V., Levitov, L. S., et al. (2018). *Nature Commun.* **9**, 4533

Bandurin, D. A., Torre, I., Kumar, R. K., et al. (2016). *Science* **351**, 1055

Bao, W., Miao, F., Chen, Z., et al. (2009). *Nature Nanotech.* **4**, 562

Bardarson, J. H., Medvedeva, M. V., Tworzydlo, J., Akhmerov, A. R., & Beenakker, C. W. J. (2010). *Phys. Rev. B* **81**, 121414

Bardarson, J. H., Tworzydlo, J., Brower, P. W., & Beenakker, C. W. J. (2007). *Phys. Rev. Lett.* **99**, 106801

Basko, D. M. (2008). *Phys. Rev.* **78**, 115432

Basov, D. N., Fogler, M. M., & García de Abajo, F. J. (2016). *Science* **354**, 195

Bassani, F., & Pastori Parravicini, G. (1975). *Electronic States and Optical Transitions in Solids*. Oxford: Pergamon

Beenakker, C. W. J. (2008). *Rev. Mod. Phys.* **80**, 1337

Beenakker, C. W. J., & Büttiker, M. (1992). *Phys. Rev. B* **46**, 1889

Beenakker, C. W. J., & van Houten, H. (1991). *Solid State Phys.* **44**, 1

Beffara, V. (2004). *Ann. Probab.* **32**, 2606

Belenkii, G. L., Salaev, E. Yu., & Suleimanov, R. A. (1988). *Usp. Fiz. Nauk.* **155**, 89

Bellissard, J., van Elst, A., & Schulz-Baldes, H. (1994). *J. Math. Phys.* **35**, 5373

Bena, C., & Kivelson, S. A. (2005). *Phys. Rev. B* **72**, 125432

Berestetskii, V. B., Lifshitz, E. M., & Pitaevskii, L. P. (1971). *Relativistic Quantum Theory, vol. 1*. Oxford: Pergamon

Berry, M. V. (1984). *Proc. R. Soc. (London) A* **392**, 45

Berry, M. V., & Mondragon, R. J. (1987). *Proc. R. Soc. (London) A* **412**, 53

Bhandary, S., Eriksson, O., Sanyal, B., & Katsnelson, M. I. (2010). *Phys. Rev. B* **82**, 165405

Binnig, G., & Rohrer, H. (1987). *Rev. Mod. Phys.* **59**, 615

Bistritzer, R., & MacDonald, A. H. (2011). *Proc. Natl. Acad. Sci. USA* **108**, 12233

Biswas, R. R., Sachdev, S., & Son, D. T. (2007). *Phys. Rev. B* **76**, 205122

Bjorken, J. D., & Drell, S. D. (1964). *Relativistic Quantum Mechanics*. New York: McGraw-Hill

Blake, P., Hill, E. W., Castro Neto, A. H., et al. (2007). *Appl. Phys. Lett.* **91**, 063124

Blanter, Ya. M., & Büttiker, M. (2000). *Phys. Rep.* **336**, 1

Blees, M. K., Barnard, A. W., Rose, P. A., et al. (2015). *Nature* **524**, 204

Blount, E. I. (1962). *Phys. Rev.* **126**, 1636

Bocquet, L., & Barrat, J.-L. (2007). *Soft Matter* **3**, 685

Bøggild, P., Caridad, J. M., Stampfer, C., et al. (2017). *Nature Commun.* **8**, 15783

Bogolyubov, N. N. (1958). *Doklady AN SSSR* **119**, 244

Bokdam, M., Amlaki, T., Brocks, G., & Kelly, P. J. (2014). *Phys. Rev. B* **89**, 201404(R)

Bohr, A., & Mottelson, B. R. (1969). *Nuclear Structure, vol. 1*. New York: Benjamin

Bolotin, K. I., Ghahari, F., Shulman, M. D., Stormer, H. L., & Kim, P. (2009). *Nature* **462**, 196

Bolotin, K. I., Sikes, K. J., Jiang, Z., et al. (2008). *Solid State Commun.* **146**, 351

Booth, T. J., Blake, P., Nair, R. R., et al. (2008). *Nano Lett.* **8**, 2442

Born, M., & Wolf, E. (1980). *Principles of Optics*. Oxford: Pergamon

Boukhvalov, D. W., & Katsnelson, M. I. (2008). *Nano Lett.* **8**, 4373

   (2009a). *J. Phys.: Condens. Matter* **21**, 344205

   (2009b). *J. Phys. Chem. C* **113**, 14176

   (2009c). *Eur. Phys. J. B* **68**, 529

   (2011). *ACS Nano.* **5**, 2440

Boukhvalov, D. W., Katsnelson, M. I., & Lichtenstein, A. I. (2008). *Phys. Rev. B* **77**, 035427

Bowick, M. J., Košmrlj, A., Nelson, D. R., & Sknepnek, R. (2017). *Phys. Rev. B* **95**, 104109

Boyda, D. L., Braguta, V. V., Katsnelson, M. I., & Ulybyshev, M. V. (2016). *Phys. Rev. B* **94**, 085421

Brandt, N. B., Chudinov, S. M., & Ponomarev, Ya. G. (1988). *Semimetals: Graphite and Its Compounds*. Amsterdam: North Holland

Brar, V. W., Decker, R., Solowan, H.-M., et al. (2011). *Nature Phys.* **1**, 43

Brau, F., Vandeparre, H., Sabbah, A., et al. (2011). *Nature Phys.* **1**, 56

Braun, O. M., & Kivshar, Y. S. (2004). *The Frenkel-Kontorova Model: Concepts, Methods and Applications*. Berlin: Springer

Bray, A. J. (1974). *Phys. Rev. Lett.* **32**, 1413

Brey, L., & Fertig, H. A. (2006). *Phys. Rev. B* **73**, 235411

   (2009). *Phys. Rev. B* **80**, 035406

Briassoulis, D. (1986). *Computers Structures* **23**, 129

Briskot, U., Schütt, M., Gornyi, I. V., et al. (2015). *Phys. Rev. B* **92**, 115426

Brouwer, P. W. (1998). *Phys. Rev. B* **58**, R10135

Brown, E. (1964). *Phys. Rev.* **133**, A1038

Burmistrov, I. S., Gornyi, I. V., Kachorovskii, V. Yu., Katsnelson, M. I., & Mirlin, A. D. (2016). *Phys. Rev. B* **94**, 195430

Burmistrov, I. S., Gornyi, I. V., Kachorovskii, V. Yu., Katsnelson, M. I., Los J. H., & Mirlin, A. D. (2018a). *Phys. Rev. B* **97**, 125402

Burmistrov, I. S., Kachorovskii, V. Yu., Gornyi, I. V., & Mirlin, A. D. (2018b). *Ann. Phys. (NY)* **396**, 119

Caldeira, A. O., & Leggett, A. J. (1983). *Ann. Phys. (NY)* **149**, 374

Calogeracos, A., & Dombey, N. (1999). *Contemp. Phys.* **40**, 313

Cangemi, D., & Dunne, G. (1996). *Ann. Phys. (NY)* **249**, 582

Cao, Y., Fatemi, V., Demir, A., et al. (2018a). *Nature* **556**, 80

Cao, Y., Fatemi, V., Fang, S., et al. (2018b). *Nature* **556**, 43

Cardy, J. (2000). *Phys. Rev. Lett.* **84**, 3507

Castro, E. V., Novoselov, K. S., Morozov, S. V., et al. (2007). *Phys. Rev. Lett.* **99**, 216802
   et al. (2010a). *J. Phys.: Condens. Matter* **22**, 175503

Castro, E. V., Ochoa, H., Katsnelson, M. I., et al. (2010b). *Phys. Rev. Lett.* **105**, 266601

Castro Neto, A. H., & Guinea, F. (2009). *Phys. Rev. Lett.* **103**, 026804

Castro Neto, A. H., Guinea, F., & Peres, N. M. R. (2006). *Phys. Rev. B* **73**, 205408

Castro Neto, A. H., Guinea, F., Peres, N. M. R., Novoselov, K. S., & Geim, A. K. (2009). *Rev. Mod. Phys.* **81**, 109

Cerda, E., & Mahadevan, L. (2003). *Phys. Rev. Lett.* **90**, 074302

Chang, M.-C., & Niu, Q. (2008). *J. Phys.: Condens. Matter* **20**, 193202

Checkelsky, J. G., Li, L., & Ong, N. P. (2008). *Phys. Rev. Lett.* **100**, 20680

Cheianov, V. V., & Fal'ko, V. I. (2006). *Phys. Rev. B* **74**, 041403

Cheianov, V. V., Fal'ko, V., & Altshuler, B. L. (2007). *Science* **315**, 1252

Cheianov, V. V., Fal'ko, V. I., Altshuler, B. L., & Aleiner, I. L. (2007). *Phys. Rev. Lett.* **99**, 176801

Cheianov, V. V., Syljuasen, O., Altshuler, B. L., & Fal'ko, V. (2009). *Phys. Rev. B* **80**, 233409

Chen, J.-H., Cullen, W. G., Jang, C., Fuhrer, M. S., & Williams, E. D. (2009). *Phys. Rev. Lett.* **102**, 236805

Chen, J.-H., Jang, C., Adam, S., et al. (2008). *Nature Phys.* **4**, 377

Chen, S., Han, Z., Elahi, M. M., et al. (2016). *Science* **353**, 1522

Chico, L., Benedict, L. X., Louie, S. G., & Cohen, M. L. (1996). *Phys. Rev. B* **54**, 2600

Chizhova, L. A., Libisch, F., & Burgdörfer, J. (2014). *Phys. Rev. B* **90**, 165404

Choi, S.-M., Jhi, S.-H., & Son, Y.-W. (2010). *Phys. Rev. B* **81**, 081407

Cocco, G., Cadelano, E., & Colombo, L. (2010). *Phys. Rev. B* **81**, 241412

Couto, N. J. G., Costanzo, D., Engels, S., et al. (2014). *Phys. Rev. B* **4**, 041019

Couto, N. J. G., Sacépé, B., & Morpurgo, A. F. (2011). *Phys. Rev. Lett.* **107**, 225501

Coxeter, H. S. M. (1989). *Introduction to Geometry.* New York: Wiley

Crassee, I., Levallois, J., Walter, A. L., et al. (2011). *Nature Phys.* **7**, 48

Creutz, M. (1983). *Quarks, Gluons and Lattices.* Cambridge: Cambridge University Press

Crossno, J., Shi, J. K., Wang, K., Liu, X., et al. (2016). *Science* **351**, 1058

Cserti, J., Csordás, A., & Dávid, G. (2007). *Phys. Rev. Lett.* **99**, 066802

Cserti, J., & Dávid, G. (2006). *Phys. Rev. B* **74**, 125419

Czycholl, G. (2008). *Theoretische Festkörperphysik von den klassischen Modellen zu modernen Forschungsthemen.* Berlin: Springer

Danneau, R., Wu, F., Craciun, M. F., et al. (2008). *Phys. Rev. Lett.* **100**, 196802

Das Sarma, S., Adam, S., Huang, E. H., & Rossi, E. (2011). *Rev. Mod. Phys.* **83**, 407

Davydov, A. S. (1976). *Quantum Mechanics.* Oxford: Pergamon

Dean, C. R., Wang, L., Maher, P., et al. (2013). *Nature* **497**, 598

Dean, C. R., Young, A. F., Meric, I., et al. (2010). *Nature Nanotech.* **5**, 722

de Andres, P. L., Guinea, F., & Katsnelson, M. I. (2012). *Phys. Rev. B* **86**, 144103

de Gennes, P.-G. (1979). *Scaling Concepts in Polymer Physics.* Ithaca, NY: Cornell University Press

de Juan, F., Grushin, A. G., & Vozmediano, M. A. H. (2010). *Phys. Rev. B* **82**, 125409

Delplace, P., & Montambaux, G. (2010). *Phys. Rev. B* **82**, 205412

De Raedt, H., & Katsnelson, M. I. (2008). *JETP Lett.* **88**, 607

De Raedt, H., & Lagendijk, A. (1985). *Phys. Rep.* **127**, 233

Diez, M., Dahlhaus, J. P., Wimmer, M., & Beenakker, C. W. J. (2014). *Phys. Rev. Lett.* **112**, 196602

Di Francesco, P., Mathieu, P., & Sénéchal, D. (1997). *Conformal Field Theory.* New York: Springer

DoCarmo, M. P. (1976). *Differential Geometry of Curves and Surfaces.* London: Prentice-Hall

Dombey, N., & Calogeracos, A. (1999). *Phys. Rep.* **315**, 41

Dresselhaus, G. (1974). *Phys. Rev. B* **10**, 3602

Dresselhaus, M. S., & Dresselhaus, G. (2002). *Adv. Phys.* **51**, 1

Drut, J. E., & Lähde, T. A. (2009a). *Phys. Rev. Lett.* **102**, 026802
(2009b). *Phys. Rev. B* **79**, 165425

Du, X., Skachko, I., Barker, A., & Andrei, E. Y. (2008). *Nature Nanotech.* **3**, 491

Du, X., Skachko, I., Duerr, F., Cucian, A., & Andrei, E. Y. (2009). *Nature* **462**, 192

Dugaev, V. K., & Katsnelson, M. I. (2012). *Phys. Rev. B* **86**, 115405
(2013). *Phys. Rev. B* **88**, 235432
(2014). *Phys. Rev. B* **90**, 035408

Edwards, D. M. (1967). *Phys. Lett. A* **24**, 350

Edwards, D. M., & Katsnelson, M. I. (2006). *J. Phys.: Condens. Matter* **18**, 7209

Efetov, K. B. (1997). *Supersymmetry in Disorder and Chaos*. Cambridge: Cambridge University Press

Elias, D. C., Gorbachev, R. V., Mayorov, A. S., et al. (2011). *Nature Phys.* **7**, 701

Elias, D. C., Nair, R. R., Mohiuddin, T. M. G., et al. (2009). *Science* **323**, 610

Elliott, R. J. (1954). *Phys. Rev.* **96**, 266

Emelyanenko, A., & Boinovich, L. (2008). *J. Phys.: Condens. Matter* **20**, 494227

Esaki, L. (1958). *Phys. Rev.* **109**, 603

Esquinazi, P., & Höhne, R. (2005). *J. Magn. Magn. Mater.* **290–291**, 20

Evers, F., & Mirlin, A. D. (2008). *Rev. Mod. Phys.* **80**, 1355

Eyring, H., Walter, J., & Kimball, G. E. (1946). *Quantum Chemistry*. Ithaca, NY: Cornell University Press

Faddeev, L. D., & Slavnov, A. A. (1980). *Gauge Fields: Introduction to a Quantum Theory*. Reading, MA: Benjamin

Fal'ko, V. I. (2008). *Phil. Trans. R. Soc. A* **366**, 205

Falkovich, G. (2011). *Fluid Mechanics: A Short Course for Physicists*. Cambridge: Cambridge University Press

Falkovsky, L. A. (1983). *Adv. Phys.* **32**, 753

Fang, S., & Kaxiras, E. (2016). *Phys. Rev. B* **93**, 235153

Fasolino, A., Los, J. H., & Katsnelson, M. I. (2007). *Nature Mater.* **6**, 858

Fedoryuk, M. V. (1977). *Method of Steepest Descent*. Moscow: Nauka

Ferrari, A. C., Meyer, J. C., Scardaci, V., et al. (2006). *Phys. Rev. Lett.* **97**, 187401

Feynman, R. P. (1955). *Phys. Rev.* **97**, 660
(1972). *Statistical Mechanics*. Reading, MA: Benjamin

Fialkovsky, I. V., & Vassilevich, D. V. (2009). *J. Phys. A* **42**, 442001

Fisher, M. E. (1964). *Am. J. Phys.* **32**, 343

Fogler, M. M. (2009). *Phys. Rev. Lett.* **103**, 236801

Fogler, M. M., Guinea, F., & Katsnelson, M. I. (2008). *Phys. Rev. Lett.* **101**, 226804

Fogler, M. M., Novikov, D. S., Glazman, L. I., & Shklovskii, B. I. (2008). *Phys. Rev. B* **77**, 075420

Foulkes, W. M. C., Mitas, L., Needs, R. J., & Rajagopal, G. (2001). *Rev. Mod. Phys.* **73**, 33

Fowler, W. A. (1984). *Rev. Mod. Phys.* **56**, 149

Fradkin, E. (1986). *Phys. Rev. B* **33**, 3263

Frank, F. C., & van der Merwe, J. H. (1949). *Proc. R. Soc. (London) A* **198**, 205

Frenkel, J., & Kontorowa, T. (1938). *Phys. Z. Sowjet.* **13**, 1

Friedel, J. (1952). *Phil. Mag.* **43**, 253

Fritz, L., Schmalian, J., Müller, M., & Sachdev, S. (2008). *Phys. Rev. B* **78**, 085416

Fuchs, J. N., Piéchon, F., Goerbig, M. O., & Montambaux, G. (2010). *Eur. Phys. J. B* **77**, 351

Fujita, M., Wakabayashi, K., Nakada, K., & Kusakabe, K. (1996). *J. Phys. Soc. Japan* **65**, 1920

Galitskii, V. M. (1958a). *Zh. Eksp. Teor. Fiz.* **34**, 151

(1958b). *Zh. Eksp. Teor. Fiz.* **34**, 1011

Gangadharaiah, S., Farid, A. M., & Mishchenko, E. G. (2008). *Phys. Rev. Lett.* **100**, 166802

Garcia-Pomar, J. L., Cortijo, A., & Nieto-Vesperinas, M. (2008). *Phys. Rev. Lett.* **100**, 236801

Gazit, D. (2009). *Phys. Rev. E* **80**, 041117

Geim, A. K. (2009). *Science* **324**, 1530

(2011). *Rev. Mod. Phys.* **83**, 851

Geim, A. K., & Novoselov, K. S. (2007). *Nature Mater.* **6**, 183

Gendenshtein, L. E., & Krive, I. V. (1985). *Sov. Phys. Usp.* **28**, 645

Georgi, A., Nemes-Incze, P., Carrillo-Bastos, R., et al. (2017). *Nano Lett.* **17**, 2240

Geringer, V., Liebmann, M., Echtermeyer, T., et al. (2009). *Phys. Rev. Lett.* **102**, 076102

Gerritsma, R., Kirchmair, G., Zahringer, F., et al. (2010). *Nature* **463**, 38

Ghosh, S., Bao, W., Nika, D. L., et al. (2010). *Nature Mater.* **9**, 555

Gibertini, M., Tomadin, A., Guinea, F., Katsnelson, M. I., & Polini, M. (2012). *Phys. Rev. B* **85**, 201405(R)

Gibertini, M., Tomadin, A., Polini, M., Fasolino, A., & Katsnelson, M. I. (2010). *Phys. Rev. B* **81**, 125437

Giesbers, A. J. M., Ponomarenko, L. A., Novoselov, K. S., et al. (2009). *Phys. Rev. B* **80**, 201403

Giesbers, A. J. M., Zeitler, U., Katsnelson, M. I. (2007). *Phys. Rev. Lett.* **99**, 206803

Giovannetti, G., Khomyakov, P. A., Brocks, G., Kelly, P. J., & van den Brink, J. (2007). *Phys. Rev. B* **76**, 073103

Giuliani, A., Mastropietro, V., & Porta, M. (2011). *Phys. Rev. B* **83**, 195401

Giuliani, G. F., & Vignale, G. (2005). *Quantum Theory of the Electron Liquid*. Cambridge: Cambridge University Press

Glazman, L. I., Lesovik, G. B., Khmelnitskii, D. E., & Shekhter, R. I. (1988). *Pis'ma ZhETF* **48**, 218

Gmitra, M., Konschuh, S., Ertler, C., Ambrosch-Draxl, C., & Fabian, J. (2009). *Phys. Rev. B* **80**, 235431

Goerbig, M. O. (2011). *Rev. Mod. Phys.* **83**, 1193

Goldman, A. I., & Kelton, R. F. (1993). *Rev. Mod. Phys.* **65**, 213

González, J., Guinea, F., & Vozmediano, M. A. H. (1994). *Nucl. Phys. B* **424**, 595

(1999). *Phys. Rev. B* **59**, 2474

Gonzáles-Herrero, H., Gómez-Rodriguez, J. M., Mallet, P., et al. (2016). *Science* **352**, 437

Gorbachev, R. V., Song, J. C. W., Yu, G. L., et al. (2014). *Science* **346**, 448

Gorbar, E. V., Gusynin, V. P., Miransky, V. A., & Shovkovy, I. A. (2002). *Phys. Rev. B* **66**, 045108

Gornyi, I. V., Kachorovskii, V. Yu., & Mirlin, A. D. (2012). *Phys. Rev. B* **86**, 165413

Greiner, W., Mueller, B., & Rafelski, J. (1985). *Quantum Electrodynamics of Strong Fields*. Berlin: Springer

Greiner, W., & Schramm, S. (2008). *Am. J. Phys.* **76**, 509

Grib, A. A., Mamaev, S. V., & Mostepanenko, V. M. (1994). *Vacuum Effects in Strong Fields*. St Petersburg: Friedmann

Grigorenko, A. N., Polini, M., & Novoselov, K. S. (2012). *Nature Photon.* **6**, 749

Gubernatis, J., Kawashima, N., & Werner, P. (2016). *Quantum Monte Carlo Methods: Algorithms for Lattice Models*. Cambridge: Cambridge University Press

Guinea, F. (1984). *Phys. Rev. Lett.* **53**, 1268

(2008). *J. Low Temp. Phys.* **153**, 359

Guinea, F., Castro Neto, A. H., & Peres, N. M. R. (2006). *Phys. Rev. B* **73**, 245426

Guinea, F., Geim, A. K., Katsnelson, M. I., & Novoselov, K. S. (2010). *Phys. Rev. B* **81**, 035408

Guinea, F., Horovitz, B., & Le Doussal, P. (2008). *Phys. Rev. B* **11**, 205421

Guinea, F., & Katsnelson, M. I. (2014). *Phys. Rev. Lett.* **112**, 116604

Guinea, F., Katsnelson, M. I., & Geim, A. K. (2010). *Nature Phys.* **6**, 30

Guinea, F., Katsnelson, M. I., & Vozmediano, M. A. H. (2008). *Phys. Rev. B* **11**, 075422

Güney, D. Ö., & Meyer, D. A. (2009). *Phys. Rev. A* **19**, 063834

Guo, H., Ilseven, E., Falkovich, G., & Levitov, L. S. (2017). *Proc. Natl. Acad. Sci. USA* **114**, 3068

Gurzhi, R. N. (1968). *Sov. Phys. Usp.* **11**, 255

Gusynin, V. P., & Sharapov, S. G. (2005). *Phys. Rev. Lett.* **95**, 146801

Gusynin, V. P., Sharapov, S. G., & Carbotte, J. P. (2006). *Phys. Rev. Lett.* **96**, 256802
   (2007a). *J. Phys.: Condens. Matter* **19**, 026222
   (2007b). *Phys. Rev. B* **75**, 165407
   (2009). *New J. Phys.* **11**, 095407

Güttinger, J., Stampfer, C., Libisch, F., et al. (2009). *Phys. Rev. Lett.* **103**, 046810

Gutzwiller, M. C. (1963). *Phys. Rev. Lett.* **10**, 159

Guyot, P., Kramer, P., & de Boissieu, M. (1991). *Rep. Prog. Phys.* **54**, 1373

Halperin, B. I. (1982). *Phys. Rev. B* **25**, 2185

Halperin, W. (1986). *Rev. Mod. Phys.* **58**, 533

Han, M. Y., Brant, J. C., & Kim, P. (2010). *Phys. Rev. Lett.* **104**, 056801

Han, M. Y., Ozyilmaz, B., Zhang, Y., & Kim, P. (2007). *Phys. Rev. Lett.* **98**, 206805

Han, W., Pi, K., Bao, W., et al. (2009a). *Appl. Phys. Lett.* **94**, 222109

Han, W., Wang, W. H., Pi, K., et al. (2009b). *Phys. Rev. Lett.* **102**, 137205

Hands, S., & Strouthos, C. (2008). *Phys. Rev. B* **78**, 165423

Hasan, M. Z., & Kane, C. L. (2010). *Rev. Mod. Phys.* **82**, 3045

Hasegawa, Y., Konno, **R.**, Nakano, H., & Kohmoto, M. (2006). *Phys. Rev. B* **74**, 033413

Hatsugai, Y. (1993). *Phys. Rev. Lett.* **71**, 3697
   (1997). *J. Phys.: Condens. Matter* **9**, 2507

Hatsugai, Y., Fukui, T., & Aoki, H. (2006). *Phys. Rev. B* **74**, 205414

Heeger, A. J., Kivelson, S., Schrieffer, J. R., & Su, W.-P. (1988). *Rev. Mod. Phys.* **60**, 781

Heikkilä, T. T., Kopnin, N. B., & Volovik, G. E. (2011). *JETP Lett.* **94**, 233

Heine, V. (1960). *Group Theory in Quantum Mechanics*. Oxford: Pergamon

Helfrich, W. (1973). *Z. Naturforsch.* **28C**, 693

Hentschel, M., & Guinea, F. (2007). *Phys. Rev. B* **76**, 115407

Herbut, I. F., Juričič, V., & Vafek, O. (2008). *Phys. Rev. Lett.* **100**, 046403

Hermann, K. (2012). *J. Phys.: Condens. Matter* **24**, 314210

Herring, C. (1966). *Exchange Interactions among Itinerant Electrons*. New York: Academic Press

Hewson, A. C. (1993). *The Kondo Problem to Heavy Fermions*. Cambridge: Cambridge University Press

Hill, A., Mikhailov, S. A., & Ziegler, K. (2009). *Europhys. Lett.* **87**, 27005

Hobson, J. P., & Nierenberg, W. A. (1953). *Phys. Rev.* **89**, 662

Hongler, C., & Smirnov, S. (2011). *Probab. Theory Relat. Fields* **151**, 735

Hostadter, D. R. (1976). *Phys. Rev. B* **14**, 2239

Huang, L., Lai, Y.-C., & Grebogi, C. (2010). *Phys. Rev. E* **81**, 055203

Hubbard, J. (1963). *Proc. R. Soc. (London) A* **276**, 238

Huefner, M., Molitor, F., Jacobsen, A., et al. (2009). *Phys. Stat. Sol. (b)* **246**, 2756

Huertas-Hernando, D., Guinea, F., & Brataas, A. (2006). *Phys. Rev. B* **74**, 155426

(2009). *Phys. Rev. Lett.* **103**, 146801

Hunt, B., Sanchez-Yamagishi, J. D., Young, A. F., et al. (2013). *Science* **240**, 1427

Hwang, E. H., & Das Sarma, S. (2007). *Phys. Rev. B* **75**, 205418

in 't Veld, Y., Schüler, M., Wehling, T., Katsnelson, M. I., & van Loon, E. G. C. P. (2019). *J. Phys.: Condens. Matter* 31, 465603

Inui, M., Trugman, S. A., & Abrahams, E. (1994). *Phys. Rev. B* **49**, 3190

Irkhin, V. Yu., Katanin, A. A., & Katsnelson, M. I. (2001). *Phys. Rev. B* **64**, 165107

(2002). *Phys. Rev. Lett.* **89**, 076401

Irkhin, V. Yu., & Katsnelson, M. I. (1985a). *J. Phys. C* **18**, 4173

(1985b). *Zh. Eksp. Teor. Fiz.* **88**, 522

(1986). *Z. Phys. B* **62**, 201

Irkhin, V. Yu., Katsnelson, M. I., & Trefilov, A. V. (1992). *J. Magn. Magn. Mater.* **117**, 210

Irkhin, V. Yu., Katsnelson, M. I., & Trefilov, A. V. (1993). *J. Phys.: Condens. Matter* **5**, 8763

(2002). *Eur. Phys. J. B* **30**, 481

Ishigami, M., Chen, J. H., Cullen, W. G., Fuhrer, M. S., & Williams, E. D. (2007). *Nano. Lett.* **1**, 1643

Ishihara, A. (1971). *Statistical Physics.* New York: Academic Press

Isichenko, M. B. (1992). *Rev. Mod. Phys.* **64**, 961

Jackiw, R., Milstein, A. L., Pi, S.-Y., & Terekhov, I. S. (2009). *Phys. Rev. B* **80**, 033413

Jackson, J. D. (1962). *Classical Electrodynamics.* New York: Wiley

Jiang, J., Saito, R., Samsonidge, G., et al. (2005). *Phys. Rev. B* **72**, 235408

Jiang, Y., Low, T., Chang, K., Katsnelson, M. I., & Guinea, F. (2013). *Phys. Rev. Lett.* **110**, 046601

Jiang, Z., Henriksen, E. A., Tung, L. C., et al. (2007a). *Phys. Rev. Lett.* **98**, 197403

Jiang, Z., Zhang, Y., Stormer, H. L., & Kim, P. (2007b). *Phys. Rev. Lett.* **99**, 106802

Jo, S., Ki, D.-K., Jeong, D., Lee, H.-J., & Kettermann, S. (2011). *Phys. Rev. B* **84**, 075453

John, D. L., Castro, L. C., & Pulfrey, D. L. (2004). *J. Appl. Phys.* **96**, 5180

Jones, R. A. (2002). *Soft Condensed Matter.* Oxford: Oxford University Press

Kadanoff, L. P., & Baym, G. (1962). *Quantum Statistical Mechanics.* New York: Benjamin

Kager, W., & Nienhuis, B. (2004). *J. Stat. Phys.* **115**, 1149

Kailasvuori, J. (2009). *Europhys. Lett.* **87**, 47008

Kailasvuori, J., & Lüffe, M. C. (2010). *J. Statist. Mech.: Theory Exp.* P06024

Kaku, M. (1988). *Introduction to Superstrings.* Berlin: Springer

Kalashnikov, V. P., & Auslender, M. (1979). *Fortschr. Phys.* **27**, 355

Kamenev, A. (2011). *Field Theory of Non-Equilibrium Systems.* Cambridge: Cambridge University Press

Kamenev, A., & Levchenko, A. (2009). *Adv. Phys.* **58**, 197

Kanamori, J. (1963). *Prog. Theor. Phys.* **30**, 276

Kane, C. L., & Mele, E. J. (1997). *Phys. Rev. Lett.* **78**, 1932

(2005a). *Phys. Rev. Lett.* **95**, 146802

(2005b). *Phys. Rev. Lett.* **95**, 226801

Karssemeijer, L. J., & Fasolino, A. (2011). *Surface Sci.* **605**, 1611

Kashuba, A. B. (2008). *Phys. Rev. B* **78**, 085415

Katanin, A. A., & Kampf, A. P. (2003). *Phys. Rev. B* **68**, 195101

Katsnelson, M. I. (1981). *Fiz. Met. Metalloved.* **52**, 436

(2006a). *Eur. Phys. J. B* **51**, 157

(2006b). *Eur. Phys. J. B* **52**, 151

(2006c). *Phys. Rev. B* **74**, 201401

(2007a). *Mater. Today* **10**, 20

(2007b). *Eur. Phys. J. B* **57**, 225

(2007c). *Phys. Rev. B* **76**, 073411

(2008). *Europhys. Lett.* **84**, 37001

(2010a). *Europhys. Lett.* **89**, 17001

(2010b). *Phys. Rev. B* **82**, 205433

Katsnelson, M. I., & Fasolino, A. (2013). *Accounts Chem. Research* **46**, 97

Katsnelson, M. I., & Geim, A. K. (2008). *Phil. Trans. R. Soc. A* **366**, 195

Katsnelson, M. I., & Guinea, F. (2008). *Phys. Rev. B* **78**, 075417

Katsnelson, M. I., Guinea, F., & Geim, A. K. (2009). *Phys. Rev. B* **79**, 195426

Katsnelson, M. I., Irkhin, V. Yu., Chioncel, L., Lichtenstein, A. I., & de Groot, R. A. (2008). *Rev. Mod. Phys.* **80**, 315

Katsnelson, M. I., Naumov, I. I., & Trefilov, A. V. (1994). *Phase Transitions* **49**, 143

Katsnelson, M. I., & Nazaikinskii, V. E. (2012). *Theor. Math. Phys.* **172**, 1263

Katsnelson, M. I., & Novoselov, K. S. (2007). *Solid State Commun.* **143**, 3

Katsnelson, M. I., Novoselov, K. S., & Geim, A. K. (2006). *Nature Phys.* **2**, 620

Katsnelson, M. I., & Prokhorova, M. F. (2008). *Phys. Rev. B* **77**, 205424

Katsnelson, M. I., & Trefilov, A. V. (2002). *Dynamics and Thermodynamics of Crystal Lattices.* Moscow: Atomizdat

Keldysh, L. V. (1964). *Zh. Eksp. Teor. Fiz.* **47**, 1515

Kellendonk, J., & Schulz-Baldes, H. (2004). *J. Fund. Anal.* **209**, 388

Kelliher, J. P. (2006). *SIAM J. Math. Anal.* **38**, 210

Kemppainen, A. (2017). *Schramm-Loewner Evolution.* Berlin: Springer

Khodel, V. A., & Shaginyan, V. R. (1990). *JETP Lett.* **51**, 553

Kim, K., DaSilva A., Huang, S., et al. (2017). *Proc. Natl. Acad. Sci. USA* **114**, 3364

Kim, K. S., Zhao, Y., Jang, H., et al. (2009). *Nature* **457**, 706

Kim, W. Y., & Kim, K. S. (2008). *Nature Nanotech.* **3**, 408

Kindermann, M., & First, P. N. (2011). *Phys. Rev. B* **83**, 045425

Kindermann, M., Uchoa, B., & Miller, D. L. (2012). *Phys. Rev. B* **86**, 115415

Kiselev, E. I., & Schmalian, J. (2019). *Phys. Rev. B* **99**, 035430

Klein, O. (1929). *Z. Phys.* **53**, 157

Kleptsyn, V., Okunev, A., Schurov, I., Zubov, D., & Katsnelson, M. I. (2015). *Phys. Rev. B* **92**, 165407

Kochan, S., Gmitra, M., & Fabian, J. (2014). *Phys. Rev. Lett.* **112**, 116602

Kochan, S., Irmer, S., & Fabian, J. (2017). *Phys. Rev. B* **95**, 165415

Kogan, E. (2011). *Phys. Rev. B* **84**, 115119

Kohmoto, M. (1985). *Ann. Phys. (NY)* **160**, 343

(1989). *Phys. Rev. B* **39**, 11943

Kohn, W. (1959). *Phys. Rev.* **115**, 1460

Kohn, W., & Luttinger, J. M. (1957). *Phys. Rev.* **108**, 590

Kondo, J. (1964). *Prog. Theor. Phys.* **32**, 37

Konschuh, S., Gmitra, M., & Fabian, J. (2010). *Phys. Rev. B* **82**, 245412

Kopnin, N. B. (2002). *Rep. Prog. Phys.* **65**, 1633

Kosevich, A. M. (1999). *Theory of Crystal Lattices.* New York: Wiley

Koshino, M., & Ando, T. (2006). *Phys. Rev. B* **73**, 245403

(2007). *Phys. Rev. B* **76**, 085425

(2010). *Phys. Rev. B* **81**, 195431

Koshino, M., & McCann, E. (2010). *Phys. Rev. B* **81**, 115315

Koskinen, P., Malola, S., & Häkkinen, H. (2008). *Phys. Rev. Lett.* **101**, 115502

Kotakoski, J., Krasheninnikov, A. V., & Nordlund, K. (2006). *Phys. Rev. B* **74**, 245420

Kotov, V. N., Uchoa, B., Pereira, V. M., Castro Neto, A. H., & Guinea, F. (2012). *Rev. Mod. Phys.* **84**, 1067

Kouwenhoven, L. P., Marcus, C. M., & McEuen, P. L. (1997). Electron transport in quantum dots. In *Mesoscopic Electron Transport* (Sohn, L. L., Kouwenhoven, L. P., & Schon, G., eds.). Dordrecht: Kluwer

Kownacki, J.-P., & Mouhanna, D. (2009). *Phys. Rev. E* **79**, 040101

Košmrlj, A., & Nelson, D. R. (2016). *Phys. Rev. B* **93**, 125431

Krekora, P., Su, Q., & Grobe, R. (2005). *Phys. Rev. A* **72**, 064103

Kretinin, A., Yu, G. L., Jalil, R., et al. (2013). *Phys. Rev. B* **88**, 165427

Krishna Kumar, R., Chen, X., Auton, G. H., et al. (2017). *Science* **357**, 181

Krishna Kumar, R., Mischenko, A., Chen, X., et al. (2018). *Proc. Natl. Acad. Sci. USA* **115**, 5135

Kroes, J. M. H., Akhukov, M. A., Los, J. H., Pineau, N., & Fasolino, A. (2011). *Phys. Rev. B* **83**, 165411

Kubo, R. (1957). *J. Phys. Soc. Japan* **12**, 570

Kubo, R., Hasegawa, H., & Hashitsume, N. (1959). *J. Phys. Soc. Japan* **14**, 56

Kumar, R. K., Bandurin, D. A., Pellegrino, F. M. D., et al. (2017). *Nature Phys.* **13**, 1182

Kuratsuji, H., & Iida, S. (1985). *Prog. Theor. Phys.* **74**, 439

Kuzemsky, A. L. (2005). *Int. J. Mod. Phys. B* **19**, 1029

Kuzmenko, A. B., van Heumen, E., Carbone, F., & van der Marel, D. (2008). *Phys. Rev. Lett.* **100**, 117401

Kuzmenko, A. B., van Heumen, E., van der Marel, D., et al. (2009). *Phys. Rev. B* **79**, 115441

Landau, L. D. (1930). *Z. Phys.* **64**, 629
    (1937). *Phys. Z. Sowjetunion* **11**, 26
    (1956). *Zh. Eksp. Teor. Fiz.* **30**, 1058

Landau, L. D., Abrikosov, A. A., & Khalatnikov, I. M. (1956). *Nuovo Cimento* **Suppl. 3**, 80

Landau, L. D., & Lifshitz, E. M. (1970). *Theory of Elasticity*. Oxford: Pergamon
    (1977). *Quantum Mechanics*. Oxford: Pergamon
    (1980). *Statistical Physics*. Oxford: Pergamon
    (1984). *Electrodynamics of Continuous Media*. Oxford: Pergamon
    (1987). *Fluid Mechanics*. Amsterdam: Elsevier

Landau, L. D., & Peierls, R. (1931). *Z. Phys.* **69**, 56

Landau, L. D., & Pomeranchuk, I. Y. (1955). *Dokl. AN SSSR* **102**, 489

Landsberg, G., & Mandelstam, L. (1928). *Naturwissenschaften* **16**, 557

Le Doussal, P., & Radzihovsky, L. (1992). *Phys. Rev. Lett.* **69**, 1209
    (2018). *Ann. Phys. (NY)* **392**, 340

Lee, C., Wei, X., Kysar, J. W., & Hone, J. (2008). *Science* **321**, 385

Lee, G.-H., Park, G.-H., & Lee, H.-J. (2015). *Nature Phys.* **11**, 925

Lee, P. A. (1993). *Phys. Rev. Lett.* **71**, 1887

Le Lay, G., Salomon, E., & Angot, T. (2017). Silicene, germanene, and stanene. In *2D Materials: Properties and Devices* (Avouris, P., Heinz, T. F., & Low, T., eds.). Cambridge: Cambridge University Press, 458–471

Levitov, L., & Falkovich, G. (2016). *Nature Phys.* **12**, 672

Levy, N., Burke, S. A., Meaker, K. L., et al. (2010). *Science* **329**, 544

Li, J., Schneider, W.-D., Berndt, R., & Delley, D. (1998). *Phys. Rev. Lett.* **80**, 2893

Libisch, F., Stampfer, C., & Burgdörfer, J. (2009). *Phys. Rev. B* **79**, 115423

Lieb, E. H. (1981). *Rev. Mod. Phys.* **53**, 603

(1989). *Phys. Rev. Leu.* **62**, 1201

Lieb, E., & Mattis, D. (1962). *J. Math. Phys.* **3**, 749

Liechtenstein, A. I., Katsnelson, M. I., & Gubanov, V. A. (1985). *Solid State Commun.* **54**, 327

Lifshitz, I. M. (1952). *Zh. Eksp. Teor. Fiz.* **22**, 475

Lifshitz, I. M., Azbel, M. Ya., & Kaganov, M. I. (1973). *Electron Theory of Metals*. New York: Plenum

Lifshitz, I. M., Gredeskul, S. A., & Pastur, L. A. (1988). *Introduction to the Theory of Disordered Systems*. New York: Wiley

Lifshitz, R. (1997). *Rev. Mod. Phys.* **69**, 1181

Lin, D.-H. (2005). *Phys. Rev. A* **72**, 012701
(2006). *Phys. Rev. A* **73**, 044701

Link, J. M., Orth, P. P., Sheehy, D. E., & Schmalian, J. (2016). *Phys. Rev. B* **93**, 235447

Liu, C.-C., Jiang, H., & Yao, Y. (2011). *Phys. Rev. B* **84**, 195430

Logemann, R., Reijnders, K. J. A., Tudorovskiy, T., Katsnelson, M. I., & Yuan, S. (2015). *Phys. Rev. B* **91**, 045420

Lopes dos Santos, J. M. B., Peres, N. M. R., & Castro Neto, A. H. (2007). *Phys. Rev. Lett.* **99**, 56802

López-Polín, G., Gómez-Navarro, C., Parente, V., et al. (2015). *Nature Phys.* **11**, 26

Los, J. H., & Fasolino, A. (2003). *Phys. Rev. B* **68**, 024107

Los, J. H., Fasolino, A., & Katsnelson, M. I. (2016). *Phys. Rev. Lett.* **116**, 015901
(2017). *NPJ 2D Mater. Appl.* **1**, 9

Los, J. H., Ghiringhelli, L. M., Meijer, E. J., & Fasolino, A. (2005). *Phys. Rev. B* **72**, 214102

Los, J. H., Katsnelson, M. I., Yazyev, O. V., Zakharchenko, K. V., & Fasolino, A. (2009). *Phys. Rev. B* **80**, 121405

Los, J. H., Zakharchenko, K. V., Katsnelson, M. I., & Fasolino, A. (2015). *Phys. Rev. B* **91**, 045415

Low, T., Chaves, A., Caldwell, J. D., et al. (2017). *Nature Mater.* **16**, 182

Low, T., Guinea, F., & Katsnelson, M. I. (2011). *Phys. Rev. B* **83**, 195436

Low, T., Jiang, Y., Katsnelson, M. I., & Guinea, F. (2012). *Nano Lett.* **12**, 850

Lucas, A., Crossno, J., Fong, K. C., Kim, P., & Sachdev, S. (2016). *Phys. Rev. B* **93**, 075426

Lucas, A., & Fong, K. C. (2018). *J. Phys.: Condens. Matter* **30**, 053001

Ludwig, A. W. W., Fisher, M. P. A., Shankar, R., & Grinstein, G. (1994). *Phys. Rev. B* **50**, 7526

Lukose, V., Shankar, R., & Baskaran, G. (2007). *Phys. Rev. Lett.* **98**, 116802

Lundeberg, M. B., Gao, Y., Asgari, R., et al. (2017). *Science* **357**, 187

Luttinger, J. M., & Kohn, W. (1958). *Phys. Rev.* **109**, 1892

Luzzi, R., Vasconcellos, A. R., & Ramos, J. G. (2000). *Int. J. Mod. Phys. B* **14**, 3189

Ma, S. K. (1976). *Modern Theory of Critical Phenomena*. Reading, MA: Benjamin

MacDonald, A. H., & Středa, P. (1984). *Phys. Rev. B* **29**, 1616

Madhavan, V., Chen, W., Jamneala, T., Crommie, M. F., & Wingreen, N. S. (1998). *Science* **280**, 567
(2001). *Phys. Rev. B* **64**, 165412

Mafra, D. L., Samsonidze, G., Malard, L. M., et al. (2007). *Phys. Rev. B* **76**, 233407

Mahan, G. (1990). *Many-Particle Physics*. New York: Plenum

Makarova, T., & Palacio, F. (eds.). (2006). *Carbon Based Magnetism: An Overview of the Metal Free Carbon-Based Compounds and Materials*. Amsterdam: Elsevier

Makhlin, Y., & Mirlin, A. D. (2001). *Phys. Rev. Lett.* **87**, 276803

Malard, L. M., Pimenta, M. A., Dresselhaus, G., & Dresselhaus, M. S. (2009). *Phys. Rep.* **473**, 51

Mañes, J. L. (2007). *Phys. Rev. B* **76**, 045430

Mañes, J. L., Guinea, F., & Vozmediano, M. A. H. (2007). *Phys. Rev. B* **75**, 155424

Manyuhina, O. V., Hertzel, J. J., Katsnelson, M. I., & Fasolino, A. (2010). *Eur. Phys. J. E* **32**, 223

Mariani, E., & von Oppen, F. (2008). *Phys. Rev. Lett.* **100**, 076801
   (2010). *Phys. Rev. B* **82**, 195403

Martin, I., & Blanter, Ya. M. (2009). *Phys. Rev. B* **79**, 235132

Martin, J., Akerman, N., Ulbricht, G., et al. (2008). *Nature Phys.* **4**, 144

Maultzsch, J., Reich, S., & Thomsen, C. (2004). *Phys. Rev. B* **70**, 155403

Mayorov, A. S., Elias, D. C., Mucha-Kruczynski, M., et al. (2011a). *Science* **333**, 860

Mayorov, A. S., Gorbachev, R. V., Morozov, S. V., et al. (2011b). *Nano Lett.* **11**, 2396

Mazurenko, V. V., Rudenko, A. N., Nikolaev, S. A., et al. (2016). *Phys. Rev. B* **94**, 214411

McCann, E., Abergel, D. S. L., & Fal'ko, V. I. (2007). *Solid State Commun.* **143**, 110

McCann, E., & Fal'ko, V. I. (2004). *J. Phys.: Condens. Matter* **16**, 2371
   (2006). *Phys. Rev. Leu.* **96**, 086805

McCann, E., Kechedzhi, K., Fal'ko, V. I., et al. (2006). *Phys. Rev. Lett.* **97**, 146805

McClure, J. W. (1956). *Phys. Rev.* **104**, 666
   (1957). *Phys. Rev.* **108**, 612

McCreary, K. M., Pi, K., Swartz, A. G., et al. (2010). *Phys. Rev. B* **81**, 115453

Mele, E. J. (2011). *Phys. Rev. B* **84**, 235439

Meric, I., Dean, C. R., Young, A. F., et al. (2010). *IEEE IEDM Tech. Dig.* 556.

Mermin, N. D. (1968). *Phys. Rev.* **176**, 250
   (1992). *Rev. Mod. Phys.* **64**, 3

Mermin, N. D., & Wagner, H. (1966). *Phys. Rev. Lett.* **17**, 22

Meyer, J. C., Geim, A. K., Katsnelson, M. I., et al. (2007a). *Nature* **446**, 60
   et al. (2007b). *Solid State Commun.* **143**, 101

Meyer, J. R., Hoffman, C. A., Bartoli, F. J., & Rammohan, L. R. (1995). *Appl. Phys. Lett.* **67**, 757

Miao, F., Wijeratne, S., Zhang, Y., et al. (2007). *Science* **317**, 1530

Migdal, A. B. (1977). *Qualitative Methods in Quantum Theory*. Reading, MA: Benjamin

Mikitik, G. P., & Sharlai, Yu. V. (1999). *Phys. Rev. Lett.* **82**, 2147
   (2008). *Phys. Rev. B* **77**, 113407

Min, H., Hill, J. E., Sinitsyn, N. A., et al. (2006). *Phys. Rev. B* **74**, 165310

Mishchenko, E. G. (2008). *Europhys. Lett.* **83**, 17005

Moldovan, D., & Golubovic, L. (1999). *Phys. Rev. E* **60**, 4377

Molitor, F., Knowles, H., Droscher, S., et al. (2010). *Europhys. Lett.* **89**, 67005

Moll, P. J. W., Kushwaha, P., Nandi, N., Schmidt, B., & Mackenzie, A. P. (2016). *Science* **351**, 1061

Moon, P., Koshino, M., & Son, Y.-W. (2019). arXiv:1901.04701

Moore, J. (2009). *Nature Phys.* **5**, 378

Mori, H. (1965). *Prog. Theor. Phys.* **34**, 399

Moriya, T. (1985). *Spin Fluctuations in Itinerant Electron Magnetism*. Berlin: Springer

Morozov, S. V., Novoselov, K. S., Katsnelson, M. I., et al. (2008). *Phys. Rev. Lett.* **100**, 016602
   et al. (2006). *Phys. Rev. Lett.* **91**, 016801

Morpurgo, A. F., & Guinea, F. (2006). *Phys. Rev. Lett.* **91**, 196804

Mott, N. F. (1974). *Metal-Insulator Transitions*. London: Taylor & Francis

Mott, N. F., & Davis, E. A. (1979). *Electron Processes in Non-Crystalline Materials.* Oxford: Clarendon

Mounet, N., & Marzari, N. (2005). *Phys. Rev. B* **71**, 205214

Muñoz-Rojas, F., Fernandez-Rossier, J., Brey, L., & Palacios, J. J. (2008). *Phys. Rev. B* **77**, 045301

Nagaev, E. L. (1983). *Physics of Magnetic Semiconductors.* Moscow: Mir (2001). *Phys. Rep.* **346**, 387

Nagaev, K. E. (1992). *Phys. Lett. A* **169**, 103

Nair, R. R., Blake, P., Grigorenko, A. N., et al. (2008). *Science* **320**, 1308

Nair, R. R., Ren, W., Jalil, R., et al. (2010). *Small* **6**, 2877

Nair, R. R., Sepioni, M., Tsai, I.-L., et al. (2012). *Nature Phys.* **8**, 199

Nair, R. R., Tsai, I.-L., Sepioni, M., et al. (2013). *Nature Commun.* **4**, 2010

Nakada, K., Fujita, M., Dresselhaus, G., & Dresselhaus, M. S. (1996). *Phys. Rev. B* **54**, 17954

Nakahara, N. (1990). *Geometry, Topology and Physics.* Bristol: IOP

Nakano, H. (1957). *Prog. Theor. Phys.* **17**, 145

Narozhny, B. N., Gornyi, I. V., Titov, M., Schütt, M., & Mirlin, A. D. (2015). *Phys. Rev. B* **91**, 035414

Nelson, D. R., & Peliti, L. (1987). *J. Physique* **48**, 1085

Nelson, D. R., Piran, T., & Weinberg, S. (eds.). (2004). *Statistical Mechanics of Membranes and Surfaces.* Singapore: World Scientific

Nemanich, R. J., & Solin, S. A. (1977). *Solid State Commun.* **23**, 417 (1979). *Phys. Rev. B* **20**, 392

Nersesyan, A. A., Tsvelik, A. M., & Wenger, F. (1994). *Phys. Rev. Lett.* **72**, 2628

Neto, C., Evans, D. R., Bonaccurso, E., Butt, H.-J., & Craig, V. S. J. (2005). *Rep. Prog. Phys.* **68**, 2859

Newton, R. G. (1966). *Scattering Theory of Waves and Particles.* New York: McGraw-Hill

Ni, Z. H., Ponomarenko, L. A., Nair, R. R., et al. (2010). *Nano Lett.* **10**, 3868

Nicholl, R. J. T., Conley, H. J., Lavrik, N. V., et al. (2015). *Nature Commun.* **6**, 8789

Nomura, K., & MacDonald, A. H. (2006). *Phys. Rev. Lett.* **96**, 256602

Novikov, D. S. (2007). *Phys. Rev. B* **76**, 245435

Novoselov, K. S. (2011). *Rev. Mod. Phys.* **83**, 837

Novoselov, K. S., Geim, A. K., Morozov, S. V., et al. (2004). *Science* **306**, 666 et al. (2005a). *Nature* **438**, 197

Novoselov, K. S., Jiang, D., Schedin, F., et al. (2005b). *Proc. Natl. Acad. Sci. USA* **102**, 10451

Novoselov, K. S., Jiang, Z., Zhang, Y., et al. (2007). *Science* **315**, 1379

Novoselov, K. S., McCann, E., Morozov, S. V., et al. (2006). *Nature Phys.* **2**, 177

Nozieres, P. (1992). *J. Phys. I (France)* **2**, 443

Ochoa, H., Castro, E. V., Katsnelson, M. I., & Guinea, F. (2011). *Phys. Rev. B* **83**,235416

Okulov, V. I., & Ustinov, V. V. (1979). *Sov. J. Low Temp. Phys.* **5**, 101

Olariu, S., & Popescu, I. (1985). *Rev. Mod. Phys.* **57**, 339

Ono, S., & Sugihara, K. (1966). *J. Phys. Soc. Japan* **21**, 861

Oostinga, K. B., Heersche, H. B., Liu, X., Morpurgo, A. F., & Vandersypen, L. M. K. (2008). *Nature Mater.* **7**, 151

Ostrovsky, P. M., Gornyi, I. V., & Mirlin, A. D. (2006). *Phys. Rev. B* **74**, 235443 (2008). *Phys. Rev. B* **77**, 195430

Ostrovsky, P. M., Titov, M., Bera, S., Gornyi, I. V., & Mirlin, A. D. (2010). *Phys. Rev. Lett.* **105**, 266803

Park, C.-H., Giustino, F., Cohen, M. L., & Louie, S. G. (2008). *Nano Lett.* **8**, 4229

Park, C.-H., & Marzari, N. (2011). *Phys. Rev. B* **84**, 205440

Park, C.-H., Yang, L., Son, Y.-W., Cohen, M. L., & Louie, S. G. (2008). *Phys. Rev. Lett.* **101**, 126804

Partoens, B., & Peeters, F. M. (2006). *Phys. Rev. B* **74**, 075404

Patashinskii, A. Z., & Pokrovskii, V. L. (1979). *Fluctuation Theory of Phase Transitions.* New York: Pergamon

Pauling, L. (1960). *The Nature of the Chemical Bond.* Ithaca, NY: Cornell University Press

Peierls, R. E. (1933). *Z. Phys.* **80**, 763

　(1934). *Helv. Phys. Acta* **7**, 81

　(1935). *Ann. Inst. Henri Poincare* **5**, 177

　(1938). *Phys. Rev.* **54**, 918

Peliti, L., & Leibler, S. (1985). *Phys. Rev. Lett.* **54**, 1690

Pellegrino, F. M. D., Angilella, G. G. N., & Pucci, R. (2010). *Phys. Rev. B* **81**, 035411

Pellegrino, F. M. D., Torre, I., Geim, A. K., & Polini, M. (2016). *Phys. Rev. B* **94**, 155414

Pellegrino, F. M. D., Torre, I., & Polini, M. (2017). *Phys. Rev. B* **96**, 195401

Pendry, J. B. (2004). *Contemp. Phys.* **45**, 191

Peng, L. X., Liew, K. M., & Kitipornchai, S. (2007). *Int. J. Mech. Science* **49**, 364

Pereira, J. M., Peeters, F. M., & Vasilopoulos, P. (2007). *Phys. Rev. B* **76**, 115419

Pereira, V. M., & Castro Neto, A. H. (2009). *Phys. Rev. Lett.* **103**, 046801

Pereira, V. M., Castro Neto, A. H., & Peres, N. M. R. (2009). *Phys. Rev. B* **80**, 045401

Pereira, V. M., Guinea, F., Lopes dos Santos, J. M. B., Peres, N. M. R., & Castro Neto, A. H. (2006). *Phys. Rev. Lett.* **96**, 036801

Pereira, V. M., Nilsson, J., & Castro Neto, A. H. (2007). *Phys. Rev. Lett.* **99**, 166802

Perenboom, J. A. A. J., Wyder, P., & Meier, F. (1981). *Phys. Rep.* **78**, 173

Peres, N. M. R. (2010). *Rev. Mod. Phys.* **82**, 2673

Peres, N. M. R., Castro Neto, A. H., & Guinea, F. (2006). *Phys. Rev. B* **73**, 195411

Peres, N. M. R., Guinea, F., & Castro Neto, A. H. (2006). *Phys. Rev. B* **73**, 125411

Platzman, P. M., & Wolf, P. A. (1973). *Waves and Interactions in Solid State Plasmas.* New York: Academic Press

Polini, M., Tomadin, A., Asgari, R., & MacDonald, A. H. (2008). *Phys. Rev. B* **78**,115426

Pomeranchuk, I., & Smorodinsky, Y. (1945). *J. Phys. (USSR)* **9**, 97

Ponomarenko, L. A., Gorbachev, R. V., Yu, G. L., et al. (2013). *Nature* **497**, 594

Ponomarenko, L. A., Schedin, F., Katsnelson, M. I., et al. (2008). *Science* **320**, 356

Ponomarenko, L. A., Yang, R., Gorbachev, R. V., et al. (2010). *Phys. Rev. Lett.* **105**, 136801

Ponomarenko, L. A., Yang, R., Mohiuddin, T. M., et al. (2009). *Phys. Rev. Lett.* **102**, 206603

Prada, E., San-Jose, P., Wunsch, B., & Guinea, F. (2007). *Phys. Rev. B* **75**, 113407

Prange, R. E., & Girvin, S. M. (eds.). (1987). *The Quantum Hall Effect.* Berlin: Springer

Principi, A., Polini, M., & Vignale, G. (2009). *Phys. Rev. B* **80**, 075418

Principi, A., Polini, M., Vignale, G., & Katsnelson, M. I. (2010). *Phys. Rev. Lett.* **104**, 225503

Principi, A., van Loon, E., Polini, M., & Katsnelson, M. I. (2018). *Phys. Rev. B* **98**, 035427

Principi, A., Vignale, G., Carrega, M., & Polini, M. (2016). *Phys. Rev. B* **93**, 125410

Prodan, E. (2009). *J. Math. Phys.* **50**, 083517

Prokhorova, M. (2013). *Comm. Math. Phys.* **322**, 385

Pyatkovskiy, P. K. (2009). *J. Phys.: Condens. Matter* **21**, 025506

Qi, X., & Zhang, S. (2010). *Phys. Today* **1**, 33

　(2011). *Rev. Mod. Phys.* **83**, 1057

Quilichini, M. (1997). *Rev. Mod. Phys.* **69**, 277

Radzig, A. A., & Smirnov, B. M. (1985). *Reference Data on Atoms, Molecules and Ions*. Berlin: Springer

Raman, C. V. (1928). *Nature* **121**, 619

Raman, C. V., & Krishnan, K. S. (1928). *Nature* **121**, 501

Rammer, J., & Smith, H. (1986). *Rev. Mod. Phys.* **58**, 323

Rashba, E. I. (2009). *Phys. Rev. B* **19**, 161409

Recher, P., Trauzettel, B., Rycerz, A., et al. (2007). *Phys. Rev. B* **76**, 235404

Reich, S., Maultzsch, J., Thomsen, C., & Ordejon, P. (2002). *Phys. Rev. B* **66**, 035412

Reijnders, K. J. A., & Katsnelson, M. I. (2017a). *Phys. Rev. B* **95**, 115310
  (2017b). *Phys. Rev. B* **96**, 045305

Reijnders, K. J. A., Tudorovskiy, T., & Katsnelson, M. I. (2013). *Ann. Phys. (NY)* **333**, 155

Robinson, J. R., Schomerus, H., Oroszlány, L., & Fal'ko, V. I. (2008). *Phys. Rev. Lett.* **101**, 196803

Roldán, R., Fasolino, A., Zakharchenko, K. V., & Katsnelson, M. I. (2011). *Phys. Rev. B* **83**, 174104

Rossi, E., & Das Sarma, S. (2008). *Phys. Rev. Lett.* **101**, 166803

Rudenko, A. N., Brener, S., & Katsnelson, M. I. (2016). *Phys. Rev. Lett.* **116**, 246401

Rudenko, A. N., Keil, F. J., Katsnelson, M. I., & Lichtenstein, A. I. (2013). *Phys. Rev. B* **88**, 081405(R)

Rudenko, A. N., Lugovskoi, A. V., Mauri, A., et al. (2019). *Phys. Rev. B* 100, 075417

Ruelle, D. (1999). *Statistical Mechanics: Rigorous Results*. London: Imperial College Press/Singapore: World Scientific

Rusin, T. M., & Zawadzki, W. (2008). *Phys. Rev. B* **78**, 125419
  (2009). *Phys. Rev. B* **80**, 045416

Russo, S., Oostinga, J. B., Wehenkel, D., et al. *Phys. Rev. B* **11**, 085413

Rycerz, A. (2010). *Phys. Rev. B* **81**, 121404

Rycerz, A., Recher, P., & Wimmer, M. (2009). *Phys. Rev. B* **80**, 125417

Ryu, S., Mudry, C., Furusaki, A., & Ludwig, A. W. W. (2007). *Phys. Rev. B* **75**, 205344

Sachs, B., Wehling, T. O., Katsnelson, M. I., & Lichtenstein, A. I. (2011). *Phys. Rev. B* **84**, 195444

Sadowski, M. L., Martinez, G., Potemski, M., Berger, C., & de Heer, W. A. (2006). *Phys. Rev. Lett. 91*, 266405

Safran, S. A., & DiSalvo, F. J. (1979). *Phys. Rev. B* **20**, 4889

Saha, K., Paul, I., & Sengupta, K. (2010). *Phys. Rev. B* **81**, 165446

Saleur, H., & Duplantier, B. (1987). *Phys. Rev. Lett.* **58**, 2325

San-José, P., González, J., & Guinea, F. (2011). *Phys. Rev. Lett.* **106**, 045502

Sasaki, K., Kawazoe, Y., & Saito, R. (2005). *Prog. Theor. Phys.* **113**, 463

Savini, G., Dappe, Y. J., Oberg, S., et al. (2011). *Carbon* **49**, 62

Schakel, A. M. J. (1991). *Phys. Rev. D* **43**, 1428

Schapere, A., & Wilczek, F. (eds.). (1989). *Geometric Phases in Physics*. Singapore: World Scientific

Schedin, F., Geim, A. K., Morozov, S. V., et al. (2007). *Nature Mater.* **6**, 652

Schoenberg, D. (1984). *Magnetic Oscillations in Metals*. Cambridge: Cambridge University Press

Schrödinger, E. (1930). *Sitz. Preufi. Akad. Wiss. Phys.-Math.* **24**, 418

Schubin, S., & Wonsowski, S. (1934). *Proc. R. Soc. (London) A* **145**, 159

Schuessler, A., Ostrovsky, P. M., Gornyi, I. V., & Mirlin, A. D. (2009). *Phys. Rev. B* **79**, 075405

Schüler, M., Rösner, M., Wehling, T. O., Lichtenstein, A. I., & Katsnelson, M. I. (2013). *Phys. Rev. Lett.* **111**, 036601

Schulman, L. S. (1981). *Techniques and Applications of Path Integration*. New York: Wiley

Sepioni, M., Nair, R. R., Rablen, S., et al. (2010). *Phys. Rev. Lett.* **105**, 207205

Shallcross, S., Sharma, S., Kandelaki, E., & Pankratov, O. A. (2010). *Phys. Rev. B* **81**, 165105

Sharapov, S. G., Gusynin, V. P., & Beck, H. (2004). *Phys. Rev. B* **69**, 075104

Sharma, M. P., Johnson, L. G., & McClure, J. W. (1974). *Phys. Rev. B* **9**, 2467

Sharon, E., Roman, B., Marder, M., Shin, G.-S., & Swinney, H. L. (2002). *Nature* **419**, 579

Shechtman, D., Blech, I., Gratias, D., & Cahn, J. W. (1984). *Phys. Rev. Lett.* **53**, 1951

Sheehy, D. E., & Schmalian, J. (2009). *Phys. Rev. B* **80**, 193411

Shklovskii, B. I., & Efros, A. L. (1984). *Electronic Properties of Doped Semiconductors*. Berlin: Springer

Shon, N. H., & Ando, T. (1998). *J. Phys. Soc. Japan* **67**, 2421

Shytov, A. V., Abanin, D. A., & Levitov, L. S. (2009). *Phys. Rev. Lett.* **103**, 016806

Shytov, A. V., Gu, N., & Levitov, L. S. (2007). arXiv:0708.3081 (unpublished)

Shytov, A. V., Katsnelson, M. I., & Levitov, L. S. (2007a). *Phys. Rev. Lett.* **99**, 236801 (2007b). *Phys. Rev. Lett.* **99**, 246802

Shytov, A., Rudner, M., Gu, N., Katsnelson, M., & Levitov, L. (2009). *Solid State Commun.* **149**, 1087

Shytov, A. V., Rudner, M. S., & Levitov, L. S. (2008). *Phys. Rev. Lett.* **101**, 156804

Simon, B. (1983). *Phys. Rev. Lett.* **51**, 2167

Slonczewski, J. S., & Weiss, P. R. (1958). *Phys. Rev.* **109**, 272

Slotman, G. J., van Wijk M. M., Zhao, P.-L., et al. (2015). *Phys. Rev. Lett.* **115**, 186801

Smirnov, S. (2001). *C. R. Acad. Sci. Paris Sér. I Math.* **333**, 239

Snyman, I., & Beenakker, C. W. J. (2007). *Phys. Rev. B* **75**, 045322

Sodemann, I., & Fogler, M. M. (2012). *Phys. Rev. B* **86**, 115408

Son, Y.-W., Cohen, M. L., & Louie, S. (2006a). *Nature* **444**, 347 (2006b). *Phys. Rev. Lett.* **97**, 216803

Song, J. C. W., Shytov, A. V., & Levitov, L. S. (2013). *Phys. Rev. Lett.* **111**, 266801

Stampfer, C., Schurtenberger, E., Molitor, F., et al. (2008). *Nano Lett.* **8**, 2378

Stander, N., Huard, B., & Goldhaber-Gordon, D. (2009). *Phys. Rev. Lett.* **102**, 026807

Stauber, T., Peres, N. M. R., & Geim, A. K. (2008). *Phys. Rev. B* **78**, 085432

Stauber, T., Peres, N. M. R., & Guinea, F. (2007). *Phys. Rev. B* **76**, 205423

Stauber, T., & Schliemann, J. (2009). *New J. Phys.* **11**, 115003

Stauber, T., Schliemann, J., & Peres, N. M. R. (2010). *Phys. Rev. B* **81**, 085409

Stefanucci, G., & van Leeuwen, R. (2013) *Nonequilibrium Many-Body Theory of Quantum Systems*. Cambridge: Cambridge University Press

Stern, F. (1967). *Phys. Rev. Lett.* **18**, 546

Steward, E. G., Cook, B. P., & Kellert, E. A. (1960). *Nature* **187**, 1015

Stöckmann, H.-J. (2000). *Quantum Chaos: An Introduction*. Cambridge: Cambridge University Press

Stolyarova, E., Rim, K. T., Ryu, S., et al. G. W. (2007). *Proc. Natl. Acad. Sci. USA* **104**, 9209

Stoner, E. C. (1936). *Proc. R. Soc. (London) A* **154**, 656

Su, R.-K., Siu, G. G., & Chou, X. (1993). *J. Phys. A* **26**, 1001

Suárez Morell, E., Correa, J. D., Vargas, P., Pacheco, M., & Barticevic, Z. (2010). *Phys. Rev. B* **82**, 121407(R)

Sugihara, K. (1983). *Phys. Rev. B* **28**, 2157

Suzuura, H., & Ando, T. (2002). *Phys. Rev. B* **65**, 235412

Taimanov, I. A. (2006). *Russ. Math. Surveys* **61**, 79

Tan, L. Z., Park, C.-H., & Louie, S. G. (2010). *Phys. Rev. B* **81**, 195426

Tang, S., Wang, H., Zhang, Y., et al. (2013). *Sci. Rep.* **3**, 2666

Tao, C., Jiao, L., Yazyev, O. V., et al. (2011). *Nature Phys.* **7**, 616

Teber, S., & Kotikov, A. V. (2014). *Europhys. Lett.* **107**, 57001

Teissier, R., Finley, J. J., Skolnick, M. S., et al. (1996). *Phys. Rev. B* **54**, 8329

Tenjinbayashi, Y., Igarashi, H., & Fujiwara, T. (2007). *Ann. Phys. (NY)* **322**, 460

Tersoff, J., & Hamann, D. R. (1985). *Phys. Rev. B* **31**, 805

Thomsen, C., & Reich, S. (2000). *Phys. Rev. Lett.* **85**, 5214

Thouless, D. J. (1983). *Phys. Rev. B* **27**, 6083

Thouless, D. J., Kohmoto, M., Nightingale, M. P., & den Nijs, M. (1982). *Phys. Rev. Lett.* **49**, 405

Tian, W., & Datta, S. (1994). *Phys. Rev. B* **49**, 5097

Tikhonenko, F. V., Horsell, D. W., Gorbachev, R. V., & Savchenko, A. K. (2008). *Phys. Rev. Lett.* **100**, 056802

Tikhonenko, F. V., Kozikov, A. A., Savchenko, A. K., & Gorbachev, R. V. (2009). *Phys. Rev. Lett.* **103**, 226801

Timoshenko, S. P., & Woinowsky-Krieger, S. (1959). *Theory of Plates and Shells*. New York: McGraw-Hill

Titov, M., & Katsnelson, M. I. (2014). *Phys. Rev. Lett.* **113**, 096801

Titov, M., Ostrovsky, P. M., Gornyi, I. V., Schuessler, A., & Mirlin, A. D. (2010). *Phys. Rev. Lett.* **104**, 076802

Tombros, N., Jozsa, C., Popinciuc, M., Jonkman, H. T., & van Wees, B. J. (2007). *Nature* **448**, 571

Tombros, N., Tanabe, S., Veligura, A., et al. (2008). *Phys. Rev. Lett.* **101**, 046601

Torre, I., Tomadin, A., Geim, A. K., & Polini, M. (2015). *Phys. Rev. B* **92**, 165433

Trushin, M., Kailasvuori, J., Schliemann, J., & MacDonald, A. H. (2010). *Phys. Rev. B* **82**, 155308

Tsidilkovskii, I. M. (1982). *Band Structure of Semiconductors*. Oxford: Pergamon (1996). *Electron Spectrum of Gapless Semiconductors*. Berlin: Springer

Tudorovskiy, T., & Katsnelson, M. I. (2012). *Phys. Rev. B* **86**, 045419

Tudorovskiy, T., & Mikhailov, S. A. (2010). *Phys. Rev. B* **82**, 073411

Tudorovskiy, T., Reijnders, K. J. A., & Katsnelson, M. I. (2012). *Phys Scr.* T146, 014010.

Tuinstra, F., & Koenig, J. L. (1970). *J. Chem. Phys.* **53**, 1126

Tworzydlo, J., Trouzettel, B., Titov, M., Rycerz, A., & Beenakker, C. W. J. (2006). *Phys. Rev. Lett.* **96**, 246802

Uchoa, B., Yang, L., Tsai, S.-W., Peres, N. M. R., & Castro Neto, A. H. (2009). *Phys. Rev. Lett.* **103**, 206804

Ugeda, M. M., Brihuega, I., Guinea, F., & Gómez-Rodríguez, J. M. (2010). *Phys. Rev. Lett.* **104**, 096804

Ukraintsev, V. A. (1996). *Phys. Rev. B* **53**, 11176

Ulybyshev, M. V., Buividovich, P. V., Katsnelson, M. I., & Polikarpov, M. I. (2013). *Phys. Rev. Lett.* **111**, 056801

Ulybyshev, M. V., & Katsnelson, M. I. (2015). *Phys. Rev. Lett.* **114**, 246801

Ustinov, V. V. (1980). *Theor. Math. Phys.* **44**, 814

Vandelli, M., Katsnelson, M. I., & Stepanov, E. A. (2019). *Phys. Rev. B* **99**, 165432

van Schilfgaarde, M., & Katsnelson, M. I. (2011). *Phys. Rev. B* **83**, 081409(R)

van Wijk, M. M., Schuring, A., Katsnelson, M. I., & Fasolino, A. (2014). *Phys. Rev. Lett.* **113**, 135504 (2015). *2D Mater.* **2**, 034010

Veselago, V. S. (1968). *Sov. Phys. Usp.* **10**, 509

Vlasov, K. B., & Ishmukhametov, B. Kh. (1964). *Zh. Eksp. Teor. Fiz.* **46**, 201

Volkov, B. A., & Pankratov, O. A. (1986). *Pis'ma Zh. Eksp. Teor. Fiz.* **43**, 99

Volovik, G. E. (1991). *JETP Lett.* **53**, 222
 (2003). *The Universe in a Helium Droplet.* Oxford: Clarendon Press
 (2018). *JETP Lett.* **107**, 516

Vonsovsky, S. V. (1946). *Zh. Eksp. Teor. Fiz.* **16**, 981
 (1974). *Magnetism.* New York: Wiley

Vonsovsky, S. V., & Katsnelson, M. I. (1979). *J. Phys. C* **12**, 2043
 (1989). *Quantum Solid State Physics.* Berlin: Springer

Vonsovsky, S. V., & Svirsky, M. S. (1993). *Usp. Fiz. Nauk* 163, 5, 115

Vonsovsky, S. V., & Turov, E. A. (1953). *Zh. Eksp. Teor. Fiz.* **24**, 419

Vozmediano, M. A. H., Katsnelson, M. I., & Guinea, F. (2010). *Phys. Rep.* **496**, 109

Wagner, M. (1991). *Phys. Rev. B* **44**, 6104

Wallace, P. R. (1947). *Phys. Rev.* **71**, 622

Wallbank, J. R., Ghazaryan, D., Misra, A., et al. (2016). *Science* **353**, 575

Wallbank, J. R., Patel, A. A., Mucha-Kruczyński, M., Geim, A. K., & Fal'ko, V. I. (2013). *Phys. Rev. B* **87**, 245408

Wang, X., Ouyang, Y., Li, X., Wang, H., Guo, J., & Dai, H. (2008). *Phys. Rev. Lett.* **100**, 206803

Wang, Y., Wong, D., Shytov, A. V., et al. (2013). *Science* **340**, 734

Wassmann, T., Seitsonen, A. P., Saitta, M., Lazzeri, M., & Mauri, F. (2008). *Phys. Rev. Lett.* **101**, 096402

Watanabe, H., Hatsugai, Y., & Aoki, H. (2010). *Phys. Rev. B* **82**, 241403

Watanabe, K., Taniguchi, T., & Kanda, H. (2004). *Nature Mater.* **3**, 404

Wehling, T. O., Balatsky, A. V., Katsnelson, M. I., et al. (2007). *Phys. Rev. B* **75**, 125425

Wehling, T. O., Balatsky, A. V., Tsvelik, A. M., Katsnelson, M. I., & Lichtenstein, A. I. (2008a). *Europhys. Lett.* **84**, 17003

Wehling, T. O., Dahal, H. P., Lichtenstein, A. I., et al. (2010b). *Phys. Rev. B* **81**, 085413

Wehling, T. O., Katsnelson, M. I., & Lichtenstein, A. I. (2009a). *Chem. Phys. Lett.* **476**, 125
 (2009b). *Phys. Rev. B* **80**, 085428

Wehling, T. O., Novoselov, K. S., Morozov, S. V., et al. (2008b). *Nano Lett.* **8**, 173

Wehling, T. O., Şaşioğlu, E., Friedrich, C., Lichtenstein, A. I., Katsnelson, M. I., & Blügel, S. (2011). *Phys. Rev. Lett.* **106**, 236805

Wehling, T. O., Yuan, S., Lichtenstein, A. I., Geim, A. K., & Katsnelson, M. I. (2010a). *Phys. Rev. Lett.* **105**, 056802

Whittaker, E. T., & Watson, G. N. (1927). *A Course of Modern Analysis.* Cambridge: Cambridge University Press

Wilson, A. H. (1965). *Theory of Metals.* Cambridge: Cambridge University Press

Wilson, K. G., & Kogut, J. (1974). *Phys. Rep.* **12**, 75

Wimmer, M., Adagideli, I., Berber, S., Tomanek, D., & Richter, K. (2008). *Phys. Rev. Lett.* **100**, 177207

Wimmer, M., Akhmerov, A. R., & Guinea, F. (2010). *Phys. Rev. B* **82**, 045409

Witowski, A. M., Orlita, M., Stepniewski, R., et al. (2010). *Phys. Rev. B* **82**,165305

Woessner, A., Lundeberg, M. B., Gao, Y., et al. (2015). *Nature Mater.* **14**, 421

Woods, C. R., Britnell, L., Eckmann, A., et al. (2014). *Nature Phys.* **10**, 451

Woods, C. R., Withers, F., Zhu, M. J., et al. (2016). *Nature Commun.* **7**, 10800

Wunsch, B., Stauber, T., Sols, F., & Guinea, F. (2006). *New J. Phys.* **8**, 318

Wurm, J., Rycerz, A., Adagideli, I., et al. (2009). *Phys. Rev. Lett.* **102**, 056806

Wurm, J., Wimmer, M., Baranger, H. U., & Richter, K. (2010). *Semicond. Sci. Technol.* **25**, 034003

Xiao, D., Chang, M.-C., & Niu, Q. (2010). *Rev. Mod. Phys.* **82**, 1959

Xing, X., Mukhopadhyay, R., Lubensky, T. C., & Radzihovsky, L. (2003). *Phys. Rev. E* **68**, 021108

Xu, Y., Yan, B., Zhang, H.-J., et al. (2013). *Phys. Rev. Lett.* **111**, 136804

Xue, J., Sanchez-Yamagishi, J., Bulmash, D., et al. (2011). *Nature Mater.* **10**, 282

Yacoby, A., & Imry, Y. (1990). *Phys. Rev. B* **41**, 5341

Yang, L., Deslippe, J., Park, C.-H., Cohen, M. L., & Louie, S. G. (2009). *Phys. Rev. Lett.* **103**, 186802

Yang, W., Chen, G., Shi, Z., et al. (2013). *Nature Mater.* **12**, 792

Yang, X., & Nayak, C. (2002). *Phys. Rev. B* **65**, 064523

Yao, W., Wang, E., Bao, C., et al. (2018). *Proc. Natl. Acad. Sci. USA* **115**, 6928

Yao, Y., Ye, F., Qi, X.-L., Zhang, S. C., & Fang, Z. (2007). *Phys. Rev. B* **75**, 041401

Yazyev, O. V. (2010). *Rep. Prog. Phys.* **73**, 056501

Yazyev, O. V., & Helm, L. (2007). *Phys. Rev. B* **75**, 125408

Yazyev, O. V., & Katsnelson, M. I. (2008). *Phys. Rev. Lett.* **100**, 047209

Yennie, D. R., Ravenhall, D. G., & Wilson, R. N. (1954). *Phys. Rev.* **95**, 500

Yoon, D., Son, Y.-W., & Cheong, H. (2011). *Nano Lett.* **11**, 3227

Yosida, K. (1996). *Theory of Magnetism*. Berlin: Springer

Young, A. F., & Kim, P. (2009). *Nature Phys.* **5**, 222

Yu, G. L., Jalil, R., Belle, B., et al. (2013). *Proc. Natl. Acad. Sci. USA* **110**, 3282

Yu, G., Wu, Z., Zhan, Z., Katsnelson, M. I., & Yuan, S. (2019). *NPJ Comput. Mater.* 5, 122

Yuan, S., De Raedt, H., & Katsnelson, M. I. (2010a). *Phys. Rev. B* **82**, 115448
  (2010b). *Phys. Rev. B* **82**, 235409

Yuan, S. Roldán, R., & Katsnelson, M. I. (2011). *Phys. Rev. B* **84**, 035439

Yudin, D., Hirschmeier, D., Hafermann, H., et al. (2014). *Phys. Rev. Lett.* **112**, 070403

Zak, J. (1964). *Phys. Rev.* **134**, A1602
  (1989). *Phys. Rev. Lett.* **62**, 2747

Zakharchenko, K. V., Fasolino, A., Los, J. H., & Katsnelson, M. I. (2011). *J. Phys.: Condens. Matter* **23**, 202202

Zakharchenko, K. V., Katsnelson, M. I., & Fasolino, A. (2009). *Phys. Rev. Lett.* **102**, 046808

Zakharchenko, K. V., Los, J. H., Katsnelson, M. I., & Fasolino, A. (2010a). *Phys. Rev. B* **81**, 235439

Zakharchenko, K. V., Roldan, R., Fasolino, A., & Katsnelson, M. I. (2010b). *Phys. Rev. B* **82**, 125435

Zarea, M., & Sandler, N. (2009). *Phys. Rev. B* **79**, 165442

Zel'dovich, Y. B., & Popov, V. S. (1972). *Sov. Phys. Usp.* **14**, 673

Zener, C. (1951a). *Phys. Rev.* **81**, 440
  (1951b). *Phys. Rev.* **82**, 403
  (1951c). *Phys. Rev.* **83**, 299

Zhang, H. G., Hu, H., Pan, Y., et al. (2010). *J. Phys.: Condens. Matter* **22**, 302001

Zhang, L., Bampoulis, P., Rudenko, A. N., et al. (2016). *Phys. Rev. Lett.* **116**, 256804

Zhang, Y., Jiang, Z., Small, J. P., et al. (2006). *Phys. Rev. Lett.* **96**, 136806

Zhang, Y., Tan, Y.-W., Stormer, H. L., & Kim, P. (2005). *Nature* **438**, 201

Zhao, P.-L., Yuan, S., Katsnelson, M. I., & De Raedt, H. (2015). *Phys. Rev. B* **92**, 045437 (2015).

Ziegler, K. (1998). *Phys. Rev. Lett.* **80**, 3113

Ziman, J. M. (2001). *Electrons and Phonons. The Theory of Transport Phenomena in Solids*. Oxford: Oxford University Press

Zubarev, D. N. (1974). *Nonequilibrium Statistical Thermodynamics*. New York: Consultants Bureau

Žutić, I., Fabian, J., & Das Sarma, S. (2004). *Rev. Mod. Phys.* **76**, 323

Zverev, M. V., & Baldo, M. (1999). *J. Phys.: Condens. Matter* **11**, 2059

Index